Airborne and Terrestrial Laser Scanning

Airborne and Terrestrial Laser Scanning

Edited by

George Vosselman
Hans-Gerd Maas

WHITTLES PUBLISHING

CRC Press
Taylor & Francis Group

Published by
Whittles Publishing,
Dunbeath,
Caithness KW6 6EY,
Scotland, UK

www.whittlespublishing.com

Distributed in North America by
CRC Press LLC,
Taylor and Francis Group,
6000 Broken Sound Parkway NW, Suite 300,
Boca Raton, FL 33487, USA

ISBN 978-1904445-87-6
USA ISBN 978-1-4398-2798-7

The publisher and authors have used their best efforts in preparing this book, but assume no responsibility for any injury and/or damage to persons or property from the use or implementation of any methods, instructions, ideas or materials contained within this book. All operations should be undertaken in accordance with existing legislation and recognized trade practice. Whilst the information and advice in this book is believed to be true and accurate at the time of going to press, the authors and publisher accept no legal responsibility or liability for errors or omissions that may have been made.

Printed by InPrint, Latvia

Contents

Preface ix
The Authors xiii
List of Abbreviations xvi

Chapter 1   Laser Scanning Technology 1
  1.1   Basic measurement principles of laser scanners 2
    1.1.1   Time-of-flight measurement 3
    1.1.2   Phase measurement techniques 5
    1.1.3   Triangulation-based measurements 8
  1.2   Components of laser scanners 11
    1.2.1   Light sources 12
    1.2.2   Laser beam propagation 12
    1.2.3   Photodetection 14
    1.2.4   Propagation medium and scene effects 14
    1.2.5   Scanning/projection mechanisms 16
  1.3   Basics of airborne laser scanning 19
    1.3.1   Principle of airborne laser scanning 21
    1.3.2   Integration of on-board systems 23
    1.3.3   Global Positioning System/Inertial Measurement
            Unit combination 24
    1.3.4   Laser scanner properties 25
    1.3.5   Pulse repetition frequency and point density 26
    1.3.6   Multiple echoes and full-waveform digitisation 28
    1.3.7   Airborne laser scanner error budget 29
  1.4   Operational aspects of airborne laser scanning 30
    1.4.1   Flight planning 30
    1.4.2   Survey flight 31
    1.4.3   Data processing 32
    1.4.4   Airborne laser scanning and cameras 34
    1.4.5   Advantages and limitations of airborne laser scanning 35
  1.5   Airborne lidar bathymetry 35

1.6   Terrestrial laser scanners 36

Acknowledgements 39

References 39

Chapter 2   Visualisation and Structuring of Point Clouds 45

2.1   Visualisation 46
    2.1.1   Conversion of point clouds to images 46
    2.1.2   Point-based rendering 53

2.2   Data structures 58
    2.2.1   Delaunay triangulation 59
    2.2.2   Octrees 60
    2.2.3   k-D tree 62

2.3   Point cloud segmentation 63
    2.3.1   3D Hough transform 63
    2.3.2   The random sample consensus algorithm 67
    2.3.3   Surface growing 72
    2.3.4   Scan line segmentation 75

2.4   Data compression 75

References 79

Chapter 3   Registration and Calibration 83

3.1   Geometric models 83
    3.1.1   Rotations 84
    3.1.2   The geometry of terrestrial laser scanning 85
    3.1.3   The geometry of airborne laser scanning 86

3.2   Systematic error sources and models 92
    3.2.1   Systematic errors and models of terrestrial laser scanning 92
    3.2.2   Errors and models for airborne laser scanning 100

3.3   Registration 109
    3.3.1   Registration of terrestrial laser scanning data 111
    3.3.2   Registration of airborne laser scanning data 119

3.4   System calibration 122
    3.4.1   Calibration of terrestrial laser scanners 122
    3.4.2   Calibration of airborne laser scanners 125

Summary 129

References 129

Chapter 4   Extraction of Digital Terrain Models 135

4.1   Filtering of point clouds 137
    4.1.1   Morphological filtering 139
    4.1.2   Progressive densification 142
    4.1.3   Surface-based filtering 144
    4.1.4   Segment-based filtering 147

    4.1.5   Filter comparison 150
    4.1.6   Potential of full-waveform information for advanced filtering 151

4.2   Structure line determination 154

4.3   Digital terrain model generation 156
    4.3.1   Digital terrain model determination from terrestrial laser
            scanning data 158
    4.3.2   Digital terrain model quality 158
    4.3.3   Digital terrain model data reduction. 161

References 163

Chapter 5   Building Extraction 169

5.1   Building detection 170

5.2   Outlining of footprints 178

5.3   Building reconstruction 182
    5.3.1   Modelling and alternatives 182
    5.3.2   Geometric modelling 183
    5.3.3   Formal grammars and procedural models. 184
    5.3.4   Building reconstruction systems 186

5.4   Issues in building reconstruction 196
    5.4.1   Topological correctness, regularisation and constraints. 196
    5.4.2   Constraint equations and their solution 197
    5.4.3   Constraints and interactivity. 199
    5.4.4   Structured introduction of constraints and weak primitives 201
    5.4.5   Reconstruction and generalisation 202
    5.4.6   Lessons learnt. 203

5.5   Data exchange and file formats for building models 206

References 207

Chapter 6   Forestry Applications 213

6.1   Introduction 213

6.2   Forest region digital terrain models 216

6.3   Canopy height model and biomass determination 217

6.4   Single-tree detection and modelling. 219

6.5   Waveform digitisation techniques. 222

6.6   Ecosystem analysis applications 225

6.7   Terrestrial laser scanning applications 227
    6.7.1   Literature overview. 228
    6.7.2   Forest inventory applications 229

References 232

Chapter 7   Engineering Applications 237

7.1   Reconstruction of industrial sites 238

7.1.1   Industrial site scanning and modelling 238
7.1.2   Industrial point cloud segmentation 239
7.1.3   Industrial object detection and parameterisation 240
7.1.4   Integrated object parameterisation and registration 242

7.2   **Structural monitoring and change detection** **243**
7.2.1   Change detection 244
7.2.2   Point-wise deformation analysis 247
7.2.3   Object-oriented deformation analysis 254
7.2.4   Outlook 260

7.3   **Corridor mapping** **261**
7.3.1   Power line monitoring 261
7.3.2   Dike and levee inspection 264

7.4   **Conclusions** **265**

References **266**

Chapter 8   Cultural Heritage Applications **271**

8.1   **Accurate site recording: 3D reconstruction of the treasury (Al-Kasneh) in Petra, Jordan** **272**

8.2   **Archaeological site: scanning the pyramids at Giza, Egypt** **275**

8.3   **Archaeological airborne laser scanning in forested areas** **279**

8.4   **Archaeological site: 3D documentation of an archaeological excavation in Mauken, Austria** **282**

8.5   **The archaeological project at the Abbey of Niedermunster, France** **284**

References **289**

Chapter 9   Mobile Mapping **293**

9.1   **Introduction** **293**

9.2   **Mobile mapping observation modes** **295**
9.2.1   Stop-and-go mode 295
9.2.2   On-the-fly mode 296

9.3   **Mobile mapping system design** **298**
9.3.2   Imaging and referencing 298
9.3.3   Indoor applications 301
9.3.4   Communication and control 301
9.3.5   Processing flow 302

9.4   **Application examples** **302**
9.4.1   Railroad-track based systems 302
9.4.2   Road-based systems 305

9.5   **Validation of mobile mapping systems** **308**

Conclusions **308**

References **309**

Index **313**

Preface

Within a time frame of only two decades airborne and terrestrial laser scanning have become well established surveying techniques for the acquisition of (geo)spatial information. A wide variety of instruments is commercially available, and a large number of companies operationally use airborne and terrestrial scanners, accompanied by many dedicated data acquisition, processing and visualisation software packages. The high quality 3D point clouds produced by laser scanners are nowadays routinely used for a diverse array of purposes including the production of digital terrain models and 3D city models, forest management and monitoring, corridor mapping, revamping of industrial installations and documentation of cultural heritage.

However, the publicly accessible knowledge on laser scanning is distributed over a very large number of scientific publications, web pages and tutorials. The objective of this book is to give a comprehensive overview of the principles of airborne and terrestrial laser scanning technology and the state of the art of processing 3D point clouds acquired by laser scanners for a wide range of applications. It will serve as a textbook for use in under- and post-graduate studies as well as a reference guide for practitioners providing laser scanning services or users processing laser scanner data.

Airborne and terrestrial laser scanning clearly differ in terms of data capture modes, typical project sizes, scanning mechanisms, and obtainable accuracy and resolution. Yet, they also share many features, especially those resulting from the laser ranging technology. In particular, when it comes to the processing of point clouds, often the same algorithms are applied to airborne as well as terrestrial laser scanning data. In this book we therefore present, as far as possible, an integral treatment of airborne and terrestrial laser scanning technology and processing.

The book starts with an introduction to the technology of airborne and terrestrial laser scanning covering topics such as range measurement principles, scanning mechanisms, GPS/IMU integration, full-waveform digitisation, error budgets as well as operational aspects of laser scanning surveys. The chapter focuses on principles rather than technical specifications of current laser scanners as the latter are rapidly outdated with ongoing technological developments.

Common to all laser scanning projects are the needs to visualise and structure the acquired 3D point clouds as well as to obtain a proper georeferencing of the data. Chapter 2 discusses techniques to visualise both original point clouds and rasterised data. Visualisation is shown to be an important tool for quality assessment. Information extraction from point clouds often starts with the grouping of points into smooth or planar surfaces of the recorded objects. The most common segmentation algorithms as well as data structures suitable for point clouds are also presented in this chapter.

Chapter 3 deals with the registration of multiple datasets and the calibration of airborne and terrestrial laser scanners. Models are elaborated that show the relation between the observations made by laser scanners and the resulting coordinates of the reflecting surface points. Based on these geometric models, the error sources of typical instrument designs are discussed. Registration of point clouds is the process of transforming a dataset to an externally defined coordinate system. For airborne surveys this normally is the national mapping coordinate system. For terrestrial laser scanning, point clouds acquired at different scan positions have to be registered with respect to each other. This chapter elaborates upon the coordinate systems involved as well as the registration methods. The chapter concludes with a discussion of calibration procedures that estimate and correct for potential systematic errors in the data acquisition.

The applications oriented part of the book starts in Chapter 4 with the extraction of digital terrain models (DTM) from airborne laser scanning data. The production of high quality DTMs has been the major driving force behind the development of airborne laser scanners. Compared to other surveying technologies, airborne laser scanning enables DTM production with higher quality at lower costs. As a consequence, in the few years since its introduction airborne laser scanning has become the preferred technology for the acquisition of DTMs. Point clouds resulting from laser scanning surveys contain points both on the terrain and also on vegetation, buildings and other objects. Whereas these points are a rich source of information for a variety of applications, non-ground points need to be removed from the point clouds for the purpose of DTM production. Chapter 4 reviews the most popular strategies and algorithms that can be applied to this so-called filtering process. In addition it also discusses the extraction of break lines, reduction of point densities and aspects of DTM quality.

With increasing sensor resolution, point clouds acquired by airborne laser scanners are nowadays dense enough to also retrieve detailed information on buildings and trees. The detection of buildings and the extraction of 3D building models in airborne laser data are the subject of Chapter 5. Detection of buildings typically makes use of the separation in ground and non-ground points by filtering algorithms, but needs to further classify the non-ground points. Following the detection of buildings, algorithms for deriving their 2D outlines are reviewed. Building outlines may be used to update traditional 2D building maps or serve as an intermediate step for a complete 3D building reconstruction procedure. Chapter 5 discusses several such procedures as well as the ongoing research questions in this field with regard to regularisation, constraints, interaction and generalisation. The chapter concludes with an overview on data exchange formats for building models.

Chapter 6 is devoted to forestry applications, not only using airborne laser scanning data, but also including analyses of terrestrially acquired data. The unique ability of airborne laser scanning to obtain both points on vegetation as well as on the ground led to the rapid acceptance of laser scanning for the purpose of forest inventory studies, forest management, carbon sink analysis, biodiversity characterisation and habitat analysis. This chapter presents an overview of techniques and studies on stand-wise as well as tree-wise monitoring of forests. Whereas stand-wise monitoring provides average values for forest inventory parameters such as tree height or timber volume for larger areas, high resolution airborne laser scanning surveys can also be used to delineate and model individual trees. Terrestrial laser scanning is exploited to obtain detailed geometric models of selected plots.

The last three chapters present applications that mainly make use of terrestrial laser scanners. A wide variety of engineering applications is presented in Chapter 7. In industry 3D CAD models of installations are required for maintenance management, safety analysis and revamping. Point clouds acquired by terrestrial laser scanners directly capture the shapes of these installations and allow an efficient modelling process. Change detection and deformation analysis play an important role in civil and geotechnical engineering applications. Terrestrial laser scanning is shown to be of considerable value for projects such as the monitoring of dams, tunnels and areas susceptible to land erosion or land slides, whereas airborne laser scanning depicts a very efficient tool for monitoring power lines and water embankments.

Applications of laser scanning to the documentation of cultural heritage are discussed in some case studies in Chapter 8. Historical buildings and sculptures are often difficult to model. The ability of terrestrial laser scanners to rapidly capture complex surfaces makes laser scanning a preferred technology for documenting cultural heritage. Terrestrial as well as airborne laser scanning have also proven their value in archaeological studies. Whereas terrestrial laser scanning is used to document excavation sites, airborne laser scanning has been used successfully to uncover land use patterns in forested areas.

Finally, Chapter 9 presents a review on vehicle-borne mobile mapping systems. The chapter discusses various modes of observation (stop-and-go, on-the-fly), design considerations and data processing flows. Present-day systems are shown together with their application to corridor mapping of road and rail environments.

The obvious need for a textbook on laser scanning has been on our minds for several years. For the realisation of this book we are extremely grateful that we could enlist renowned scientists to share their knowledge and invest their precious time in contributing dedicated chapters.

We would also like to express our gratitude to Keith Whittles, who gave us a push to accept the challenge of editing this book. Sincere thanks also go to the staff of Whittles Publishing for their professional handling of the manuscript and marketing efforts.

We hope that this book may serve many students, researchers and practitioners for their work in the exciting field of airborne and terrestrial laser scanning and 3D point cloud processing.

George Vosselman
International Institute for Geo-Information Science
and Earth Observation (ITC)

Hans-Gerd Maas
Dresden University of Technology

The Authors

Jean-Angelo Beraldin has been Senior Researcher with the National Research Council (NRC), Canada since 1986. He obtained his BEng and MASc (Electrical Engineering) in 1984 and 1986, respectively. His research activities in digital 3D imaging have produced technologies that have been licensed and are now being commercialised in various sectors, including automotive, space, cultural heritage and entertainment. From 2004 to 2009 Angelo led a research initiative aimed at the development of cultural heritage applications of 3D imaging and modelling technologies. Since 2009 Angelo has been leading the research activities in 3D imaging and modelling metrology at the NRC, Canada.

François Blais is Principal Research Officer and Group Leader at the National Research Council (NRC), Canada. He obtained his BSc and MSc in Electrical Engineering from Laval University, Quebec City. Since 1984 his research has resulted in the development of a number of innovative 3D sensing technologies licensed to various industries for various applications, including the in-orbit 3D laser inspection of NASA's space shuttles as well as the scanning of important sites and objects of art, including the *Mona Lisa* by Leonardo da Vinci. In September 2005 he was recognised among the top 10 Canadian Innovators by *The Globe and Mail*, an important newspaper in Canada. During his career he has received several awards of excellence for his work.

Claus Brenner is with the Institute of Cartography and Geoinformatics at the Leibniz Universität Hannover, Germany where he leads a junior research group on automatic methods for the fusion, reduction and consistent combination of complex heterogeneous geoinformation. His current research interests are automated object extraction from images and laser scanning data, modelling and generalisation, as well as navigation and driver assistance systems.

Christian Briese studied Geodetic Engineering at the Vienna University of Technology, Austria. He received his Diploma (2000) and PhD (2004) with honours at the

Institute of Photogrammetry and Remote Sensing. His main research interests are laser scanning, covering the whole data acquisition process up to the generation of final geometric models.

Klaus Hanke graduated in Surveying Engineering from the University of Technology in Graz, Austria, and obtained his PhD and habilitation from the University of Innsbruck, Austria. He has been a professor at the Surveying and Geoinformation Unit of the University of Innsbruck since 1997. Together with Pierre Grussenmeyer he chaired an ISPRS working group on Cultural Heritage Documentation 2004–2008. He has been Vice-President of CIPA, the ICOMOS International Scientific Committee on Heritage Documentation, since 2006.

Reinhard Klein received his MSc in Mathematics and his PhD in Computer Science from the University of Tübingen, Germany. In September 1999 he became an associate professor at the University of Darmstadt, Germany and head of the Animation and Image Communication research group at the Fraunhofer Institute for Computer Graphics. He has been a professor at the University of Bonn and Director of the Institute of Computer Science II since October 2005.

Hansjörg Kutterer received a Diploma in Geodesy from the University of Karlsruhe, Germany in 1990, a PhD in 1993 and habilitation in 2001. From 1990 to 2000 he worked at the Geodetic Institute of the University of Karlsruhe as a research associate. From 2000 to 2003 he was a senior scientist at Deutsches Geodätisches Forschungsinstitut (DGFI) in Munich. Since 2004 he has been a full professor of Geodetic Engineering and Data Analysis at the Leibniz Universität Hannover, Germany. Hansjörg's current fields of scientific interest are geodetic multi-sensor systems, geodetic monitoring, terrestrial laser scanning, mobile mapping, and data analysis with emphasis on interval and fuzzy techniques. He is a regular member of the German Geodetic Commission (DGK), Fellow of the International Association of Geodesy (IAG), and a member of the editorial board of the *Journal of Applied Geodesy*.

Derek Lichti received his Bachelor's degree in Survey Engineering from Ryerson University, Canada and his MSc and PhD from the University of Calgary, Canada. From 1999 to 2007 he was a faculty member at Curtin University of Technology in Perth, Australia. He joined the Department of Geomatics Engineering at the University of Calgary in 2008, where he is currently Associate Professor. He was the 2004–2008 Chair of the ISPRS (International Society for Photogrammetry and Remote Sensing) Working Group on Terrestrial Laser Scanning and is the 2008–2012 Chair of the Working Group on Terrestrial Laser Scanning and 3D Imaging.

Roderik Lindenbergh studied Mathematics at the University of Amsterdam, the Netherlands. He obtained a PhD in Mathematics from the University of Utrecht, the

Netherlands for his work on Voronoi diagrams. After his PhD he joined Delft University of Technology to work in the field of Mathematical Geodesy and Positioning, where he is now employed as an assistant professor. His research interests include deformation analysis, spatiotemporal interpolation and quality control of large spatial data sets. Roderik works on the integration of GPS and MERIS (Medium Resolution Imaging Spectrometer) water vapour data and is involved in the analysis of ICES at full waveform laser altimetry data and in the quality aspects of terrestrial laser scanning.

Uwe Lohr holds a PhD in Physics from the Technical University of Braunschweig, Germany. His experience in airborne laser scanning started in 1995 as CEO and sales manager of TopoSys® GmbH, Germany. For the last three years he has been responsible for international business development and sales with ASTEC GmbH, Germany.

Hans-Gerd Maas graduated in Geodesy at the University of Bonn, Germany, and obtained his PhD and habilitation in Photogrammetry from ETH Zurich, Switzerland. He was associate professor for Remote Sensing at Delft University of Technology, the Netherlands from 1998 to 2000. Since 2000 he has held the post of full professor of Photogrammetry at Dresden University of Technology, Germany. He has chaired several ISPRS (International Society for Photogrammetry and Remote Sensing) working groups dealing with laser scanning and was ISPRS Technical Commission V President 2004–2008.

George Vosselman graduated in Geodetic Engineering from the Delft University of Technology, the Netherlands and obtained his PhD from the University of Bonn, Germany. He was appointed full professor of Photogrammetry and Remote Sensing at the Delft University of Technology in 1993. Since 2004 he has been full professor of Geo-Information Extraction with Sensor Systems at the International Institute for Geo-Information Science and Earth Observation (ITC), Enschede, the Netherlands. He has been chairing ISPRS working groups on laser scanning since 2000. He has been Editor-in-Chief of the *ISPRS Journal of Photogrammetry and Remote Sensing*, since 2005.

List of Abbreviations

AM	amplitude modulation
APD	avalanche photodiode
API	application programming interface
B-rep	boundary representation
CHM	canopy height model
CIR	colour infrared
CSG	constructive solid geometry
CW	continuous wave
DBA	diameter at breast height
DCM	digital canopy model
DEM	digital elevation model
DGPS	differential global positioning system
DSM	digital surface model
DTM	digital terrain model
ECEF	Earth-centred, Earth-fixed
EDM	electronic distance measurement
ENU	east-, north- and up-axes
ET	empirical terms
FM	frequency modulation
FOV	field of view
FWF	full-waveform
GBSAR	ground-based synthetic radar aperture
GIS	geographic information system
GNSS	global navigation satellite systems
GPS	global positioning system
HNF	Hesse normal function
ICP	iterative closest point

IGS	International GPS Service
IMU	inertial measurement unit
INS	inertial navigation system
InSAr	interferometric synthetic aperture radar
ITRF	international terrestrial reference frame
LAI	leaf area index
lidar	light detection and ranging
LOD	level of detail
LULUCF	land use, land-use change and forestry
LVIS	laser vegetation imaging sensor
MDL	minimum description length
MER	minimum enclosing rectangle
MMS	mobile mapping system
MSAC	M-estimator sample consensus
nDSM	normalised digital surface model
NDVI	normalised digital vegetative index
NED	north-, east- and down-axes
OGC	Open Geospatial Consortium
PAI	plant area index
PN	pseudo-noise
PPP	precise point positioning
PPS	pulse per second
PRF	pulse repetition frequency
RANSAC	random sample consensus algorithm
RFID	radiofrequency identification
R-G-B	red-green-blue
RJMCMC	reversible jump Markov chain Monte Carlo
RMSE	root mean square error
RTK	real-time kinematic
SAR	synthetic aperture radar
SBET	smoothed best estimated trajectory
SLAM	simultaneous localisation and mapping
SNR	signal-to-noise ratio
SVD	singular value decomposition
sw	swath width
TIN	triangulated irregular network
UWB	ultra wide band
VLMS	vehicle-borne laser mapping system
WAI	woody area index

Laser Scanning Technology | 1

J.-Angelo Beraldin, François Blais and Uwe Lohr

In the last 50 years, many advances in the fields of solid-state electronics, photonics, and computer vision and graphics have made it possible to construct reliable, high resolution and accurate terrestrial and airborne laser scanners. Furthermore, the possibility of processing dense point clouds in an efficient and cost-effective way has opened up a multitude of applications in fields such as topographic, environmental, industrial and cultural heritage 3D data acquisition.

Airborne and terrestrial laser scanners capture and record the geometry and sometimes textural information of visible surfaces of objects and sites. These systems are by their nature non-contact measurement instruments and produce a quantitative 3D digital representation (e.g. point cloud or range map) of a surface in a given field of view with a certain measurement uncertainty. Full-field optical 3D measurement systems in general can be divided into several categories (Figure 1.1). Airborne and terrestrial laser scanners are usually part of what are classified as time-of-flight based optical 3D measurement systems. These systems use a laser source to scan a surface in order to acquire dense range data. Triangulation systems using light sheet or strip projection techniques and passive systems exploiting surface texture (stereo image processing) will be covered only briefly here.

This chapter reviews the key elements that compose airborne and terrestrial laser scanners from basic physical principles to operational quality parameters. The chapter is divided into the following sections: Section 1.1 gives an introduction to the basic principles of time-of-flight measurement and optical triangulation. A description of the elements that are common to most laser scanning systems is given in Section 1.2. Airborne laser scanner system components are reviewed in Section 1.3, followed by operational aspects for use of airborne laser scanners in Section 1.4. The principle of airborne lidar (light detection and ranging) bathymetry is briefly outlined in Chapter 1.5. Some principles and examples of terrestrial laser scanners are presented in Section 1.6. As market evolution has a great impact on the number, type and performance parameters

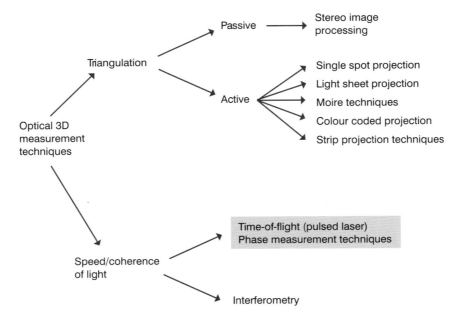

Figure 1.1 Classification of optical 3D measurement systems.

of laser scanners commercially available at a given time, no effort has been made to list and compare commercial systems in detail. Some recent publications survey commercial systems [Blais, 2004; Fröhlich and Mettenleiter, 2004; Pfeifer and Briese, 2007; Lemmens, 2009].

1.1 Basic measurement principles of laser scanners

There are two basic active methods for optically measuring a 3D surface: light transit time estimation and triangulation. As illustrated in Figure 1.2(a), light waves travel with a known velocity in a given medium. Thus, the measurement of the time delay created by light travelling from a source to a reflective target surface and back to a light detector offers a very convenient method of evaluating distance. Such systems are also known as time-of-flight or lidar (light detection and ranging) systems. Time-of-flight measurement may also be realised indirectly via phase measurement in continuous wave (CW) lasers. Triangulation exploits the cosine law by constructing a triangle using an illumination direction (angle) aimed at a reflective surface and an observation direction (angle) at a known distance (base distance or baseline) from the illumination source (Figure 1.2b). Interferometry (which is not covered here) can be classified separately as a third method or included with time-of-flight methods depending on how the metric used to measure shape is seen [Seitz, 2007].

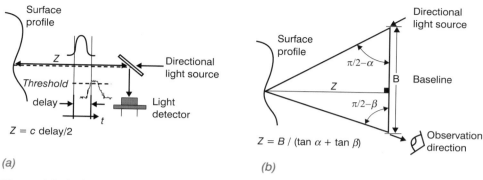

Figure 1.2 Active methods for optically measuring a 3D surface: (a) light transit time; and (b) triangulation. (NRC Crown copyright)

1.1.1 Time-of-flight measurement

Early work on time-of-flight ranging systems used radio waves in what became known as radar (radio detection and ranging) [e.g. Skolnik, 1970]. The fundamental work can be traced back to experiments conducted by Hertz in the late 1880s. With the advent of lasers in the late 1950s it became possible to image a surface with angular and range resolutions much higher than that obtained with radio waves. The theoretical principles are the same for all systems that use radiated electromagnetic energy for ranging except for their implementations, the differences in their performance and obviously their use.

A fundamental property of a light wave is its propagation velocity. In a given medium, light waves travel with a finite and constant velocity. Thus, the measurement of time delays (also known as time-of-flight) created by light travelling in a medium from a source to a reflective target surface and back to the source (round trip, τ) offers a very convenient way to evaluate the range ρ

$$\rho = \frac{c}{n}\frac{\tau}{2} \tag{1.1}$$

The current accepted value for the speed of light in a vacuum is $c = 299\ 792\ 458$ m/s. If the light waves travel in air then a correction factor equal to the refractive index, which depends on the air temperature, pressure and humidity, must be applied to c, $n \approx 1.00025$. Let us assume $c = 3 \times 10^8$ m/s and $n = 1$ in the rest of the discussion.

More than one pulse echo can be measured due to multiple returns that are caused by the site characteristics, especially when vegetation (the canopy) is scanned [Hofton *et al.*, 2000; Wagner *et al.*, 2006; Pfeifer and Briese, 2007]. Airborne systems capture at least the first returned pulse or echo and also the last echo for each emitted pulse. Most airborne systems are capable of capturing four to five separate echoes. Multiple echo measurements are also becoming available for terrestrial time-of-flight scanners.

It is important to have a more detailed look at pulse shape and pulse repetition time. The characteristics of a transmitted pulse are the pulse width t_p and the pulse rise time t_r (Figure 1.3). A typical pulse width of $t_p = 5$ ns corresponds to a length of about 1.5 m at the speed of light, while a pulse rise time $t_r = 1$ ns corresponds to a length of 0.3 m.

According to Equation (1.1), the relationship between ρ (the distance between the laser scanner and the illuminated spot) and time-of-flight τ is given by

$$\tau = n\frac{2\rho}{c}$$

With $n = 1$, the time-of-flight is 6.7 µs at a distance (the survey height in an airborne system) of 1000 m. When assuming that from a distance ρ two or more echoes are generated from only one pulse P_1, different ranges can only be discriminated if the echoes E_{11} and E_{12} are separated, i.e. do not have an overlap. This means (see also Figure 1.3)

$$\tau^{12} \geq \tau^{11} + t_p$$

With the above relation between time-of-flight and distance and the pulse length l_p, which is given by $t_p = \frac{n}{c}l_p$, this leads to

$$2\frac{n}{c}\rho^{12} - 2\frac{n}{c}\rho^{11} \geq \frac{n}{c}l_p$$
$$\rho^{12} - \rho^{11} \geq \frac{1}{2}l_p \tag{1.2}$$

Two echoes (for example related to objects heights) can only be discriminated if their distance is larger than half of the pulse length l_p. This means that for a pulse

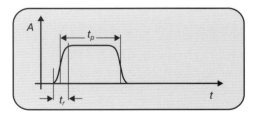

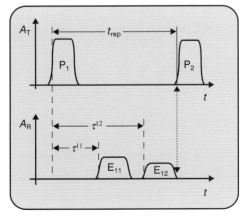

t_p : pulse width
t_r : pulse rise time
t_{rep} : pulse repetition time
τ : time of flight

Figure 1.3 Pulse characteristics and measurement principle.

width of 5 ns objects can be detected as separate objects if their distance is at least larger than 0.75 m.

An essential part of the time-of-flight measurement is the detection method for determining the time-of-flight (and thus the range). The detector will generate a time-tagged trigger pulse depending on the implemented criterion. Some detection methods take characteristic points of the path of the pulse as the decisive factor.

- **Peak detection:** The detector generates a trigger pulse at the maximum (amplitude) of the echo. Time-of-flight is the time delay given by the time span from the maximum of the emitted pulse to the maximum of the echo. Correct detection can become problematic if the echo provides more than one peak.

- **Threshold or leading edge detection:** Here the trigger pulse is actuated when the rising edge of the echo exceeds a predefined threshold. The disadvantage of this method is that the time-of-flight strongly depends on the echo's amplitude.

- **Constant fraction detection:** This method produces a trigger pulse at the time an echo reaches a preset fraction (typically 50%) of its maximum amplitude. The advantage of this method is that it is relatively independent of an echo's amplitude.

Each detector has its pros and cons, however constant fraction has proven to be a good compromise [Wagner *et al.*, 2004; Jutzi, 2007].

The range uncertainty for a single pulse buried in additive white noise is approximately given by

$$\delta_{r-p} \approx \frac{c}{2} \frac{t_r}{\sqrt{SNR}} \tag{1.3}$$

where t_r is the rise time of the laser pulse leading edge and the SNR (signal-to-noise ratio) is the power ratio of signal over noise. Assuming $SNR = 100$, $t_r = 1$ ns, and a time interval counter with adequate resolution then the range uncertainty will be about 15 mm. Most commercial time-of-flight based laser scanner systems provide a range uncertainty in the order of 5–10 mm (with a few at 50 mm) as long as a high signal-to-noise ratio is maintained. In the case of uncorrelated 3D samples, averaging N independent pulse measurements will reduce δ_{r-p} by a factor proportional to the square root of N. Obviously this technique reduces the data rate by N and is of limited applicability in scanning modes.

1.1.2 Phase measurement techniques

Besides using repetitive short laser pulses as described above, time-of-flight may also be realised by amplitude modulation (AM) using phase difference, frequency modulation (FM) exploiting beat frequencies, phase-coded compression, which includes binary sequences, linear recursive sequences and pseudo-noise (PN) sequences, and

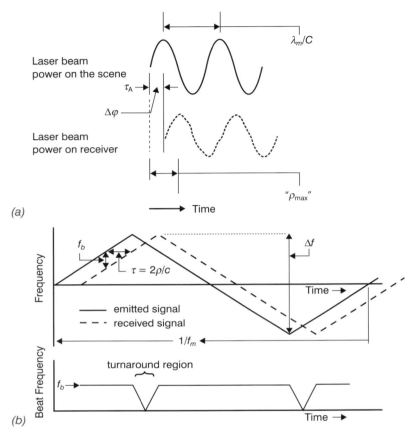

Figure 1.4 Measurement principle of systems based on CW modulation: (a) AM using the phase difference; and (b) FM exploiting beat frequencies. (NRC Crown copyright)

chaotic lidar based on chaotic waveforms generated by a semiconductor laser [Bergstrand, 1960; Takeuchi *et al.*, 1983; Wehr and Lohr, 1999; Amann *et al.*, 2001; Lin and Liu, 2004].

Continuous wave (CW) modulation avoids the measurement of short pulses by modulating the power or the wavelength of the laser beam. For AM, the intensity of the laser beam is amplitude modulated, e.g. a sine wave. The projected (emitted) laser beam and collected laser light are compared. The phase difference ($\Delta\varphi$) between the two waveforms yields the time delay ($\tau = \Delta\varphi/2\pi \times \lambda_m/c$) and the range is found using Equation (1.1). Figure 1.4(a) shows an illustration of the principle.

The range uncertainty is given approximately by

$$\delta_{r-AM} \approx \frac{1}{4\pi}\frac{\lambda_m}{\sqrt{SNR}} \tag{1.4}$$

where λ_m = wavelength of the amplitude modulation (or c/f_m). A low frequency f_m, makes the phase detection less precise. Because the returned wave cannot be associated with a specific part of the original signal, it is not possible to derive the absolute distance information, as multiples of the wavelength cannot be resolved by the pure phase measurement. This is known as the ambiguity interval and is given by

$$\rho_{\max-AM} = \frac{c}{2}\frac{1}{f_m} = \frac{\lambda_m}{2} \tag{1.5}$$

To get around the range ambiguity interval, multiple frequency waveforms can be used. For instance, assuming a two-tone AM system (10 MHz and 150 MHz) and a $SNR = 1000$, the range uncertainty is about 5 mm and the ambiguity is 15 m. Note that Equation (1.4) is applied to the highest frequency and Equation (1.5) to the lowest frequency. Typical commercial systems can scan a scene at data rates ranging from 10 kHz to about 625 kHz and offer a maximum operating range of about 100 m. While this is sufficient for many terrestrial applications, airborne laser scanners rarely use phase measurement techniques due to their range limitation. Some optimised AM systems can achieve sub-100 μm uncertainty values but over very shallow depths of field.

The second most popular continuous wave system covered in this section is based on frequency modulation (FM) with either coherent or direct detection [e.g. Bosch and Lescure, 1995; Amann *et al.*, 2001; Ruff *et al.*, 2005]. In one implementation, the frequency (wavelength) of the laser beam is linearly modulated either directly at the laser diode or with an acousto-optic modulator. The linear modulation is usually shaped as a triangular or sawtooth wave (chirp). The chirp duration, T_m *(1/f_m)*, can last several milliseconds. The important aspects of this method are determined by the coherent detection taking place at the optical detector and the fact that the beat frequency (f_b) resulting from this optical mixing encodes the round trip time delay as $4\rho f_m \Delta f / c$ (for a stationary target object; for a moving object see Schneider *et al.*, 2001). Figure 1.4(b) shows an illustration of the principle. In general, the range uncertainty is given approximately by

$$\delta_{r-FM} \approx \frac{\sqrt{3}}{2\pi}\frac{c}{\Delta f}\frac{1}{\sqrt{SNR}} \tag{1.6}$$

where Δf is the tuning range or frequency excursion [Skolnik, 1980; Stone *et al.*, 2004]. The ambiguity range depends on the chirp duration and is given by

$$\rho_{\max-FM} = \frac{c}{4}T_m \tag{1.7}$$

With coherent detection, assuming a tuning range of 100 GHz, a chirp time of 1 ms, $\sqrt{SNR} = 400$ and a low data rate, a system can achieve a theoretical measurement uncertainty of about 2 μm on a cooperative (that yields the best SNR) surface located at 10 m [Stone *et al.*, 2004]. Furthermore, because of the coherent detection, the dynamic range is typically 10^9. In practice, a typical commercial system can provide a measurement uncertainty of about 30 μm at a data rate of 40 points/s and 300 μm at about 250 points/s [Guidi *et al.*, 2005]. In direct detection (heterodyne), the mixing takes place in the electronic circuits [Dupuy and Lescure, 2002]. Stable and linear frequency modulation of the laser source is highly critical for 3D laser scanners based on time-of-flight systems and in particular for systems based on FM modulation [Amann *et al.*, 2001].

Generally, amplitude modulated systems tend to have very high data rates (up to 1 million 3D points per second) but shorter operating ranges (less than 100 m). Pulse-based time-of-flight systems on the other hand are characterised by a much longer range (e.g. 800 m) but lower data rates (usually < 50 000 points/s in terrestrial systems and < 200 000 points/s in airborne systems).

1.1.3 Triangulation-based measurements

While mid- and long-range laser scanners are usually based on time-of-flight technology, systems designed for measuring distances smaller than about 5 m often use the triangulation principle. Triangles are the basis for many measurement techniques, from basic geodetic measurements performed in the Ancient World to more sophisticated land surveying using theodolites. Though the Greek philosopher Thales (sixth century BC) has been credited with the discovery of five theorems in geometry, it is Gemma Frisius, who lived in the sixteenth century, who presented the mathematics of locating the coordinates of a remote point using this method [Singer *et al.*, 1954]. One of the earliest papers on triangulation-based spot scanning for the capture and recording of 3D data was published by Forsen in 1968. Kanade (1987) presents a collection of chapters from different authors which describe a number of active close-range 3D vision systems.

The basic geometric principle of optical triangulation for a single static laser beam is shown in Figure 1.5. A laser source projects a beam of light on a surface of interest. The scattered light from that surface is collected from a vantage point distinct from the projected light beam. This light is focused (imaged) onto a position-sensitive detector. The knowledge of both projection and collection angles (α and β) relative to a baseline (B) determines the dimensions of a triangle (cosine law) and hence the (X, Z) coordinate of a point on a surface.

The angle $\Delta\beta$ which corresponds to a displacement of a surface profile by Δz is measured by an equivalent shift on a position-sensitive detector that acts as an angle sensor and provides signals that are interpreted as a position p. Many laser spot position detectors have been proposed in the past for 3D imaging [e.g. Beraldin *et al.*, 2003]. By simple trigonometry, the (X, Z) coordinate of the laser spot on the object is

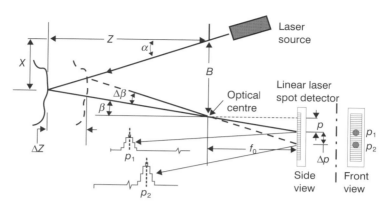

Figure 1.5 Single point optical triangulation based on a single laser beam and a position-sensitive detector. (NRC Crown copyright)

$$X = Z \tan(\alpha) \tag{1.8}$$

and

$$Z = \frac{B f_0}{p + f_0 \tan(\alpha)} \tag{1.9}$$

where p is the position of the imaged spot on the spot position detector and f_0 is the effective distance between the laser spot detector and the collection lens (simplified as a pin hole camera model).

With reference to Figure 1.5, the errors with a triangulation-based laser scanner come mainly from the estimate of the laser spot position p, through the uncertainty of laser spot position δ_p. Application of the error propagation principle to the range equation (Equation 1.9) gives the uncertainty in Z

$$\delta_z \approx \frac{Z^2}{f B} \delta_p \tag{1.10}$$

where Z is the distance to the surface, f is the effective focal distance and length B is the triangulation baseline. The value of δ_p depends on the type of laser spot sensor used, the peak detector algorithm, the signal-to-noise ratio (*SNR*: power ratio of signal over noise) and the imaged laser spot shape. In the case of discrete response laser spot sensors and assuming both a high *SNR* and a centroid-based method for peak detection, the limiting factor will be speckle noise [Baribeau and Rioux, 1991; Dorsch *et al.*, 1994]. The spot location uncertainty is approximately given by

$$\delta_p \approx \frac{1}{\pi \sqrt{2}} \lambda \, fn \tag{1.11}$$

where *fn* is the receiving lens f-number (i.e. f/Φ, where Φ is the aperture diameter) and λ is the laser wavelength. Substitution of Equation (1.11) back into Equation (1.10) shows that δ_z does not depend on the effective focal distance f.

$$\delta_z \approx \frac{\lambda}{\pi\sqrt{2}} \frac{Z^2}{B\Phi} \tag{1.12}$$

In fact, a low uncertainty is achieved by the use of a short laser wavelength, e.g. blue laser as opposed to a near-infrared laser. Further, the last term in the equation above can be decomposed into two terms, i.e. a term for the baseline–distance ratio B/Z (related to the triangulation angle) and the observation angle, Φ/Z. In other words, when these are large, δ_z will be small. A large lens diameter will limit the focusing range. Many triangulation-based 3D cameras will use the Scheimpflug condition when mounting the position sensor with the collecting lens [Blais, 2004]. Figure 1.6(a) shows an optical geometry based on the Scheimpflug condition (note the angle on the position sensor with respect to the collecting lens). This photodetector orientation ensures that all the points on the laser beam will be in focus on the detector. Speckle noise affects time-of-flight systems differently than it does triangulation systems. The main effect is seen in the intensity image with less in the 3D image, especially when the *SNR* is high [Cramer and Peterson, 1996].

Range measurement accuracy also depends on the *SNR* in laser spot detection, which is inversely proportional to the square of the scanner–surface distance. Therefore, for triangulation-based systems the range uncertainty will in theory depend on the fourth power of the range (range equation geometry and *SNR* depend on the

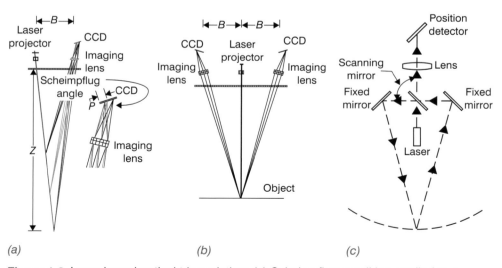

(a)   (b)   (c)

Figure 1.6 Laser-based optical triangulation: (a) Scheimpflug condition applied to a photodetector; (b) dual-view triangulation probe; and (c) autosynchronised scanning. (NRC Crown copyright)

distance squared), which make them poor candidates for long-range applications! Interestingly, the range between 2 m and 10 m represents a transition between triangulation and time-delay based systems, while triangulation systems very clearly dominate in the sub-metre range. Several phase-based (AM) systems are now covering that zone [Fröhlich and Mettenleiter, 2004]. Triangulation-based systems are increasingly confined to applications, e.g. industrial components inspection, where operating distances are less than 1 m and the instantaneous fields of view are small, e.g. 40° × 40°. Larger scan areas are sometimes accomplished with ancillary systems such as articulated arms, laser trackers or photogrammetric techniques.

The scanning of a laser spot can be achieved by a suitable combination of mirrors, lenses and galvanometers as described in Section 1.2.5. Redundancy can improve the quality and the completeness of the 3D data (Figure 1.6b). An actual implementation of a line scanner based on a galvanometer is shown schematically on Figure 1.6(c). The diagram demonstrates that both the projection and the collection of the laser light can be synchronised, i.e. autosynchronised scanning [Rioux *et al.*, 1984].

1.2 Components of laser scanners

The single point spot distance measuring devices discussed in Section 1.1 need to be coupled to optomechanical, solid-state scanners or, more recently, integrated into semiconductor devices to capture and record surfaces of objects and sites. The different implementations of these systems share a similar structure where the building blocks are part of what is called the detection chain of the system. This structure is shown schematically on Figure 1.7. The following subsections will describe the different elements.

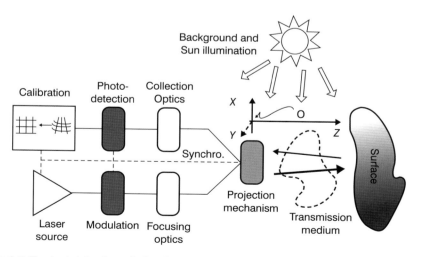

Figure 1.7 Typical detection chain of a laser scanning system. (NRC Crown copyright)

1.2.1 Light sources

Two types of light sources have been used in 3D imaging: incoherent sources of light (incandescent, luminescent or sunlight) and lasers (gas, solid-state or semiconductor). In general, the light produced by lasers (light amplification by stimulated emission of radiation) is far more monochromatic, directional, bright and spatially coherent than that from any other light sources. Spatial coherence allows the laser beam to stay in focus when projected on a surface. The coherence length is given by the speed of light *c* divided by the spectral width of the laser source (a function of its spectral purity). Unfortunately, the high spatial coherence means that speckle is produced when a rough surface is illuminated with a coherent laser source. Speckle appears as a granular pattern attached to the surface of the object and can be considered a source of multiplicative noise. Linear filtering techniques based on temporal operators will not reduce speckle-induced noise. Compact lasers for 3D measurement systems are now commercially available with a choice of a multitude of wavelengths in the range of 0.4 to 1.6 µm.

Many applications require eye safety and hence it is necessary to calculate the maximum permissible exposure, which varies with wavelength and operating conditions. The maximum permissible exposure is defined as the level of laser radiation to which a person may be exposed without hazardous effect or adverse biological changes to the eye or skin [ANSI Z136.1-2007]. This standard also contains many provisions for the safe use of laser systems and for the preparation of the environment surrounding laser systems for each of the four laser classes. For example, fractions of a milliwatt in the visible spectrum are sufficient to cause eye damage, since the laser light entering the pupil is focused at the retina through the lens. At longer wavelengths, e.g. 1.55 µm, however, strong absorption by water present in the eye serves to reduce the optical power reaching the retina, thus increasing the damage threshold.

1.2.2 Laser beam propagation

In Section 1.1, laser beams were represented by a ray with no physical dimensions. Though very convenient for explaining basic principles, it is far from the real situation in laser scanners. The manipulation and use of laser beams with a Gaussian shaped transversal profile is fundamental in understanding many optical systems and their optical limitations. Known as Gaussian beams, they are the solutions of the wave equation derived from Maxwell's equations in free space under certain conditions. Because of diffraction, and even in the best laser-emitting conditions, these laser beams do not maintain collimation with distance. Using the Gaussian beam propagation formula, it is found that the dependence of the beam radius $w(z)$ at the e^{-2} irradiance contour (13.5% of peak intensity) in the direction of propagation z is governed by

$$w(z) = \omega_0 \left[1 + \left(\frac{\lambda(z - z_0)}{n \pi \omega_0^2} \right)^2 \right]^{1/2} \tag{1.13}$$

where the distance z is taken from the focusing lens to the beam waist location at its minimum size (radius) ω_o, λ is the wavelength of the laser source and n is the refractive index of the propagation medium [Born and Wolf, 1999]. One can relate the RMS width of the intensity distribution to the beam radius, i.e. $\sigma(z) = w(z)/2$. In accordance with the Rayleigh criterion which is the usual convention for beam propagation (when the beam radius increases by a factor of $\sqrt{2}$), the depth of focus (or depth of field) for a Gaussian beam in the focused case (see Figure 1.8) is

$$D_f = \frac{2\pi n}{\lambda} \omega_0^2 \qquad (1.14)$$

For both collimated, d_o (see Figure 1.9), and focused cases, ω_o (see Figure 1.8), the far-field beam divergence is given by

$$\theta = \frac{\lambda}{\pi n \, \omega_0} \qquad (1.15)$$

As an example, assume $\lambda = 532$ nm (green laser), $n = 1$ (vacuum); $\omega_o = 1$ mm ($d_o/2$) will yield an ideal divergence of about 0.17 mrad, corresponding to a laser spot diameter (at the e^{-2} contour) of 34 cm at a distance of 1000 m. To summarise, diffraction will limit the lateral (x-y plane) spatial resolution. This laser beam once it intersects with a surface will produce a spot diameter with a finite dimension (laser footprint) that will vary according to the local surface slope and the material characteristics.

It is important to note that the equations presented in this section apply to both time-of-flight and triangulation flying spot laser scanners. Geometrical optics can be used for analysing a laser scanner but the Gaussian beam assumption must be used to complete that analysis [Self, 1983]. Finally, the spatial resolution will depend on the beam quality and characteristics. Even with the high sampling densities (over-sampling) achievable with optical scanning mechanisms, the resolution will ultimately be limited by diffraction. In the case of non-Gaussian beam propagation,

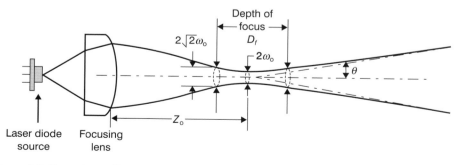

Figure 1.8 Focusing a Gaussian laser beam. (NRC Crown copyright)

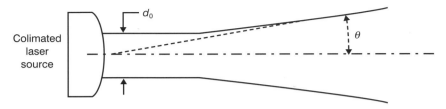

Figure 1.9 Gaussian beam in the collimated case (θ also known as half-angle divergence). (NRC Crown copyright)

more sophisticated methods may be required to fully understand a system, e.g. using the beam propagation ratio M^2; see the definition in ISO-11146-1, 2005.

1.2.3 Photodetection

Each implementation of a laser scanner requires a specific sensor to collect the laser light reflected from the scene. Traditional photosensors for time-of-flight systems are p-n photodiodes, avalanche photodiodes (APD) [Koskinen *et al.*, 1992] and photo-multipliers. The first two are photovoltaic detectors. Avalanche photodiodes and photo-multipliers are characterised by an internal gain that generates excess noise that is not found with p-n photodiodes. They can also be used in the so-called photon counting mode when the collected light levels are expected to be very low. Recent advances in CMOS (complementary metal oxide semiconductor) architecture and device technology have created a new wave of sensors that are arranged as an array of photosensors coupled to signal processing on a per pixel basis [Hosticka *et al.*, 2006]. These constitute the basic sensor for flash lidar, where range images are captured simultaneously rather than in a scanning mode.

1.2.4 Propagation medium and scene effects

With reference to Figure 1.7, laser light has to travel to the object and back through a transmission medium (air, water, vacuum, etc.). For the case of time-of-flight systems, a correction for the refractive index with respect to a vacuum has to be made, which is typically in the order of 300 ppm. Such corrections can be hidden in the calibration procedure. In high altitude flights, one may also want to consider the temperature gradient between flying height and the ground; for the accuracy demands of airborne laser scanning, averaging aircraft and ground temperature and pressure measurements will usually be sufficient.

The strength of a pulse echo characterises the reflectance of the illuminated spot on the ground. Advanced systems record the amplitude (often called the "intensity") of each of the incoming echoes as 8 bit or 16 bit values. Intensity images look like black and white photographs, although they only show the relative reflectivity of objects at the specific wavelength of the laser (monochromatic image). An amplitude image, as complementary information to pure 3D data, may, for example, help to identify objects which are difficult to recognise by elevation data only. Depending on the target, there will either be specular reflection or diffuse (Lambertian) reflection or a mixture of both. Reflectivity of materials is wavelength dependent and varies from close to 0% (e.g. black rubber, coal dust) to nearly 100% (e.g. snow in the visible wavelength range).

A return echo coming from a low reflecting target will have a lower amplitude compared to that from a high reflecting target. Due to this fact, simple pulse detectors in time-of-flight systems tend to put out longer ranges for low reflecting targets. As a consequence, for instance, white markings on a dark asphalt runway will hover above the runway in a range image if this effect is not corrected. Correction may be done with help of tables providing time (= range) corrections for low reflecting targets or by the use of constant-fraction detection techniques.

In addition to reflectivity, the form of the target has an effect on the shape of a return echo. A tilted surface (or, more generally, a non-perpendicular beam incidence angle) will lead to an elliptical footprint with a varying travel time, broadening the shape of the return echo. Local surface roughness within the laser spot diameter will further contribute to smearing the echo shape.

The reflectivity of the illuminated spot on the ground determines not only the precision and reliability of the range measurement, but also the maximum operating range. A diffuse Lambertian scatterer obeys the cosine law by distributing reflected light as a function of the cosine of the reflected angle. For a nearly perpendicular beam incidence angle, the amplitude collected at the receiver decreases with the square of the range. This means that the laser power requirement grows with the square of the range. Usually manufacturers of laser scanners specify for which targets (reflectivity, diffuse/specular reflection) the stated maximum range is valid. Artificial and natural sources of light will also affect scanners. For instance, reflections of sunlight over objects will be collected by the photodetector in the laser scanner and reduce the signal-to-noise ratio.

The underlying hypothesis of active optical geometric measurements is that the measured surface is opaque, diffusely reflecting and uniform, and that the surroundings next to the surface being measured do not create spurious signals (unwanted scattering). Hence, not all materials and scenes can be measured well. Marble, for instance, shows translucency and a non-homogeneity at the scale of the measurement process, leading to a bias in the distance measurement as well as an increase in noise level (uncertainty) that varies with the laser spot size [Godin *et al.*, 2001; El-Hakim *et al.*, 2008]. The quality of 3D data may also drop substantially when a laser beam crosses objects with sharp discontinuities such as edges or holes. The errors are biases that can be removed in some applications [Blais *et al.*, 2005].

1.2.5 Scanning/projection mechanisms

The acquisition of a densely sampled 3D surface with a single laser beam requires a scanning mechanism to move the laser beam over the surface of the object. Some basic scanning modes can be distinguished:

- Scanning is performed by two mirrors mounted orthogonally. This is typical for terrestrial scanners with a window-like field of view.

- The laser beam is scanned in one direction with a scanning mirror and the device assembly is rotated with a mechanical stage. This is typical for terrestrial scanners with a panoramic or hemispheric field of view. A similar assembly can be used where the rotation motion is replaced by a translation, e.g. in rail or tunnel inspection [Langer *et al.*, 2000].

- The laser beam is scanned in one direction and mounted in an aircraft or on a vehicle. In this case the second scanning direction is provided by aircraft or vehicle motion, and a GPS (global positioning system)/IMU (inertial measurement unit) combined unit is used to measure the position and orientation of the platform.

- In the case of triangulation scanners, a line may be projected rather than a single spot. In this case, scanning is limited to one direction. Projection of multiple lines or strip patterns enables full-field techniques without the necessity of scanning.

- Scanning can also be avoided by flash lidar techniques, which are enabled by recent advances in CMOS technology where time-of-flight sensors are arranged as an array of photosensors coupled to signal processing on a per pixel basis [Hosticka *et al.*, 2006]. Such systems are based on a floodlight device projecting light on the area of interest and a bi-dimensional device capturing time-of-flight data. They are often reported as flash LADAR (light detecting and ranging) or focal plane time-of-flight systems.

Figure 1.10 shows a number of different scan mechanisms which are in use in airborne and/or terrestrial laser scanners.

Oscillating mirror: Many commercial airborne systems utilise the oscillating mirror technique, in which a swivelling mirror directs the laser pulse across the swath. Data points are generated in both directions of the scan, resulting in a zigzag pattern on the ground. The distance of laser points within a scan line (across track) varies because the mirror is constantly accelerating and slowing down: larger point distances are found in the middle of the swath, smaller point distances at its end, where the mirror direction turns back. This behaviour can partly be compensated for by controlling the mirror movement with a specific electrical actuation. At the end of the swath, the distance dy of scan lines on the ground (see also Figure 1.10a) is given by

$$dy = v * t = v * \frac{1}{f_s} \tag{1.16}$$

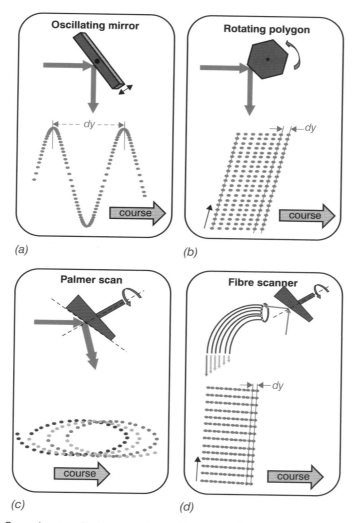

Figure 1.10 Scanning mechanisms and resulting ground patterns.

where v is the cruising speed and fs the scan frequency or scan rate (t is the back and forth travel time of the mirror). At a cruising speed of 70 m/s and a scan rate of 40 Hz, laser spots in the middle of the swath have an along-track distance of 0.88 m, while the along-track distance is up to 1.75 m at the edges as a consequence of the zigzag pattern. One of the advantages of oscillating mirrors is that the scan angle is variable (e.g. 0° to 75°) and the scan rate is also variable (e.g. 0 to 100 Hz). Scan angle, scan rate, flying height and laser pulse repetition frequency determine the maximum point distance in across- and along-track directions. So, the scanner can be configured in such a way that a predefined distance of laser spots on the ground will be kept. In most applications a point density which is homogeneous across and along the track will be chosen, with some variations caused by the zigzag pattern, the non-linear motion of the mirror and the topography itself. In general, oscillating mirror systems operate from 100 m to 6000 m above ground. Due to their flexibility, airborne laser scanners with an oscillating mirror technique can be very well configured to meet different mission requirements.

Rotating polygonal mirror: In rotating mirror systems a rotating polygon mirror is used for beam deflection. Data points are generated in only one scan direction. The scan lines are parallel and, compared to oscillating mirror systems, the measurement pattern shows a more uniform spread of points on the ground. The scanner can even be configured to produce laser points equally spaced in directions both along and across the track. Systems with rotating mirror techniques provide scan angles between about 30° and 60°.

Palmer scanner: The mirror device deflecting the laser beam is constructed in such a way that the mirror surface and rotation axis form an angle which is not equal to 90°. Palmer scan based systems are mainly used in terrestrial laser scanners. In the case of airborne systems, the resulting pattern on the ground is elliptical. Due to the scanning mechanism, objects might be illuminated twice (once in the forward scan, and again in the backward scan), reducing shadowed areas in the elevation data in airborne systems. More details of this technology may be found in Baltsavias (1999) and Wehr and Lohr (1999).

Glass fibre scanner: In the case of a glass fibre scanner, single laser pulses are fed subsequently into neighbouring glass fibres by a scanning mirror. A type of mechanical scan is achieved by a number of glass fibres arranged in a linear array directed down at ground [Lohr and Eibert, 1995]. An advantage of this technology is that the scan mechanism is extremely stable because the glass fibres are glued during the manufacturing process. The scan angle will also be fixed in the manufacturing phase. A typical configuration provides 128 glass fibres and a scan angle of 14°. This configuration will usually produce different point spacing across and along the track. The across-track distance, dx, of laser spots on the ground is

$$dx = \frac{h * \theta}{n - 1} \tag{1.17}$$

where h is the height above the ground, θ is the full scan angle (in radians) and n is the number of fibres. At a height of 1000 m, a scan angle of 14° and 128 fibres, the across distance of laser spots is 1.9 m at a swath width of 246 m. Taking a scan rate of 630 Hz and a cruising speed of 70 m/s results in a distance along the track of adjacent scan lines of 0.11 m. Assuming a beam divergence (full angle) of 0.5 mrad, the footprint will be 0.5 m from this survey height. So, there will be a high overlap of laser spots in the direction along the track.

It is important to mention the limitations of optomechanical and purely mechanical scanning devices. Optomechanical devices include mainly low inertia galvanometers/resonators or polygonal scanners (from mono-facet to *n*-facet mirrors; Marshall, 1985). Other methods have used acousto-optic deflectors, diffraction grating based, micromirror device scanners and holographic scanners. Galvanometers in oscillating mirror systems require the mounting of a mirror on the device shaft. The optical projection angle is twice the mechanical angular motion of the mirror. Galvanometers are fully programmable, i.e. sinusoidal, sawtooth, triangular, and randomly defined waveforms are possible (<100 Hz for large angles). Limitations will come from wobble (perpendicular to the scan axis), jitter (in the scan direction), dynamic deformation of the mirror, angular position transducer uncertainty and drive electronics limitations. Resonant scanners are fixed frequency devices. They can achieve 10–20 kHz scanning frequencies for small angles and mirrors. Polygonal scanners on the other hand can achieve very high scan rates (e.g. 40 000 rpm). Because of air drag, the limiting speed is about Mach 1 on the periphery of the mirror. They are composed of one mirror or multiple facets. Typical scanning errors with polygonal scanners are wobble, jitter, facet-to-facet irregularities and drive electronics limitations. Absolute angular position is usually accomplished with an external transducer, which will yield some uncertainty. Rotation stages are limited by both wobble and eccentricity. Translation stages are limited by flatness, straightness and roll wandering.

Flash lidar: As an extension of sequential scanning approaches, one can also project modulated light with either an array of LEDs (light-emitting diodes) or laser diodes in a full-field approach. The demodulation of the returned light is done with an external electro-optical device, which may be a gain controlled microchannel plate [Muguira *et al.*, 1995], or using specially designed CMOS chips [Lange and Seitz, 2001; Juberts and Barbera, 2004; Ruff *et al.*, 2005; Hosticka *et al.*, 2006] (Figure 1.11). Many of the devices described in these references are based on the direct detection of light as opposed to interferometric methods, which are based on optical coherent detection of light [Bouma and Tearney, 2002]. With the latter method, the mixing and correlation of the returned signal with a reference signal occur directly on the photodetector by what is known as coherent field superposition. As opposed to scanning systems, here a complete 3D image is generated in one shot delivering a 3D image without mechanical parts. Disadvantages of flash lidar techniques, compared to scanning systems, are limited spatial resolution and the range limitation caused by the spreading of the pulse energy over a larger field of view.

1.3 Basics of airborne laser scanning

In the early 1970s, it was shown that airborne lidar systems were able to measure distances between aircraft and ground targets to a precision of less than 1 m. However, laser altimeter systems did not come into widespread use for precise topographic mapping

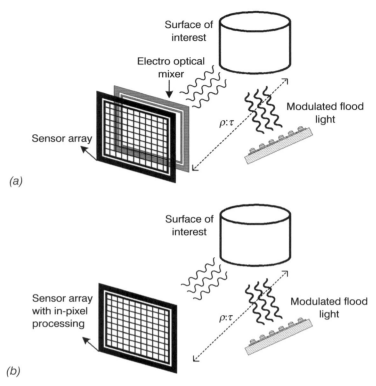

Figure 1.11 Flash 3D system with floodlight: (a) detection using an external mixer and correlator; and (b) detection on a specially designed sensor or focal plane array. (NRC Crown copyright)

mainly for two reasons. First, for precise mapping, the vertical position of the aircraft had to be known to a level of accuracy comparable to the measurement capability of the lidar system. Second, the horizontal position of the illuminated spot on the ground (laser footprint) had also to be known. Although the second requirement is less stringent, the means to determine both aspects for larger areas with sufficient quality were not available at that time. Trials were done to determine aircraft altitude by recording pressure data with precise aircraft aneroid barometric altimeters [Arp, 1982] and vertical accelerometers [Walsh, 1984; Krabill *et al.*, 1984]. Horizontal control was a tedious process as it was done post-flight by means of time-tagged photographs and rarely by IMU. At the end of the 1980s with the availability of GPS, a method was developed allowing precise registration of position and orientation over larger areas. With the introduction of differential GPS (DGPS) the scanner position became known in horizontal and vertical coordinates in the sub-decimetre range [Krabill, 1989; Friess, 1989]. Enhancements in DGPS technology along with Kalman filtering and use of IMUs provided sufficient accuracy from the beginning of the 1990s [Lindenberger,

1989; Schwarz *et al.*, 1994; Ackermann, 1994]. The standard accuracy of elevation data became ±10 cm in height, and ±50 cm in position. Until the end of the 1980s range measurements were done by laser profilers [Ackermann, 1988; Lindenberger, 1993] providing laser pulses but no scanning mechanism. In the early 1990s profilers were replaced by scanning devices which generated 5000 to 10 000 laser pulses per second at that time. Nowadays, laser pulse rates reach 300 kHz, however, depending on the kind of scanning mechanism 100% of the instrument's pulse rate may not really be available on the ground.

Airborne laser scanning is now a common technique for generating high quality 3D presentations (digital elevation models) of the landscape (Figure 1.12). The following will outline the principle and system components of airborne laser scanning. Characteristics of this measurement technique as well as accuracies and limitations will be discussed. The specifics of calibration, processing of gathered data and product generation will not be discussed as these are covered in other chapters. Several airborne laser scanners are commercially available and in operational service. An overview of the specifications of various available airborne laser scanners is given by Lemmens (2009).

1.3.1 Principle of airborne laser scanning

Airborne laser scanning is done either from a fixed wing aircraft or a helicopter. The technique is based on two main components: a laser scanner system which measures

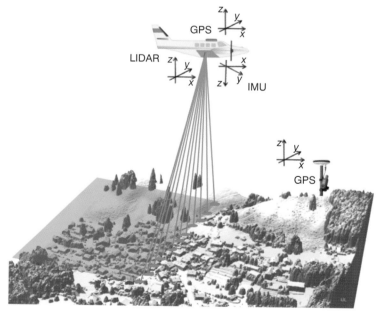

Figure 1.12 Airborne laser scanning principle.

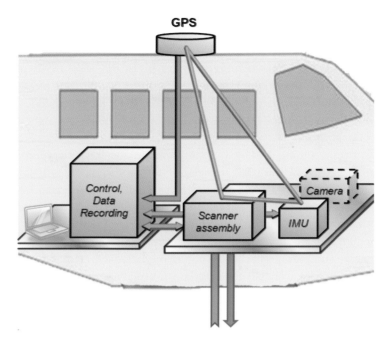

Figure 1.13 On-board components of an airborne laser scanner.

the distance to a spot on the ground illuminated by the laser and a GPS/IMU combination to measure exactly the position and orientation of the system. Active systems based on laser scanning are relatively independent of sunlight. They may be operated during the day or at night. This characteristic is a considerable advantage of airborne laser scanning compared to other methods of surveying landscapes.

The basic components of an airborne laser scanner (Figure 1.13) are:

- **Scanner assembly** comprising laser, scanning mechanics and optics. The laser system (mostly a pulsed time-of-flight measurement system), mounted over a hole in the aircraft's fuselage, continuously sends laser pulses towards the terrain as the aircraft flies. Depending on aircraft velocity and survey height, current technology allows measurement densities between 0.2 and about 50 points/m^2. Modern scanner assemblies provide a roll compensation to compensate for the roll of the aircraft. This feature helps to avoid gaps in coverage which might occur between adjacent swaths due to roll. Roll compensation allows the overlap between flight lines to be planned to be smaller and therefore gives an economic advantage.

- **Airborne GPS antenna:** the standard is a dual frequency antenna recording GPS signals at a sampling rate of 2 Hz. The antenna is mounted at an exposed position on top of the aircraft, providing an undisturbed view to GPS satellites.

- **Inertial measurement unit (IMU):** the IMU is either fixed directly to the laser scanner or close to it on a stable survey platform. Typically it records acceleration data and rotation rates at a sampling rate of 200 Hz. Acceleration data can be used to support the interpolation of the platform position on the GPS trajectory, while rotation rates are used to determine platform orientation. The combination of GPS and IMU data allows one to reconstruct the flight path (air trajectory) to an accuracy of better than 10 cm.

- **Control and data recording unit:** this device is responsible for time synchronisation and control of the whole system. It stores ranging and positioning data gathered by the scanner, IMU and GPS. Modern laser scanners, which generate up to 300 000 laser pulses per second, produce about 20 Gbyte of ranging data per hour, while GPS and IMU data only sum up to about 0.1 Gbyte per hour.

- **Operator laptop:** this serves as a means of communications with the control and data recording unit, to set up mission parameters, and to monitor the system's performance during the survey.

- **Flight management system:** this is a means for the pilot to display the pre-planned flight lines, which provides support for him in completing the mission.

An airborne laser scanner is completed by a GPS ground station. The ground station serves as a reference station for off-line differential GPS (DGPS) calculation. Differential GPS is crucial for compensating atmospheric effects disturbing precise position determination and for achieving decimetre accuracy. In order to cope with varying atmospheric conditions, the distance between the aircraft and the GPS ground station should not exceed 30 km (although sometimes adequate accuracy is reached for longer distances). Nowadays, several countries operate a network of permanent GPS stations, so there is often no need to set up one's own station.

Airborne laser scanners are often complemented by a medium-sized digital camera system. Image data taken simultaneously with range data may support data interpretation significantly in cases where it is difficult to recognise objects only from range data. Image data will usually offer a better spatial resolution and form a basis for integrated 3D point cloud and image data processing. The camera head and the recording unit are separate. The optimum location for the camera is on the scanner assembly's ground plate, because then the existing IMU registrations can be shared for georeferencing. A separate IMU will be needed if the dimensions of the scanner assembly do not allow accommodation of a camera and a scanner assembly on the same survey platform.

1.3.2 Integration of on-board systems

In order to integrate/transform accurately GPS, IMU and laser data, the spatial relationship of scanner assembly, IMU and GPS antenna must be known. Of course, the time dependencies of all three systems must also be known accurately. The time synchronisation is achieved by the PPS signal (pulse per second) of the GPS which triggers the internal clocks of the laser scanner and the IMU. Time stamps are used to link range measurements and trajectory positions.

In order to integrate the spatial relationships, a coordinate system will be defined for the survey platform. The translation parameters (Δx, Δy, Δz) between the various coordinate systems of the GPS antenna, IMU and scanner assembly are determined by surveying exactly the vectors between the origins of the various units and those with respect to the survey platform. This task can be done by a terrestrial survey as an accuracy of ±0.01 m is sufficient. After this the spatial relations of all units are known and the coordinate system of the survey platform can be taken as the origin for all future measurements. The bore-sight calibration to determine misalignment angles between the coordinate systems of the IMU and the scanner assembly is carried out by specific calibration flights and procedures (see also Chapter 3). Last but not least, the quality of calibration can be checked during the data processing (generation of a point cloud) and may be corrected if necessary.

1.3.3 Global Positioning System/Inertial Measurement Unit combination

As discussed above, GPS and IMU are used to reconstruct the flight trajectory. GPS data are processed by differential GPS methods to get precise information on positions for the whole survey flight. So, for example, a cruising speed of 70 m/s and a 2 Hz GPS recording rate result in a spacing of 35 m at which positions are available.

For complete restitution of the flight trajectory an IMU is needed which provides position data as well as roll, pitch and yaw angles of the survey platform, typically at a 200 Hz sampling rate. As IMU systems suffer from systematic drift effects, IMU data may not be used over a longer period of time without losing accuracy. The IMU drift in positioning is compensated for by updating its position measurements by more reliable GPS positions. This technique delivers exact positions for the whole survey flight and additionally is also a means of checking calculated GPS positions for errors.

During the survey flight the aircraft will follow linear flight lines by only minor movements in roll, pitch and yaw. Such a practice is necessary to maintain the planned measurement density and to guarantee coverage without gaps. However, due to the

Table 1.1 Performance data of a high-end IMU

Position	0.05–0.3 m
Velocity	0.005 m/s
Roll, pitch, yaw	0.005°
Noise	0.005°/√h
Systematic drift	0.01°/h

IMU's drift characteristic, this practice runs the risk of gathering erroneous orientation data (see Table 1.1). The problem is that IMU gyros will "fall asleep", if not nudged for a long time. Depending on the IMU's characteristics, the critical period of time may be about 10 minutes. The problem may be bypassed by planning flight lines no longer than those given by the critical time period. The movement (turn) of the aircraft, when approaching the next flight line, is sufficient to "wake up" the gyros again.

Merging GPS and IMU measurement results in an optimum set of altitude and attitude data, sometimes called "smoothed best estimated trajectory" (SBET). If altitude and attitude data have been interpolated to 800 Hz, position and orientation are available at approximately 0.09 m intervals (at an aircraft speed of 70 m/s). Using common GPS/IMU processing, the platform coordinates can be restituted to an accuracy of about 10 cm. An offset, still included in these data, can be removed during data processing with strip-adjustment procedures and control data.

1.3.4 Laser scanner properties

Commercial airborne laser systems for land applications operate at wavelengths between 800 nm and 1550 nm. The spectral width is typically between 0.1 and 0.5 nm. As the reflectivity of an object depends on the wavelength, different laser systems show different pros and cons when scanning the Earth's surface: at wavelengths close to the visible part of the spectrum, the absorption of water is high. Therefore, water surfaces will rarely be seen by laser scanners operating in that part of the spectrum. At about 1550 nm the reflectivity of ice and snow is low, therefore scanners operating at 1550 nm will not be

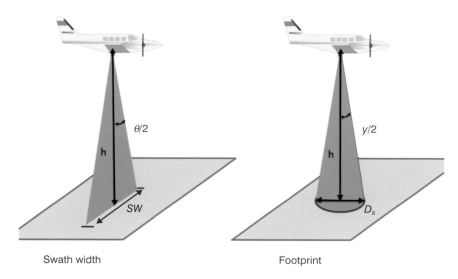

Figure 1.14 Swath and laser footprint (footprint is much smaller in reality).

the optimum choice when surveying snow fields. Attention must also be paid to eye safety which means that the human eye must not be damaged when looking into a laser beam (Section 1.2.1).

The swath width sw of a scanner is given by

$$sw = 2h \tan\left(\frac{\theta}{2}\right) \tag{1.18}$$

where θ is the full scan angle and h is the height above ground (Figure 1.14). Airborne scanners allow scan angles between about 5° and 75°. For example, the swath width will be 574 m at a height of 1000 m and a scan angle of 32°.

The laser beam widens with the distance from the laser scanner. A relationship, similar to the previous one, describes the diameter D_s of the illuminated footprint on the ground

$$D_s = 2h \tan\left(\frac{\gamma}{2}\right) \tag{1.19}$$

where γ (full angle) is the beam divergence and h the height above the ground assuming the spot shape to be a circle (Figure 1.14). Typically, beam divergences are between 0.1 mrad and 1 mrad. So, the footprint will be 0.2 m from a survey height of 1000 m and a divergence of 0.2 mrad. The laser beam profile does not have sharp edges, but the irradiance falls off gradually away from the centre of the beam. Usually, the diameter D_s is determined at a position at which the irradiance drops to $1/e^2$ times the total irradiance (see also Section 1.2.2).

1.3.5 Pulse repetition frequency and point density

In the 1990s most airborne lidar systems were characterised by pulse repetition frequencies (PRF) in the low kilohertz range. With increasing PRF, the laser spot density on the ground also increased. The pulse repetition frequencies of current airborne laser scanners reach up to 300 kHz. The mean point density P_M on the ground can be calculated as follows. The area covered in a certain time t at a cruising speed v and swath width sw (Equation 1.18) is given by

$$F = v \times t \times sw \tag{1.20}$$

The number of laser pulses P generated during a certain time interval t is $P = PRF \times t$. Now the mean point density P_M can be calculated by means of P and the area covered during t

$$P_M = \frac{PRF \times t}{F} = \frac{PRF}{v \times sw} = \frac{PRF}{2vh \tan(\theta/2)} \tag{1.21}$$

So, the highest point density is reached with high pulse repetition frequencies, low cruising speed v, low survey height h and small scan angle θ. In the previous section the example for the oscillating mirror showed that from a survey height of 1000 m and a scan angle of 32°, the swath width is 574 m. A pulse repetition frequency of 80 kHz and a cruising speed of 70 m/s results in a mean point density of about 2 points/m².

With respect to formula (1.21), two facts should be noted:

- P_M is only a mathematical description suggesting a homogeneous distribution of measurements over the area F. In reality, the point distribution is not necessarily homogeneous, but depends on several items such as scan pattern (i.e. scan mechanism) and surface topography.

- In general, the maximum available pulse repetition frequency is not a fixed figure but depends on survey height h. The reason is the ambiguity constraint as well as the balance between pulse energy and repetition rate.

It should be noted that time-of-flight receivers also record the amplitude of an echo. The value may be stored for each of the recorded echoes and gives information on the reflectance properties of the target illuminated by the laser spot.

In standard systems, range measurements are acquired at equally spaced time intervals given by $1/PRF$. Obviously, when the observed pulse time delay τ is larger than $1/PRF$, an ambiguity occurs because two pulses can be received in the same time interval. Hence, time-of-flight systems based on this principle have an inherent ambiguity interval $\rho_{\text{max-}p}$ given by

$$\rho_{\text{max-}p} = \frac{c}{2} \frac{1}{PRF} \tag{1.22}$$

For $PRF = 10\,000$ Hz, $\rho_{\text{max-}p}$ is equivalent to an interval of 15 km, which covers virtually all terrestrial and airborne applications. In the case of airborne laser scanners and a pulse repetition frequency of 200 000 Hz, $\rho_{\text{max-}p}$ is equivalent to only 0.75 km, which means that ranges beyond 750 m may lead to ambiguities. This relationship between maximum pulse rate and survey height has to be taken into account when planning a survey flight. Until recently, this critical pulse frequency determined the maximum pulse rate, as airborne receivers could only handle one pulse in the air at a time. Now, new multipulse receivers are in operation which record echoes that arrive after a subsequent pulse has been emitted, which means that the laser may send pulses even while other pulses are still in the air. Compared to the past, higher pulse rates can now be used from the same survey altitude. A look at the performance parameters of current lidar systems shows that systems operating in multipulse mode send double the number of pulses as in single-pulse mode.

1.3.6 Multiple echoes and full-waveform digitisation

The number and the form factor (signal shape) of return echoes depend on the type and orientation of the illuminated surface. Figure 1.15 shows different situations in which a laser spot illuminates an object. The picture shows for each situation the location of the return echo (pulse on the P axis) as generated by the receiver of a time-of-flight measuring system. A time-of-flight receiver provides only a stop signal at a certain rise time of the echo. Such systems are also called discrete echo systems. They do not provide information on an echo's shape. In contrast, full waveform recording allows analysis of the echo's full shape by digitising the return echo at gigahertz resolution. In Figure 1.15 the waveform of return pulses is sketched on the FW axis.

In Figure 1.15(a) the laser spot hits flat terrain. The result is only one return echo. The shape of the waveform is similar to that of the emitted signal. In Figure 1.15(b) the laser beam hits the roof of a building. The echo coming from the roof is slightly broadened, as the laser beam hits a sloping surface. As before, the pulse-measuring system only produces a stop signal. However, due to the slowly rising edge of the echo, the associated range will depend on the type of detector (and its strategy) used for stop pulse generation. In Figure 1.15(c) the laser spot hits the ground and also illuminates a part of the roof. This situation gives two separate return echoes, one from the roof and another one from the ground. The time-of-flight detector will generate two stop signals. However, the two echoes have different waveforms which may be stored by the full waveform receivers. As the information on an echo's shape is conserved in waveform recording, only this technique maintains information on the nature of a target [Wagner *et al.*, 2004].

In a more complex situation, as shown in Figure 1.16, multiple echoes will be generated. The first hit on a branch of the tree produces a first echo as well as an associated value for echo amplitude (Figure 1.16a). There will be some intermediate echoes coming from branches at different heights and finally a last echo from the ground, if the tree is transparent enough. Figure 1.16(b) depicts the full waveform

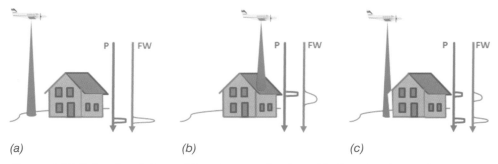

(a)   *(b)*   *(c)*

Figure 1.15 Echo signals from different target situations (P: discrete return pulses; FW: waveforms).

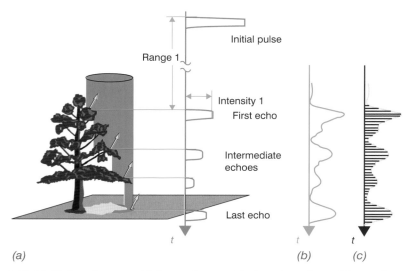

Figure 1.16 Multiple echoes and full waveforms: (a) discrete pulses; (b) waveform; and (c) digitised waveform.

while Figure 1.16(c) shows the waveform after digitisation. Section 6.5 will treat the analysis and use of waveform data in more detail.

Typically, time-of-flight receivers can capture up to four echoes per pulse: first, second, third and last. However, it should be kept in mind that objects (e.g. branches in Figure 1.16) can only be detected as separate objects if their distance is larger than the pulse width (see also Section 1.1.1). This distance is 0.75 m for a pulse width of 5 ns. Waveform receivers have digitisation intervals typically between 0.5 ns and 1 ns, which correspond respectively to range intervals of 0.075 m and 0.15 m.

1.3.7 Airborne laser scanner error budget

The various components of an airborne laser scanner contribute to the final accuracy of elevation data [Schenk, 2001]. In practice, standard vertical and horizontal accuracies (at 1σ) in the final elevation data are usually between 0.05 m and 0.2 m in height and between 0.2 m and 1.0 m in position (when considering flight heights up to 2000 m above the ground).

The main sources which contribute to the final accuracy are:

- errors due to erroneous calibration of the GPS, IMU and scanner assembly;
- errors due to limited accuracy of the flight path restitution;
- errors due to the complexity of the target; it should be noted that in the case of a sloping terrain an error in position on the ground also leads to an error in elevation;

- errors due to multipath reflections (i.e. if the laser beam is reflected by another object on the ground before reaching the detector);
- errors due to coordinate transformation and geoid correction.

Chapter 3 gives more details on error budget and describes how to minimise systematic errors.

1.4 Operational aspects of airborne laser scanning

Independent of the specific type of airborne laser scanning system, there are several common aspects of project planning and processing. The following might serve as a general outline of phases within an airborne laser scanning project. In principle, each project can be divided into three phases:

- flight planning;
- a survey campaign which includes the survey flight, the operation of GPS ground station(s) and the collection of ground reference data;
- data processing and quality control.

1.4.1 Flight planning

The project requirements in terms of point density and characteristics of the project area (extension, location and topography) form the basis for planning of the survey flight. Based on the project demands on the one hand, and the performance parameters of the laser scanner in use on the other, a laser scanner setup has first to be defined. Usually, the system manufacturer delivers a configuration program for this purpose, which needs information such as survey height, cruising speed and scan angle to be input in order to output parameters such as average point density and swath width. Appropriate choice of a scan angle is of essential importance. Seen from the economic point of view, the aim is to find the largest scan angle at which the project requirements are still met. In flat terrain, larger scan angles may be the best choice. In urban areas or in the presence of dense vegetation (forest), shadowing effects must be taken into account. Therefore, smaller scan angles should be chosen here.

Bearing in mind the IMU drift characteristics, the maximum length of a flight line will typically be 30 to 40 km (depending on the chosen cruising speed). Cross lines are planned at regular distances, at least one for each flight block. During data processing these cross lines will help checking the internal accuracy of merged scan lines and so be part of the quality control.

Flight planning is supported by specific software. The aim is to cover the project area by parallel flight lines including a certain overlap between consecutive lines. Typically, an overlap of 20% is taken to avoid gaps and to compensate for minor difficulties likely to occur in aircraft navigation during the survey flight. Figure 1.17 shows on the

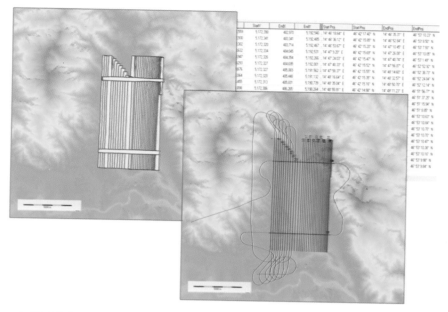

Figure 1.17 Flight plan.

left the planned swaths over an elevation model of the project area. The table on the right lists all flight lines and the start and stop coordinates for each line. The image in the lower right depicts the planned flight lines in blue and the part of the flight mission which has already been carried out in red. It should be noted that some of the flight tracks belong to an in-flight manoeuvre carried out to initialise the IMU.

High-end flight planning software allows integration of a digital terrain model (DTM) of the project area. This is of particular help in mountainous areas: at a constant flying altitude the swath width varies with terrain height and the swath might become too narrow for complete coverage if the terrain rises. Planning flight lines over a DTM avoids such gaps in coverage and so increases considerably the reliability and efficiency of the planning. The flight plan is stored and later transferred to the aircraft's flight management and guidance system.

1.4.2 Survey flight

The aim of the actual survey flight is to collect data along the previously set up flight plan. The ideal time for a survey mission depends on project needs (leaf-on/leaf-off conditions, free of snow, etc.) as well as favourable weather condition (no rain, no fog, no storms). As mentioned earlier, laser scanning is an active measurement method that allows survey flights either in the daytime or at night.

During the flight mission the control and data recording unit (see Section 1.3.1) provides continuous feedback of scanner functions. Frequently, the system operator also keeps a log including remarks on unusual equipment and/or environment events. This log may facilitate analysis at some stage of the data processing.

Usually two GPS ground stations are operated during the survey flights, one as the base station, another one as back-up. The location of the base station is chosen so that the distance between aircraft and station will not exceed 30 km. For larger project areas several base stations may be installed, which record GPS data simultaneously and allow the critical 30 km barrier to be exceeded. Over the years several countries have installed a network of permanent ground stations and offer GPS data as a service. So, frequently there is no longer the need to operate their own GPS ground stations.

The final quality of an elevation model is checked by means of ground truth data. Flat areas like tennis courts, road crossings and parking lots (if they are really flat) serve to control height accuracy, while roof planes of buildings turn out to be the best choice for position control. For the purpose of quality control some of these ground truth objects located in the project area will have to be terrestrially surveyed. Usually, it is sufficient for each survey flight to have two or three objects for height control and some more for position control.

In the case of longer survey campaigns and/or project areas which also might be far away from the processing facility, coverage control and check of data quality should be done on-site. The advantage of such on-site activities is to recognise the need of reflying flight lines at a time when equipment and operators are still on-site.

1.4.3 Data processing

After a survey mission there are three datasets present:

- GPS ground station data;
- navigation data (airborne GPS and IMU data);
- ranging data (including time tags), amplitude data, scanner parameters (e.g. scan angle) and echo counts.

Processing of these data can be divided into two parts: first, the common processing of ground GPS and airborne GPS/IMU data, and second, common processing of these data with range measurements (Figure 1.18). For both parts of the process there are several software packages available on the market.

The first task in GPS/IMU processing is to correct airborne GPS measurements by means of GPS ground station data. Then, the derived DGPS dataset is integrated with IMU measurements to obtain the best restitution of the flight path, optimised altitude and attitude data.

Flight path data and ranging measurements are now merged with the help of time tags. In this phase of the processing all kinds of correction data (calibration data, mounting parameters, atmospheric correction) are also taken into account. At this stage calibration parameters might be slightly adapted, if it turns out that they are no longer optimal. Generally, this phase also includes the transformation of computed X,

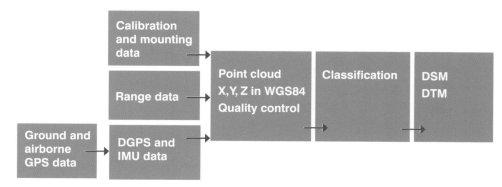

Figure 1.18 Processing scheme for airborne laser scanning.

Y, Z coordinates from WGS84 into the final coordinate system. *X, Y, Z* data at this processing level is named 3D point cloud. These data are normally stored in the LAS format which became a standard for lidar data in recent years [ASPRS, 2008]. Non-interpolated point cloud data are elevation data showing a measurement distribution reflecting the scan geometry.

As a first quality control, the dataset can be checked for completeness and achieved measurement density. Next step in the processing may be to control/adjust flight lines in height to receive a best-fitting point cloud set. Now, the position and height of the complete point cloud may be checked by means of ground truth data, and corrected if necessary. More details on these aspects are given in Chapters 2 and 3.

Point cloud data in LAS format contain information on several echoes and also amplitude information. As an example, Figure 1.19 shows a point cloud in which

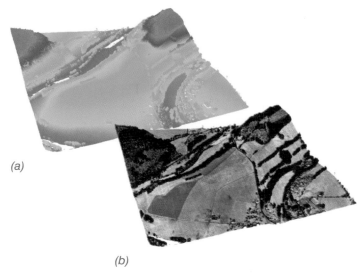

(a)

(b)

Figure 1.19 (a) Colour-coded point cloud; and (b) amplitude image.

different colours stand for different elevations. The figure shows the associated amplitude image in the lower part. Obviously the amplitude information helps to distinguish details in areas with only minor differences in elevation. Echoes may be classified as "only echo", "first of many", "second of many", …, "last of many". In further processing, appropriate echo classes will be chosen depending on whether a digital surface model (DSM), a digital terrain model (DTM) or some other product has to be generated.

1.4.4 Airborne laser scanning and cameras

Airborne laser scanning systems are frequently supplemented by medium-sized digital cameras allowing collection of elevation and image data during the same survey flight. Table 1.2 lists typical performance parameters of a medium-sized digital camera used with airborne laser scanners. In general, these cameras do not allow gathering of red–green–blue (RGB) and colour infrared (CIR) images simultaneously.

Of course, lasers scanners may also be operated simultaneously with a large format digital camera. However, there are some restrictions which have to be taken into account.

- **Accommodation:** Operating a large format digital camera and a laser scanning system simultaneously needs an aircraft with two holes in the fuselage. High weight and power consumption might need use of a large (and thus more expensive) aircraft.

- **Survey height:** It has to be taken into account that large format digital cameras have a fixed field of view (FOV), which means that a certain survey height stands for a certain ground sampling distance and swath width. On the other hand, for a specific laser scanner there is a characteristic relationship between FOV and point density for a certain height above the ground. So, for common operation of both the systems, a trade-off must be made between image ground sampling distance, swath width and point density.

Such a constraint does not exist for medium-sized digital cameras as they can be equipped with various objectives.

In any case, it has to be noted that the basic advantage of the active laser scanning system of being relatively weather independent is lost when commonly operated with a camera. Moreover, data collection is limited to daylight.

Table 1.2 Typical performance parameter of a medium-sized digital camera

Array size	7200 × 5400
Ground sampling distance (GSD)	0.05–0.4 m
Spectral channels	RGB or CIR
Max. frame rate	2.5 s
Radiometric resolution	12 bit

1.4.5 Advantages and limitations of airborne laser scanning

As an active system operating with light, a laser scanner needs a clear view to the ground. It cannot penetrate clouds, fog and dense vegetation. The laser beam will easily go through the canopy of deciduous trees, especially in winter time when the leaves are off. However, it will not see the ground below dense conifers and in multistorey rainforest. Total reflection might be the reason if a water surface is seen only close to the nadir direction, and a wet street or a wet roof may also lead to drop-outs.

Despite its limitations, airborne laser scanning has turned out to be a very effective means for producing high quality digital elevation models. Laser scanning technology has some advantages compared to other methods of generating elevation data. Some of its pros are:

- **High measurement density and high data accuracy:** The highest measurement densities (about 30 measurements/m²) are reached from a helicopter. The standard accuracy of elevation data in the local coordinate system is 0.05–0.20 m for height and 0.2–1.0 m for position.

- **Fast data acquisition:** For point densities of 1 point/m² and higher airborne laser scanning is accepted to be a very fast means of generating accurate elevation models. There is an example in Section 1.3.5 in which 2 points/m² are collected from 1000 m above ground at a cruising speed of 70 m/s. Taking these figures and in addition considering the turn of the aircraft at each end of a flight line leads to coverage of about 50 km²/h. Fast acquisition is supported by the fact that scanning can be done at any time, day and night.

- **Canopy penetration:** If the canopy is not too dense, part of the laser light beam may penetrate to the ground enabling production of an elevation model of the forest floor. High penetration rates are reached in the case of deciduous trees in winter time, when the leaves are off.

- **Minimum amount of ground truth data:** Terrestrial work is minimised because even for large flight blocks only few ground references are needed.

1.5 Airborne lidar bathymetry

Airborne lidar bathymetry describes a lidar-based technique for acquiring digital bathymetric (water depth) data. Airborne lidar bathymetry is highly efficient in shallow water where traditional survey methods cannot be applied [Guenther, 2000; Francis *et al.*, 2003]. Penetration depth is a function of water clarity and will decrease with increased water turbidity. Under favourable conditions mapping can be conducted in depths up to 50 m. Compared to conventional bathymetry, the methodology offers significant savings both in operational cost and increase in productivity. For relief in low waters and complex areas the survey speed may be up to 20 times quicker than traditional soundings executed with hydrographic ships.

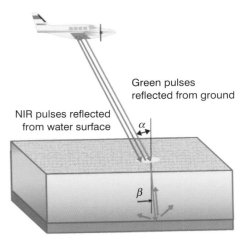

Figure 1.20 Principle of airborne lidar bathymetry.

Bathymetry systems work with two lidar systems: one in the infrared (e.g. 1064 nm) and another in the green part (e.g. 532 nm) of the spectrum [Wozencraft and Millar, 2005]. The green laser light is refracted as it moves into the water and reflected by the seabed. The longer infrared wavelength is reflected by the water surface and is used to determine the distance between survey platform and water surface. Water depth and location of the illuminated spot on the seabed can be calculated by means of scan angle, distance to water surface, time-of-flight and refraction angle of the green laser pulse obtained via Snell's Law (Figure 1.20).

Compared to airborne laser scanning, lidar bathymetry needs higher energy as the green laser beam undergoes refraction, absorption and scattering during its passage through the water. The high energy results in longer pulses (about 250 ns) and lower pulse repetition frequency. At a higher pulse rate, lidar bathymetry may also work in topographic mode, scanning the land surface. By using both modes simultaneously, land surface and seabed may be surveyed by the same system.

Tables 1.3 and 1.4 show examples of airborne lidar bathymetry performance parameters. It should be noted that accuracy parameters are frequently given by assuming a flat water surface. In the case of rough sea with high waves, accuracy will deteriorate significantly. White water due to breaking waves on shoals, in the surf zone or the track of a ship will lead to no depths being measured.

1.6 Terrestrial laser scanners

The fundamentals of laser distance measurement and scanning, as outlined in Sections 1.1 and 1.2, apply to airborne systems as well as to terrestrial systems. Terrestrial laser scanners are usually operated on a tripod. Mobile mapping systems, wherein a scanner is mounted on a (terrestrial) vehicle, are treated separately in Chapter 9. Trian-

Table 1.3 Performance parameters of an airborne lidar bathymetry system in hydrographic mode

Operating altitude	200–400 m
Laser pulse rate	3000–4000 Hz
Aircraft speed	>60 m/s
Sounding density	2 m × 2 m to 5 m × 5 m
Typical swath width	300 m (@ 4 m × 4 m)
Horizontal accuracy	±2.5 m
Depth accuracy	±0.25 m
Sounding depth	0.2–50 m

Table 1.4 Performance parameters of an airborne lidar bathymetry system in topographic mode

Operating altitude	300–1000 m
Laser pulse rate	20 000–64 000 Hz
Horizontal accuracy	2 m
Height accuracy	0.25 m

gulation scanners, which are mounted on a robot arm or an assembly line in industrial applications, are not treated here.

While an airborne laser scanner needs only one scanning direction (the other one is accomplished by the moving aircraft), a terrestrial laser scanner will usually come with a 2D scanning device. In practice, one can distinguish window-like, panoramic and (hemi)spherical scanners. Window-like scanners have a rectangular field of view similar to that of a conventional camera, which is often realised by two combined galvanometer mirrors. Panoramic scanners have a horizontal 360° field of view and a vertical opening angle of typically 80–90°. This can for instance be realised by a polygonal wheel for the vertical scanning direction (see Section 1.2.5) combined with a horizontal 360° rotation of the whole instrument. Replacing the polygonal wheel by a Palmer scan system (Figure 1.21) allows generation of a field of view which is basically only limited by self-occlusions of the instrument itself and its tripod.

Section 1.1 discussed the advantages and disadvantages of different range measurement principles: direct time-of-flight measurement delivers a larger range capacity, while phase-measurement based CW systems provide better accuracy and a higher data rate. Time-of-flight measurement is clearly dominant in airborne laser scanner systems, which are usually employed at flying altitudes between 200 m and 4000 m. As many applications in terrestrial laser scanning have range requirements less than 100 m, phase-measurement systems are more common in terrestrial laser scanning. In fact, there is a group of phase-measurement systems on the market, which provide a

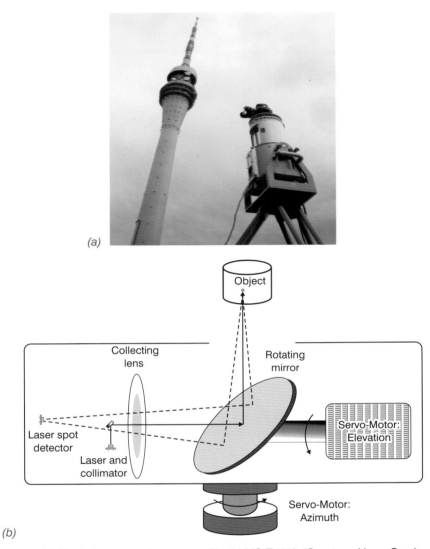

Figure 1.21 (a) Typical panoramic scanner Riegl LMS Z420i. (Courtesy Hans-Gerd Maas). (b) Hemispherical laser scanner principle with a Palmer mirror and a rotation stage. (NRC Crown copyright)

non-ambiguous range of 20–80 m and a range measurement accuracy level of 1–3 mm, while time-of-flight based terrestrial systems have a maximum range of typically 250–1000 m at a range measurement accuracy of 5–10 mm, with one instrument even reaching up to 6000 m. Triangulation scanners are only of interest at ranges of less than 2–5 m due to their accuracy characteristics (Section 1.1.3). Short-range industrial grade triangulation scanners deliver data rates up to 150 00 000 points/s.

A typical value for the angular precision of rotation stage, polygonal wheel and Palmer mirror rotation devices of high quality instruments is of the order of 0.04 mrad. This corresponds to a planimetric error of 4 mm at 100 m distance. The depth and planimetric accuracy of all time-of-flight and phase-measurement systems show different characteristics. With increasing range, the range accuracy will decrease slightly, while the planimetric accuracy will decrease linearly.

Like airborne systems, many terrestrial laser scanners offer an integrated camera, which may be used for colourising point clouds and 3D models or to facilitate 3D point cloud interpretation. In an integrated self-calibrating photogrammetric bundle adjustment, a digital camera may also support the self-calibration of a laser scanner and optimise its accuracy potential [Schneider and Maas, 2007].

Acknowledgements

The authors wish to acknowledge the various collaborators who have participated in the results discussed in this paper: L. Cournoyer, J. Domey, G. Godin, S.F. El-Hakim, M. Picard, and M. Rioux from the NRC Canada and D. Silbermann from ASTEC GmbH, Germany. The sustained and professional services of the NRC Canada Institute for Scientific and Technical Information (CISTI) were instrumental in gathering many of the references listed in this chapter.

References

Ackermann, F., 1988. Digital terrain models of forest areas by airborne profiling. In: *High Precision Navigation,* Linkwitz, K. and Hangleiter, U. (Eds.), 239–250, Springer-Verlag, New York, NY, USA.

Ackermann, F., 1994. On the status and accuracy performance of GPS photogrammetry. *Proceedings of Mapping and Remote Sensing Tools for the 21st Century*, 26–29 August, Washington, DC, 80–90.

Amann, M.-C., Bosch, T., Lescure, M., Myllylä, R. and Rioux, M., 2001. Laser ranging: a critical review of usual techniques for distance measurement. *Optical Engineering*, 40(1), 10–19.

ANSI Z136.1-2007, 2007. Safe Use of Lasers, American National Standards Institute, March 16, 2007, 276 p.

Arp, H., Griesbach, J.C. and Burns, J.P., 1982. Mapping in tropical forests: a new approach using the laser APR. *Photogrammetric Engineering and Remote Sensing*, 48(1), 91–100.

ASPRS, 2008. LASer (LAS) File Format Exchange Activities. http://www.asprs.org/society/committees/standards/lidar_exchange_format.html (accessed 20 March 2009).

Baribeau, R. and Rioux, M., 1991. Influence of speckle on laser range finders. *Applied Optics*, 30(20), 2873–2878.

Baltsavias, E.P., 1999. Airborne laser scanning: basic relations and formulas. *ISPRS Journal of Photogrammetry and Remote Sensing*, 54(2–3), 199–214.

Beraldin, J.-A., Blais, F., Rioux, M., Domey, J., Gonzo, L., De Nisi, F., Comper, F., Stoppa, D., Gottardi, M. and Simoni, A., 2003. Optimized position sensors for flying-spot active triangulation systems. *Proceedings of the Fourth International Conference on 3-D Digital Imaging and Modeling*, 6–10 October 2003, Banff, Canada, 29–36.

Bergstrand, E., 1960. The geometer: a short discussion of its principal function and future development. *Journal of Geophysics Research*, 65(2), 404–409.

Blais, F., 2004. Review of 20 years of range sensor development. *Journal of Electronic Imaging*, 13(1), 232–240.

Blais, F., Taylor, J., Cournoyer, L., Picard, M., Borgeat, L., Dicaire, L.-G., Rioux, M., Beraldin, J.-A., Godin, G., Lahnanier, C. and Aitken, G., 2005. Ultra-high resolution imaging at 50 μm using a portable XYZ-RGB color laser scanner. *International Workshop on Recording, Modelling and Visualization of Cultural Heritage*, 22–27 May 2005, Ascona, Switzerland.

Born, M. and Wolf, E., 1999. *Principles of optics: electromagnetic theory of propagation, interference and diffraction of light*, with contribution by A. J. Bhatai *et al.*, 7th edn. Cambridge University Press, Cambridge, UK.

Bosch, T. and Lescure, M. (Eds.), 1995. *Selected Papers on Laser Distance Measurements*, vol. MS 115, Thompson, B.J. (General ed.), SPIE Milestone Series, SPIE, Bellingham, WA, USA,

Bouma, B.E. and Tearney, G.J., 2002. *Handbook of Optical Coherent Tomography*, Marcel Dekker Inc., New York, NY, USA and Basel, Switzerland.

Carmer, D.C. and Peterson, L.M., 1996. Laser radar in robotics. *Proceedings of the IEEE*, 84(2), 299–320.

Dorsch, R.G., Hausler, G. and Herrmann, J.M., 1994. Laser triangulation: fundamental uncertainty in distance measurement. *Applied Optics*, 33(7), 1306–1314.

Dupuy, D. and Lescure, M., 2002. Improvement of the FMCW laser range-finder by an APD working as an optoelectronic mixer. *IEEE Transactions on Instrumentation and Measurement*, 51(5), 1010–1014.

El-Hakim, S.F., Beraldin, J.-A., Picard, M. and Cournoyer, L., 2008. Surface reconstruction of large complex structures from mixed range data – the Erechtheion experience. *International Archives of Photogrammetry, Remote Sensing and Spatial Information Sciences*, 37(Part B5), 1077–1082.

Francis, K., LaRocquel, P., Gee, L. and Paton, M., 2003. Hydrographic lidar processing moves into the next dimension. *Proceedings of the US Hydro Conference*, 24–27 March 2003, Biloxi, MS, USA, The Hydrographic Society of America, Rockville, MD, USA; http://www.thsoa.org/hy03/5_3.pdf (accessed 15 June 2009).

Forsen, G.E., 1968. Processing visual data with an automaton eye. *Pictorial Pattern Recognition*, Cheng, G.C., Ledley, R.S., Pollock, D.K. and Rosenfeld, A. (Eds.) Thompson Book Co., Washington, DC, 471–502.

Friess, P., 1989. Empirical accuracy of positions computed from airborne GPS data. In *High Precision Navigation*, Linkwitz, K. and U. Hangleiter (Eds.) Springer Verlag, Berlin and Heidelberg, Germany, 163–175.

Fröhlich, C. and Mettenleiter, M., 2004. Terrestrial laser scanning – new perspectives in 3d surveying. *International Archives of Photogrammetry, Remote Sensing and Spatial Information Sciences*, 36(Part 8/W2), 7–13.

Godin, G., Rioux, M., Beraldin, J.-A., Levoy, M., Cournoyer, L. and Blais, F., 2001. An assessment of laser range measurement on marble surfaces. *5th Conference on Optical 3D Measurement Techniques*, 1–3 October 2001, Vienna, Austria, Wichmann Verlag, Heidelberg, Germany, 49–56.

Guenther, G., 2000. Meeting the accuracy challenge in airborne LIDAR bathymetry. *Proceedings of EARSeL–SIG-Workshop LIDAR*, 14–16 June 2000, Dresden, Germany.

Guidi, G., Frischer B., De Simone, M., Cioci, A., Spinetti, A., Carosso, L., Micoli, L.L., Russo, M. and Grasso, T., 2005. Virtualizing ancient Rome: 3D acquisition and modelling of a large plaster-of-Paris model of imperial Rome. *Videometrics VIII, SPIE*, 5665, 119–133.

Hofton, M.A., Minster, J.B. and Blair, J.B., 2000. Decomposition of laser altimeter waveforms. *IEEE Transactions on Geoscience and Remote Sensing*, 38(4), 1989–1996.

Hosticka, B., Seitz, P. and Simoni, A, 2006. Optical time-of-flight sensors for solid-state 3D-vision. *Encyclopedia of Sensors*, Grimes, C.A., Dickey, E.C. and Pishko, M.V. (Eds.) ASP, Valencia, CA, USA, vol. 7, 259–289.

ISO 11146-1, 2005. Lasers and laser-related equipment—Test methods for laser beam widths, divergence angles and beam propagation ratios—Part 1: Stigmatic and simple astigmatic beams.

Juberts, M. and Barbera, A.J., 2004. Status report on next-generation LADAR for driving unmanned ground vehicles, *Mobile Robots XVII, SPIE*, 5609, Gage D.W. (Ed.), 1–12.

Jutzi, B., 2007. Analyse der zeitlichen Signalform von rückgestreuten Laserpulsen. PhD thesis, Verlag der Bayerischen Akademie der Wissenschaften, Series C, no. 611, 99 p.

Kanade, T., 1987. Three-dimensional machine vision. *Kluwer International Series in Engineering and Computer Science*, Kluwer, Dordrecht, The Netherlands.

Koskinen, M., Kostamovaara, J. and Myllylä, R., 1992. Comparison of the continuous wave and pulsed time-of-flight laser rangefinding techniques. *Optics, Illumination and Image Sensing for Machine Vision VI, SPIE*, 1614, 296–305.

Krabill, W.B., Collins, G.J., Link, L.E., Swift, R.N. and Butler, M.L., 1984. Airborne laser topographic mapping results. *Photogrammetric Engineering and Remote Sensing*, 50(6), 685–694.

Krabill, W.B., 1989. GPS applications to laser profiling and laser scanning for digital terrain models. In *Photogrammetric Week 1989*, Wichmann Verlag, Heidelberg, Germany.

Langer, D., Mettenleiter, M., Härtl, F. and Fröhlich, C., 2000. Imaging ladar for 3-D surveying and CAD modeling of real-world environments. *The International Journal of Robotics Research*, 19(11), 1075–1088.

Lange, R. and Seitz, P., 2001. Solid-state time-of-flight range camera. *IEEE Journal of Quantum Electronics*, 37(3), 390–397.

Lemmens, M., 2009. Airborne lidar sensor: product survey. *GIM International*, 23(2), 16–19.

Lin, Fan-Y. and Liu J.-M., 2004. Chaotic lidar. *IEEE Journal of Selected Topics In Quantum Electronics*, 10(5), 991–997.

Lindenberger, J., 1989. Test results of laser profiling for topographic terrain survey. *Proceedings 42nd Photogrammetric Week*, Stuttgart, Germany, Wichmann Verlag, Heidelberg, Germany, 25–39.

Lindenberger, J., 1993. Laser-Profilmessung zur topographischen Geländeaufnahme. PhD dissertation, Deutsche Geodätische Kommission bei der Bayerischen Akademie der Wissenschaften.

Lohr, U. and Eibert, M., 1995. The TopoSys laser scanner system. *Proceedings Photogrammetric Week*, Stuttgart, Germany, Wichmann Verlag, Heidelberg, Germany, 263–267.

Marshall, G.F., 1985. Laser beam scanning. *Optical Engineering Series*, vol. 8, CRC Press, Boca Raton, FL, USA.

Muguira, M.R., Sackos, J.T., and Bradley, B.D., 1995. Scannerless range imaging with a square wave. *Proceedings SPIE*, 2472, 106–113.

Pfeifer, N. and Briese, C., 2007. Geometrical aspects of airborne laser scanning and terrestrial laser scanning. *International Archives of Photogrammetry, Remote Sensing and Spatial Information Sciences*, 36 (Part 3/W52), 311–319.

Rioux, M., 1984. Laser range finder based on synchronized scanners. *Applied Optics*, 23(21), 3837–3844.

Ruff, W.C., Aliberti, K., Giza, M., Hongen Shen, Stann, B. and Stead, M., 2005. Characterization of a 1X32 element metal-semiconductor-metal optoelectronic mixer array for FM/cw LADAR. *IEEE Sensors Journal*, 5(3), 439–445.

Schenk, T., 2001. Modeling and analyzing systematic errors in airborne laser scanners. *Technical Notes in Photogrammetry*, No 19.

Schneider, D. and Maas, H.-G., 2007. Integrated bundle adjustment with variance component estimation – fusion of terrestrial laser scanner data, panoramic and central perspective image data. *International Archives of Photogrammetry, Remote Sensing and Spatial Information Sciences*, 36(Part 3/W52) (on CD-ROM).

Schneider, R., Thürmel, P. and Stockmann, M., 2001. Distance measurement of moving objects by frequency modulated laser radar. *Optical Engineering*, 40(1), 33–37.

Schwarz, K.P., Chapman, M., Cannon, M., Gong, P. and Cosandier, D., 1994. A precise positioning/attitude system in support of airborne remote sensing. *Proceedings of the GIS/ISPRS Conference*, 6–10 June, Ottawa, Canada.

Seitz P., 2007. Photon-noise limited distance resolution of optical metrology methods. *Optical Measurement Systems for Industrial Inspection V, Proceedings of SPIE*, 6616, Wolfgang Osten, Christophe Gorecki, Erik L. Novak (Eds.), SPIE, Bellingham, WA, USA, 66160D.

Self, S.A., 1983. Focusing of spherical Gaussian beams. *Applied Optics*, 22(5), 658–661.

Singer, C., Holmyard, E.J., Hall, A.R. and Williams, T.I. (Eds.), 1954. *A History of Technology*, Oxford University Press, Oxford, UK and New York, NY, USA.

Skolnik, M.L., 1970. *Radar Handbook*, McGraw Hill, New York, NY, USA, Chapter 37.

Skolnik, M.L. 1980. *Introduction to radar systems*, 2nd edn., McGraw-Hill, New York, NY, USA,

Stone, W. C., Juberts, M., Dagalakis, N., Stone, J. and Gorman, J., 2004. *Performance Analysis of Next-Generation LADAR for Manufacturing, Construction, and Mobility*, National Institute of Standards and Technology, Gaithersburg, MD, USA, NISTIR 7117.

Takeuchi, N., Sugimoto, N., Baba, H. and Sakurai, K., 1983. Random modulation cw lidar. *Applied Optics*, 22(9), 1382–1386.

Wagner, W., Ullrich, A., Melzer, T., Briese, C. and Kraus, K., 2004. From single-pulse to full-waveform airborne laser scanners: Potential and practical challenges. *International Archives of Photogrammetry, Remote Sensing and Spatial Information Sciences*, 35(Part B3), 201–206.

Wagner, W., Ullrich, A., Ducic, V., Melzer, T. and Studnicka, N., 2006. Gaussian decomposition and calibration of a novel small-footprint full-waveform digitising airborne laser scanner. *ISPRS Journal of Photogrammetry and Remote Sensing*, 60(2), 100–112.

Walsh, E.J., 1984. Electromagnetic bias of 36 GHz RADAR altimeter measurements of MSL. *Marine Geodesy*, 8(1–4), 265–296.

Wehr, A. and Lohr, U., 1999. Airborne laser scanning – an introduction and overview. *ISPRS Journal of Photogrammetry and Remote Sensing*, 54(2–3), 68–82.

Wozencraft, J.M. and Millar, D., 2005. Airborne LIDAR and integrated technologies for coastal mapping and nautical charting. *Marine Technology Society Journal*, 39(3), 27–35.

BLOM
IMAGING THE WORLD

Visualisation and Structuring of Point Clouds | 2

George Vosselman and Reinhard Klein

Once a point cloud has been derived from the laser scanner observations, visualisation is commonly used to inspect whether the complete object or area has been recorded and to obtain a first impression of the data quality. Images of point clouds efficiently present the large amount of information that is captured by a laser scanner. Various techniques like colouring or shading can be used to obtain an easy to interpret overview or to emphasise local phenomena. Although visualisations of point clouds are useful for many purposes, most laser scanning projects will require further processing of the point clouds to extract information like surface shapes, tree heights and density, building or plant models, or distances between power lines and vegetation. To extract this kind of information, further structure has to be given to the point cloud. Point clouds are delivered as long lists of X-, Y- and Z-coordinates, possibly with attributes such as amplitude or echo count. Although points are usually listed in scan line order, this does not yet enable one to, for example, quickly determine all neighbouring points within some radius. To make such queries data structures such as a Delaunay triangulation, k-D tree or octree have to be introduced. Once a data structure is available, further structure can be given to a point cloud by determining sets of nearby points that belong to the same planar or smooth surface. Features extracted by such a point cloud segmentation can be used for many purposes, such as registration of multiple point clouds, extraction of terrain points from an airborne laser scanning survey, or modelling of buildings.

Before covering these topics, we therefore first discuss the visualisation and structuring of point clouds in this chapter. Various techniques for visualisation are discussed in Section 2.1. Section 2.2 describes the most common data structures used for organising point clouds. It also discusses the advantages and disadvantages of these data structures compared to storing airborne laser data in raster height images. Methods for segmenting point clouds and recognising shapes like planes and cylinders are presented in Section 2.3. As datasets of multiple billion points are becoming more and more common, techniques for data compression and reduction may be required if data cannot be processed sequentially. The last section of this chapter therefore deals with the topic of data compression. Methods for data reduction are addressed in Chapter 4, which covers the extraction of digital terrain models from airborne laser scanner data. These methods intend to preserve important details of the surveyed terrain surface, such as break lines, while thinning the point cloud.

2.1   Visualisation

Visualisation can be achieved by converting the point cloud to various types of raster images, but it is also possible to browse through 3D point clouds and visualise them using point-based rendering techniques like point splatting. In Section 2.1.1 we will first discuss visualisation in raster images. Section 2.1.2 will then elaborate on point splatting.

2.1.1   Conversion of point clouds to images

Visualisation of laser scanner data in imagery is primarily used for airborne laser scanning or terrestrial scans taken from a single position. Data resulting from the combination of multiple terrestrial scans are often truly three-dimensional and therefore less suitable for visualisation in an image. Airborne laser scanning data or data from a single terrestrial scan can, however, be considered as 2.5D and be converted to height or range imagery. In the following sections it is assumed that the data are acquired by an airborne laser scanner.

2.1.1.1 Height images

The creation of a height image from a point cloud consists of three steps:

- definition of the image grid;
- determination of the height for each pixel; and
- conversion of the height to a grey value or colour.

Definition of the image grid

The image grid defines the area that should be imaged as well as the pixel size. For smaller datasets the pixel size can be set to a value close to the point spacing in the point cloud. In this way most pixels will contain one point and all height variations will be visible in the image. To generate an overview image of larger projects, the pixel size can be defined by the size of the project area divided by the desired number of pixels of the overview image.

Determination of pixel heights

The easiest and quickest way to determine the height of a pixel is often referred to as binning. Binning only makes use of the laser points within the bounds of a pixel. The height of a pixel is defined as a function of the height of the points within the pixel. This can be, for example, the average of all point heights. Alternatively, one can also take the height of the point that is nearest to the pixel centre or the median height as the pixel height value.

If a pixel contains no point, its height value is undefined. This may be due to bad surface reflectance properties in the area of the pixel, occlusions next to buildings or other objects, or irregularities in the point spacing. The latter occurs in particular when the pixel size chosen is close to the point spacing. To avoid many small gaps in the height image, the height of pixels without a point is often inferred by interpolation or extrapolation of the heights of the neighbouring pixels.

A more advanced way to determine the height of the pixels is to first define a triangular irregular network (TIN) of the original laser points and then to derive the height of the pixel centres by interpolation in the TIN. To avoid interpolation over large gaps without laser points, a maximum size of the TIN mesh used for interpolation can be set. Interpolation in a TIN is more accurate than binning, but also more time consuming. For most visualisation purposes the result of binning suffices.

Conversion to grey value or colour

Once a height has been determined for every pixel, it needs to be converted to a grey value or colour. The most straightforward method is to determine the minimum and maximum height in a dataset, assign grey value 0 to the minimum height and grey value 255 to the maximum height and perform a linear interpolation for all heights in between. This will, in many cases, however, not result in a good height image. In most datasets there will be a few relatively high objects, whereas most of the other objects and the terrain are low. The image created by linear interpolation between the minimum and maximum height will therefore appear rather dark and low objects will be hard to separate from the terrain. Better results can be obtained assigning grey value 255 to a value that is lower than the maximum height. All heights equal to or above this value will then be imaged white. Alternatively, one can apply a non-linear transformation from heights to grey values that stretch the height differences of the lower pixels, e.g. a square root or logarithmic function followed by a linear contrast stretch.

Similar considerations have to be made when converting the heights to colours. A linear transformation between the height and the hue will often result in an image with large areas in very similar colours.

Figure 2.1 shows two visualisations of a high density point cloud over the town of Enschede, the Netherlands. The dataset was recorded with the FLI-MAP 400 system (Fugro-Inpark 2007) with 20 points/m^2. The pixel size in this image is 0.5 m. The grey value image was created by binning. Because the pixel size is more than twice the average point spacing (of 0.22 m), pixels without points, displayed in black, are hardly caused by irregularities in the point spacing. They are only seen next to buildings (due to occlusions) as well as on some flat roofs that were most likely covered with water

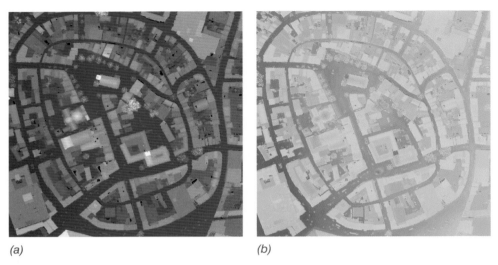

(a)   *(b)*

Figure 2.1 Heights converted to (a) grey value image; and (b) colour image.

and did not sufficiently reflect the laser beam towards the receiver. The conversion of heights to grey values was a linear transformation with a lower than maximum height value set to 255. As a result all points on the two towers in the centre obtained grey value 255 and do not show the height variations of the tower roofs. The conversion from grey values to colours was a piecewise linear transformation emphasising variations in the lower heights.

2.1.1.2 Shaded images

Height images are useful for obtaining a global impression of the surveyed area, but less useful for seeing the shapes of buildings and other objects. In particular, in the case of large height differences in the area, height variations within a building roof are relatively small and will not result in a visible variation of grey tones within the imaged roof. For visualisation of such local height variations, it is useful to derive a shaded image.

A shaded image can be produced by a gradient operator in combination with a transformation to map the gradients to the range of 0 to 255. The gradients g' can be computed by convolving the height image g as in

$$g' = \begin{pmatrix} -1 & 0 \\ 0 & 1 \end{pmatrix} \otimes g \tag{2.1}$$

Note that the height image should contain the original heights. Using an image with heights already converted to grey values reduces the quality of the gradient computation. At locations of height jump edges, like building edges, the gradients will be very large compared to the gradients of roof slopes or terrain slopes. A linear stretch of the gradient range would therefore hardly show the moderate slopes. As in the case of the non-linear transformation of heights to grey values as discussed in the previous section, one should also apply a non-linear transformation to emphasise the moderate slopes. Strong slopes are effectively reduced by applying a logarithmic function such as

$$g'' = \frac{g'}{|g'|}\log\left(|g'|+1\right) \qquad (2.2)$$

After this transformation, the gradients can be mapped to the range of 0 to 255 by

$$g''' = \frac{g''}{\max\left(|g''|\right)} * 127.5 + 127.5 \qquad (2.3)$$

The shaded image in Figure 2.2(a) was produced with the described procedure. The image clearly shows both roof edges as well as variations in roof slopes. The latter can not be seen well in the height image in Figure 2.1(a). The height image, however, gives a better impression of the global height variations in a scene. An effective method for visualising both the global and local height variations in one image is to multiply the coloured height image with the shaded image. Figure 2.2(b) shows the result of such a multiplication after a contrast enhancement. In this image, the colour is used to obtain an impression of the global height distribution, whereas the gradients reveal the local height variations.

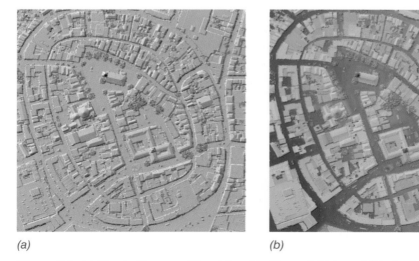

(a) (b)

Figure 2.2 (a) Shaded image of the dataset in Figure 2.1(a); and (b) shaded colour image derived by pixel-wise multiplication of the shaded image with Figure 2.1(b) and a contrast enhancement.

2.1.1.3 Other imagery

Most laser scanners record the amplitude of the reflected laser pulse. As lasers operate in the near infrared part of the spectrum, this reflectance strength can be used to create a monochromatic near infrared image. An example is shown in Figure 2.3(b). These reflectance strength images are sometimes also called intensity images. They are valuable for scene interpretation and can support the understanding of the objects recorded in the point cloud, in particular if no other optical imagery is recorded simultaneously.

For various reasons, the images can, however, be noisy. The amount of energy emitted by laser scanners may vary from pulse to pulse by 15–20%. This, of course, also leads to variations in the strength of the reflected energy. Furthermore, laser scanners only register the maximum amplitude of the returned signal. In comparison, an optical sensor integrates the amount of incoming energy over the exposure time. This integration reduces the noise, but is currently not applied in laser scanners. Finally, the small footprint compared to the distance between points, as well as the distribution of the energy over multiple reflections in the case of scatterers at different heights within the footprint, further contribute to the noisy appearance of reflectance strengths.

Much information can also be retrieved from point density images. These images are derived by simply counting the number of points within a pixel. The image in Figure 2.3(c) shows variations in point density due to small rotations of the aircraft. It also shows higher than average point densities in areas with vegetation as well as along many building edges. This increased number of points is caused by multiple reflections of the pulses as well as by points on the building walls. Fewer points than average are recorded on water surfaces due to absorption of the pulse energy by the water.

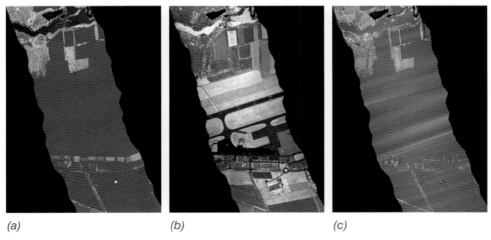

(a)   (b)   (c)

Figure 2.3 (a) Height image; (b) reflectance strength image; and (c) point density image of a strip over Münster Airport, Germany. (Data courtesy TopScan GmbH)

2.1.1.4 Visualisation for quality control

Visualisation is a powerful tool for checking various aspects of the quality of a laser scanning survey. To inspect whether the complete project area has been scanned, Figure 2.4 displays the areas covered by the strips as well as their overlaps. For every pixel in this image, the number of strips with data in the pixel was counted. In the case of Figure 2.4 the image immediately shows several large gaps between strips due to an unfortunate mistake in the flight planning. Note that some of the areas without data are caused by non-reflecting water surfaces.

Most project requirements specify a minimum number of points per square metre. By thresholding a point density image, like Figure 2.3(c), areas with a lower point density can easily be identified. From the shapes of the areas in Figure 2.5 it can be seen that some areas have lower point density due to water surfaces, whereas other areas are due to instabilities of the platform. Small gaps between the strips are also visible.

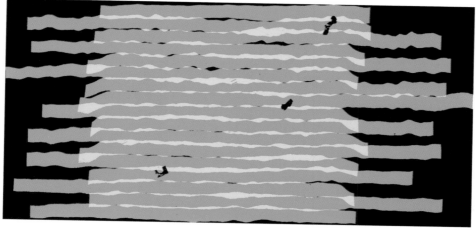

Figure 2.4 Coverage of a project area. Black: area not covered in survey. Green: area covered by one strip. Yellow: area covered by two strips. (Data courtesy EuroSDR)

0   1-7   > 8 pts/m^2

Figure 2.5 Varying point densities due to platform instability and water surfaces.

Images of height differences between strips are useful as a first check on the accuracy of a survey. Calibration errors, such as incorrectly estimated bore-sight misalignments, often result in systematic planimetric offsets between strips. These offsets result in height differences on sloped surfaces. This is shown in Figure 2.6. In the flat areas the height differences below 10 cm are imaged in grey. On gable roofs, however, clearly systematic positive or negative height differences are shown in colour. Height differences may also occur due to vegetation. Whereas the laser beam may reflect from leaves in one strip, it may penetrate to the ground in the overlapping strip due to the different scan direction. These height differences are generally large. To enable the detection of small systematic height differences, the large height differences are ignored and their locations are imaged in grey. The remaining small height differences in low vegetation show up as salt and pepper noise.

Figure 2.7 displays height differences between overlapping strips in a flat area. These height differences were explained by improper functioning of the inertial measurement unit. Height differences larger than 0.3 m are mapped to grey value 127, which is also used to show areas without significant height differences. In this way the large height differences due to vegetation are ignored and the images clearly show the systematic pattern of small height differences.

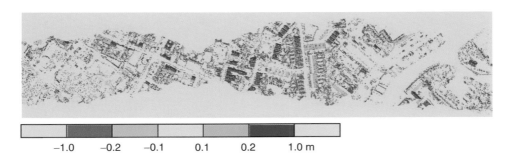

Figure 2.6 Height differences between two strips due to planimetric offsets.

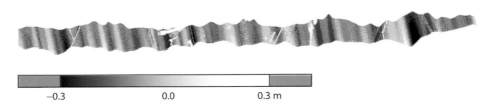

Figure 2.7 Height differences between two strips due to problems with the inertial measurement unit.

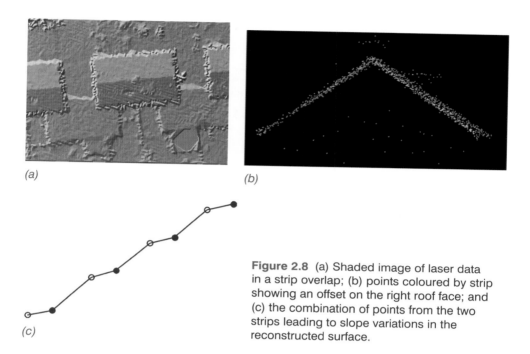

Figure 2.8 (a) Shaded image of laser data in a strip overlap; (b) points coloured by strip showing an offset on the right roof face; and (c) the combination of points from the two strips leading to slope variations in the reconstructed surface.

Problems with the accuracy of the recorded data may also show up in shaded images of overlaps between strips. In the case of a height error or planimetric error the point clouds of both strips will show an offset. The combination of points from these strips results in a systematic pattern in which the scan line direction and the extent of the overlap can be recognised. An example is shown in Figure 2.8(a). The sloped roof of the middle building shows a striped pattern in the shaded image as if the roof surface had many local variations in slope. In reality this is caused by the combination of points from two strips (Figure 2.8b).

The above examples show that visualisation can be a powerful tool for detecting various kinds of errors in the laser scanning data. Techniques to model and eliminate systematic errors will be addressed in Chapter 3.

2.1.2 Point-based rendering

Whereas the above visualisation methods convert the point clouds to raster images, point clouds can also be browsed directly with point cloud viewers. Nowadays, standard graphics APIs such as OpenGL or DirectX support points as a primitive. In addition to the x, y, z coordinates of the point they allow specification of the size of the point in image space (in pixels), material properties like colour and the normal in a point.

Depending on the graphics hardware the point is rendered as a quad or a circle, with uniform depth in the image space coordinate system. Textures cannot be specified.

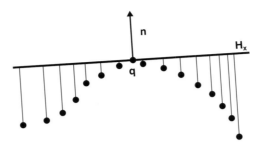

Figure 2.9 Estimating the normal direction in a point q. The non-linear least-squares problem is solved by an eigenvector computation.

Occlusion is resolved using the standard z-buffer algorithm [Shirley, 2005]. The big advantage of this kind of rendering is that only simple preprocessing of the point cloud is required and the rendering is fast due to hardware support so that currently about 50 million points can be rendered per second.

Note that without specifying normals no shading can be performed and since scanning techniques do not deliver points and normals, for each point the normals have to be estimated in a preprocessing step. This can be done by formulating the task as a least-squares problem (see Figure 2.9).

Given the point cloud $P = \{\mathbf{p}_i\}_{i=1,...,N}$ and a point $\mathbf{q} \in P$, we search a plane $H(\mathbf{x}): \mathbf{n}'\mathbf{x} - \mathbf{n}'\mathbf{q} = 0, \|\mathbf{n}\| = 1$ passing through $\mathbf{q}$ that minimises $\sum_i \left(\mathbf{n}'\left(\mathbf{q} - \mathbf{p}_i\right)\right)^2$

where the points $\mathbf{p}_i$ should be taken from a local neighbourhood of $\mathbf{q}$. This can be done by considering only the k-nearest neighbours of $\mathbf{q}$ or by weighting close points with a locally supported weight function θ. Using such a weight function, the normal is defined as

$$\mathbf{n} = \min_{\|\mathbf{n}\|=1} \sum_i \left(\mathbf{n}'\left(\mathbf{q} - \mathbf{p}_i\right)\right)^2 \theta\left(\|\mathbf{p}_i - \mathbf{q}\|\right) \qquad (2.4)$$

This non-linear optimisation problem can be solved by rewriting the functional

$$\sum_i \left(\mathbf{n}'\left(\mathbf{q} - \mathbf{p}_i\right)\right)^2 \theta\left(\|\mathbf{p}_i - \mathbf{q}\|\right) = \mathbf{n}'\underbrace{\left(\sum_i \left(\mathbf{q} - \mathbf{p}_i\right)\left(\mathbf{q} - \mathbf{p}_i\right)' \theta\left(\|\mathbf{p}_i - \mathbf{q}\|\right)\right)}_{\text{weighted variance matrix}} \mathbf{n} \qquad (2.5)$$

Performing a singular value decomposition of the weighted variance matrix $\sum_i \left(\mathbf{q} - \mathbf{p}_i\right)\left(\mathbf{q} - \mathbf{p}_i\right)' \theta\left(\|\mathbf{p}_i - \mathbf{q}\|\right) = \mathbf{U}\text{diag}\left(\lambda_0, \lambda_1, \lambda_2\right)\mathbf{U}'$ the functional evaluates to

$$\mathbf{n}'\mathbf{U}\text{diag}\left(\lambda_0, \lambda_1, \lambda_2\right)\mathbf{U}'\mathbf{n} = \lambda_0\left(\mathbf{U}'\mathbf{n}\right)_x^2 + \lambda_1\left(\mathbf{U}'\mathbf{n}\right)_y^2 + \lambda_2\left(\mathbf{U}'\mathbf{n}\right)_z^2 \qquad (2.6)$$

If we order the eigenvalues in such a way that $\lambda_0 \leq \lambda_1 \leq \lambda_2$ the functional takes its minimum under all unit vectors $\mathbf{n}$ for $\mathbf{U}^t\mathbf{n} = (1,0,0)$,which in turn means that $\mathbf{n}$ is the eigenvector of the weighted variance matrix to the smallest eigenvalue. Note that the other two eigenvectors of the weighted variance matrix can be used to define a natural orthogonal coordinate system in the tangent space T_q of the surface at $\mathbf{q}$.

Unfortunately, simple point rendering only delivers very restricted rendering quality. Small distances between point cloud and observer lead to blocky artefacts and large distances to aliasing. In the first case upsampling of the point cloud is omitted; in the second case appropriate downsampling is neglected (see Figure 2.10).

One simple way to perform downsampling and to overcome the aliasing problems is to organise the point cloud in a hierarchical manner, as in QSplat [Rusinkiewitz and Levoy, 2000], which generates a hierarchy of bounding spheres. This hierarchy is used for visibility culling, level-of-detail control, and rendering (see Figure 2.11). Each node of the tree contains a sphere centre and radius, a normal and, optionally, a colour. For optimisation purposes, in the inner nodes the maximum angular deviation of the point normals of the subtree from the average normal of an inner node is stored (width of a normal cone). During rendering the appropriate spheres are selected from the hierarchy depending on the distance between observer and sphere and then rendered as simple points with material and normal. The selection of the spheres is done by traversing the hierarchy from the root to the leaves.

A more general high quality rendering approach for points incorporating upsampling of the data for close-up views and appropriate downsampling for distant views is surface splatting, the most widely used technique for advanced point rendering

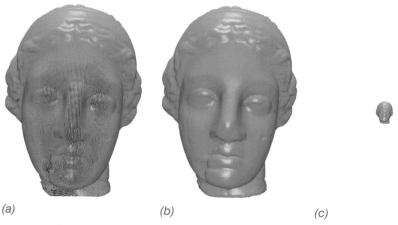

(a)　　　　　　　　*(b)*　　　　　　　　*(c)*

Figure 2.10 Problems of simple point rendering. (a) If the observer is close to the objects, the points become visible. (b) Increasing the point size leads to blocky artefacts, e.g. at the eyes or the tip of the nose. (c) If the viewer is far away, aliasing artefacts occur, e.g. the nose appears overly large.

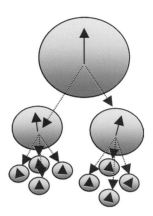

Figure 2.11 Bounding sphere hierarchy.

[Zwicker *et al.*, 2001] (see Figure 2.12). In a splat representation of a point cloud the original surface S around the points is locally approximated by a small disk or ellipse with normal, called a splat, in order to fill gaps between the sample points efficiently. The approximation error of this representation depends on second-order derivatives. The sampling density can be low in flat areas of the surface where large splats are used and is higher with smaller splats in curved areas (see Figure 2.13). In this way the approximation error can be decreased by a factor of two by approximately doubling the number of splats.

Given a point set $P = \{\mathbf{p}_i\}_{i=1,\dots,N}$ with normals $\{\mathbf{n}_i\}_{i=1,\dots,N}$ and colour values $\{c_i = (r_i, g_i, b_i)\}_{i=1,\dots,N}$ as attributes $\{a_i\}_{i=1,\dots,N}$, we define a continuous attribute function f_a on the surface S represented by the set of points as follows

$$f_a(\mathbf{q}) = \sum_k a_k r_k \left(\Pi_k(\mathbf{q}) - \Pi_k(\mathbf{p}_k) \right) \tag{2.7}$$

where $\mathbf{q}$ is a point on S, a_k are the attributes of point $\mathbf{p}_k$, Π_k the projection operator onto the tangent plane of S at $\mathbf{p}_k$ and $r_k : T_{\mathbf{p}_k} \to \mathfrak{R}$, a basis function with local support around $\mathbf{p}_k$. In general a perspective projection m is used to project a surface point $\mathbf{q}$ onto a point $\mathbf{q}' = m(\mathbf{q})$ in the image plane. The attributes of the projected point $\mathbf{q}'$ should be exactly the same as the attributes of

$$g_a(\mathbf{q}') = f_a\left(m^{-1}(\mathbf{q}')\right) = \sum_k a_k r_k \left(\Pi_k\left(m^{-1}(\mathbf{q}')\right) - \Pi_k(\mathbf{p}_k) \right)$$
$$= \sum_k a_k \underbrace{r_k \circ \Pi_k \circ m^{-1}}_{r'_k} \left(\mathbf{q}' - m(\mathbf{p}_k)\right) = \sum_k a_k r'_k \left(\mathbf{q}' - m(\mathbf{p}_k)\right) \tag{2.8}$$

This equation states that we can first warp each basis function r_k defined on the tangent plane $T_{\mathbf{p}_k}$ individually into a basis function $r'_k = r_k \circ \Pi_k \circ m^{-1}$ in the image plane and

(a)

(b)

(c)

Figure 2.12 Splatting solves some of the problems of simple point rendering. (a) Splatted model; (b) close up view does not show blocking artefacts; and (c) distant views are anti-aliased.

then sum up the contribution of these kernels in image space. This approach is called surface splatting [Zwicker *et al.*, 2001].

To simplify computations, instead of the mapping $\Pi_k \circ m^{-1}$, which maps a point in image space first onto a point on the surface by the inverse of the projective mapping m^{-1} and then onto the tangent plane $T_{\mathbf{p}_k}$, a mapping m_k^{-1} is used, which maps a point in image space directly onto the tangent plane $T_{\mathbf{p}_k}$ by m^{-1}. Although this approximation leads to a view-dependent weighting of the attributes in most cases the resulting artefacts can be neglected.

So far the surface splatting algorithm directly rasterises the object space reconstruction filter r_k without taking into account the pixel sampling grid in image space. Thus, in the case of minification and high frequencies, resampling filters may miss pixels leading to aliasing artefacts. In order to overcome this problem, in the original splatting algorithm the reconstruction kernels r_k are band-limited with a screen-space prefilter. This can be done efficiently by using an affine mapping approximation instead of the projective mapping and Gaussians for both the reconstruction kernel and the low-pass filter.

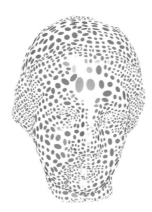

Figure 2.13 An illustration of splats on downsampled point clouds using the model in Figure 2.10. Flat areas are represented by larger splats.

To get the final reconstruction kernel $\hat{r}_k$ in the image a Gaussian low-pass filter in image space can then easily be convolved with the reconstruction kernel r'_k.

For the evaluation of the actual normal and colour attributes these individual reconstruction kernels $\hat{r}_k$ of the points $\mathbf{p}_k$ have to be evaluated and their contributions accumulated for each pixel while taking into account visibility information. To eliminate hidden surfaces, a slightly modified z-buffer can be used. Instead of discarding pixel contributions from splats with a greater depth than already contained in the z-buffer, the z-test checks if the depth of the current splat is greater than the depth value of the z-buffer plus some user-defined threshold epsilon. In order to deal with the infinite support of the Gaussian reconstruction kernel and to come up with a computationally efficient approximation, the kernels are only evaluated in an area where they are greater than a given threshold.

Beyond purely point-based rendering techniques several approaches have already shown that mixing splats with polygons achieves outstanding performance without deficiencies with respect to rendering quality [Guthe *et al.*, 2004; Wahl *et al.*, 2005; Guennebaud *et al.*, 2006]. Indeed, even though points are particularly efficient for handling and rendering complex geometries, they are also significantly less efficient for representing flat surfaces. Therefore, hybrid rendering approaches dynamically select for each part of the model the best representation at the best level-of-detail. A detailed representation and overview of current point rendering techniques, including reconstruction and animation techniques, is given in a monograph on point-based graphics [Gross and Pfister, 2007].

2.2 Data structures

Images like those generated in Section 2.1.1 are a very convenient data structure for 2.5D data. They allow one to make use of a large collection of commercial and public domain image-processing tools for information extraction. The image raster implicitly

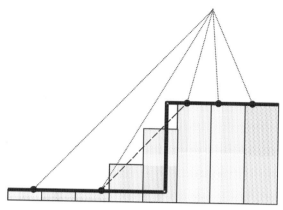

Figure 2.14 Interpolation of heights over an occluded area next to a building.

defines four- or eight-connected neighbours of a pixel, requires relatively little memory, and can be processed quickly. Yet, there are several reasons why images are often not an optimal data structure for processing laser scanning data. In order not to lose resolution, the pixel size of an image has to be chosen close to the point spacing of a point cloud. The resulting pixels without laser points then have to be filled by interpolation. These interpolated heights are, for example, incorrect in the case of areas that are occluded by a building (Figure 2.14). Because the heights are somewhere between the ground and the building roof, these pixels are likely to lead to classification errors. Storing the laser points in a raster image also complicates the recognition of multiple surfaces. This is in particular true for merged point clouds of various terrestrial scans around an object, but also holds for the recognition of a ground surface below vegetation.

For these reasons other data structures may be preferred for processing the laser data, even though this may require a larger programming effort and result in slower algorithms. The next sections discuss the suitability of Delaunay triangulations, k-D trees and octrees for structuring laser scanner data.

2.2.1 Delaunay triangulation

A triangulation of a set of points divides the convex hull of the points into a set of triangles such that the nodes of the triangles correspond to the point set. The Delaunay triangulation is a very popular triangulation and has the property that a circle through the three points of any triangle does not contain any other point of the point set [e.g. Okabe *et al.*, 1992]. The Delaunay triangulation also creates compact triangles with the largest minimum angle. An example of a triangulated laser scanning point cloud is shown in Figure 2.15.

Within a triangulated point cloud the edges of the triangles define the neighbourhood relationships between the points. The so-called first neighbours of a point are directly connected by an edge in the triangulation. Most of these edges will be short.

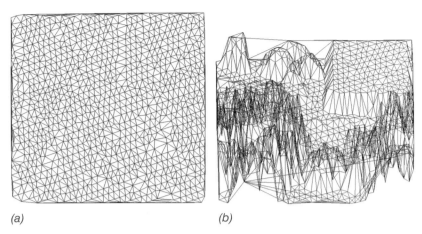

(a)   *(b)*

Figure 2.15 (a) Top view; and (b) slanted view of a triangulated point cloud with a building, terrain and trees[1].

Only near the convex hull will very elongated triangles with long edges occur. Second neighbours can be defined as points that are indirectly connected through one other point (which has to be a first neighbour). For algorithms that require a set of neighbouring points around a selected point, one can select the neighbours connected through the triangulation that are within a specified distance. Many algorithms that have been designed for raster data (e.g. image segmentation, mathematical morphology or image matching) can also be applied to data organised in a Delaunay triangulation.

It should, however, be noted that a triangulation is defined in a plane, typically the *XY*-plane. The height distribution of points has no influence on a triangulation in the *XY*-plane. As a consequence, points that are very close in the *XY*-plane, and therefore share a triangle edge, may not be close in 3D. Because the Delaunay triangulation is still a 2D data structure, it cannot solve all raster data structure problems. In particular, in the presence of multiple surfaces above each other, the triangulation will generate many edges between points of these surfaces that may be far apart. In Figure 2.15(b) this is visible as the many edges between points on the trees and points on the terrain surface. For datasets with multiple layers, the data structures described in the next sections are more suitable.

2.2.2 Octrees

Octrees are efficient tree data structures for handling large point sets. In addition to spatial indexing, common uses of octrees in the context of point cloud processing

---

1 The triangulation was calculated with the software package Triangle [Shewchuk, 1996]. The software contains implementations of robust and efficient algorithms for both standard and constrained Delaunay triangulations. It can be downloaded from http://www.cs.cmu.edu/~quake/triangle.html.

are compression, streaming of levels of detail and view frustum culling. Octrees are the 3D analogue of quadtrees. Each node of the tree represents a cuboidal volume, also called a cell. Starting with a 3D bounding box cell enclosing the entire data space as the root, each internal cell containing data is recursively subdivided into up to eight non-empty octants. The *region octree* uses a regular binary subdivision of all dimensions. Note that in this way the location of the internal cells is implicitly defined by their level and position in the octree (see Figure 2.16). A *point octree* shares the same features of all octrees but the centre of subdivision is always at an arbitrary point inside a cell. While the point octree adapts better to the data than the region octree, it needs to store additional data for the split positions.

A node of an octree is similar to a node of a binary tree, with the major difference being that it has eight pointers, one for each octant, instead of two for "left" and "right" as in an ordinary binary tree. A point octree additionally contains a key which is usually decomposed into three parts, referring to x, y and z coordinates. Therefore a node contains the following information:

- eight pointers: oct[1],…,oct[8];
- (optionally) a pointer to parent: oct;
- pointer to list of data points (commonly stored in leaves only, oct[i] can be used); and
- split coordinates x, y, z (optionally for the region octree).

An octree is said to be complete if every internal node has exactly eight child nodes. The number of internal nodes (n_{nodes}) in a complete octree with resolution s is given by

$$n_{\text{nodes}} = \sum_{i=0}^{(\log_2 s)-1} 8^i = \frac{s^3 - 1}{7} \approx 0.14\, n_{\text{data points}} \tag{2.9}$$

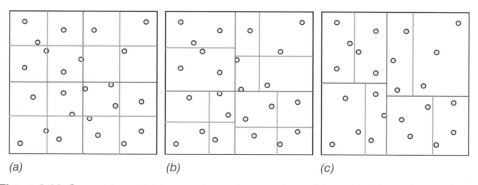

Figure 2.16 Comparison of (a) a regular region quadtree, (b) an adaptive point quadtree, and (c) a 2D tree.

Therefore, storing the hierarchy of a complete octree instead of the leaves only requires an overhead of about 0.14 but has the advantage that it allows for a simple array-like data structure and organisation. No pointers are needed. The array element at index 1 is the root, and element i has the elements $8i$-6, $8i$-5,...,$8i$, $8i + 1$ as child octants. Such a representation is referred to as a linear octree. An example of the use of octrees for compression can be found in Section 2.4.

2.2.3 k-D tree

In contrast to octrees, k-D trees can guarantee a fully balanced hierarchical data structure for point sets sampled from a k-dimensional manifold. Originally, the name stood for k-dimensional tree. k-D trees are a compact spatial data structure well suited for important tasks like the location of nearest neighbours. The k-D tree is a k-dimensional binary search tree in which each node is used to partition one of the dimensions. Therefore, to organise a 3D point set a 3D tree is used. Each node in the tree contains pointers to the left and right subtrees. All nodes except for the leaf nodes have one axis-orthogonal plane that cuts one of the dimensions (x, y or z) into two pieces. All points in the left subtree are below this plane and all points in the right subtree are above the plane. Each leaf node contains a pointer to the list of points located in the corresponding cell defined by the intersection of all half-spaces given by nodes in the path of the root node to the leaf node itself. This data structure makes it possible to locate a point in the k-D tree with N points in time $O(\log N)$ on average but at worst in time $O(N)$ if the tree is very skewed. If the tree is balanced, the worst case time becomes $O(\log N)$. The average time to locate the k nearest neighbours is in the order of $O(k + \log N)$.

Similar to the linear octree a left balanced k-D tree can be represented very compactly by using a heap-like data structure [Sedgewick, 1992]. In this structure it is not

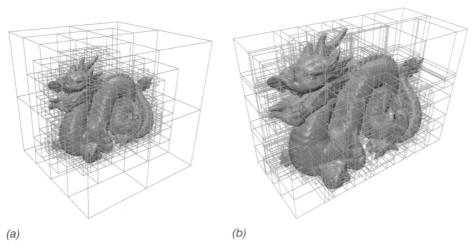

(a)   *(b)*

Figure 2.17 (a) An octree, and (b) a 3D tree. The 3D tree adapts better to the given data than the octree.

necessary to store pointers to the subtrees explicitly. In the heap structure the array element at index 1 is the root, and element i has element $2i$ as left child and element $2i + 1$ as right child. Left-balancing the tree ensures that the heap does not contain empty slots. Not storing pointers saves eight bytes (on a 32-bit architecture), which leads to substantial savings when a large number of points has to be processed.

The construction of a balanced k-D tree starts with the bounding box enclosing all points and recursively splits the current cell along one dimension into two subregions enclosing an equal number of elements by calculating the median of the corresponding coordinate of the point set. An efficient median search algorithm can be found in most textbooks on algorithms [for example Sedgewick, 1992]. Instead of repeatedly cycling through the split dimensions, the best results are achieved by splitting the dimension of largest spatial extent at each recursion step. A comparison of the 2D analogue of an octree with a k-D tree is shown in Figure 2.16. While a point octree may avoid empty cells and improve data distribution over a region octree, the k-D tree can achieve a better fill rate using a minimal number of subdivisions. A 3D tree is shown in Figure 2.17.

A generic k nearest-neighbours search algorithm begins at the root of the k-D tree and adds points to a list if they are within a certain distance. For the k nearest neighbours, the list is sorted such that the point that is furthest away can be deleted if the list contains k points and a new closer one is found. Instead of naïve sorting of the full list it is better to use a max heap [Sedgewick, 1992], also called a priority queue. When the max heap contains k elements, we can use the distance of the point in the heap that is furthest away to adjust the range of the query [Samet, 1990].

2.3 Point cloud segmentation

Segmentation is often the first step in extracting information from a point cloud. A segmentation algorithm groups points that belong together according to some criterion. The most common segmentations of point clouds are those that group points that fit to the same plane, cylinder or smooth surface. Segmentation is then equivalent to the recognition of simple shapes in a point cloud [Vosselman *et al.*, 2004]. These segments can be used for a variety of purposes. Planar segments are required for modelling buildings (Chapters 5 and 8), but are also applied to register terrestrially scanned point clouds (see Section 3.3.1.3 on feature-based registration). Recognition of cylinders and tori is required for modelling industrial installations (see Section 7.1) and smooth surface segments can be used to extract the terrain surface from airborne laser scanning point clouds (see Section 4.1.4 on segment-based filtering). This section discusses several frequently used segmentation methods.

2.3.1 3D Hough transform

The Hough transform is a well-known method for the detection of lines in a 2D image [Hough, 1962; Ballard and Brown, 1982]. After a brief explanation of this classical method, it is extended to a method for detection of planes in a 3D point cloud. Furthermore, a method for the detection of cylinders is presented.

2.3.1.1 Detection of lines in a 2D space

A line in 2D can be described by

$$d = X\cos\alpha + Y\sin\alpha \tag{2.10}$$

with α as the angle between the line and the Y-axis and d as the distance between the line and the origin. The Hough transform is based on the duality between points and lines in 2D. For a fixed α and d, Equation (2.10) defines the infinite set of positions (X, Y) on the line. If, however, a fixed X and Y are chosen, the equation can also be interpreted as the infinite set of lines described by (α, d) that pass through the position (X, Y). In order to detect the lines through the points in an object space, the Hough transform maps all points to the lines in a parameter space spanned by the line parameters α and d. An example is given in Figure 2.18. Inserting the coordinates $(-2, -1)$ of the thick point of the Figure 2. 18(a) into Equation (2.10), results in the thick curve in the parameter space shown in Figure 2.18(b). This thick curve defines the set of all possible lines in object space that pass through position $(-2, -1)$. When doing so for all points in the object space, the parameter spaces shows a set of curves. Most of these curves intersect at the location $(135°, \frac{1}{2}\sqrt{2})$. This implies that most points in the object space lie on the line defined by these values of α and d. The Hough transform therefore detects lines in the object space by determining locations in the parameter space with a high number of crossing curves.

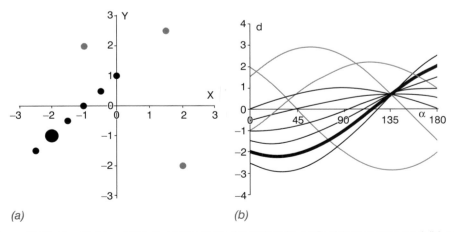

(a)   (b)

Figure 2.18 Hough transform in a 2D space: (a) points in a 2D object space; and (b) corresponding curves in the 2D parameter space.

To do this by computer, the parameter space, also called a Hough space, needs to be discretised. For each bin the number of curves passing through the bin is counted. The bin with the highest count yields the line parameters of the most likely line in the object space. The bin sizes have to be chosen carefully. Due to noise in the point coordinates, the curves in the Hough space may not exactly intersect at one point. When the bin size is very small, the curves may pass through different nearby bins and the maximum in the Hough space will become less pronounced. On the other hand, choosing a large bin size will reduce the accuracy of the estimated line parameters. A way in between may be to choose a relatively small bin size and to convolve the Hough space with a box filter before searching for the maximum.

Once the maximum of the Hough space has been selected, it has to be verified whether all points on the determined line indeed belong to one and the same line segment. In Figure 2.18 seven points are on a line. All corresponding curves intersect in the Hough space. One point is, however, further away from the other six points. This can not be deduced from the Hough space, but needs to be verified with a connected component analysis in the object space.

Due to the discretisation of the Hough space, the line parameters may not be very accurate. A least-squares fit to the selected points is therefore often applied to improve the line parameter values.

After the first line has been extracted, the contributions of the points of that line to the Hough space are removed by decrementing the corresponding counts in the Hough space. The process of extracting the best line from the Hough space continues until no more lines with a minimum number of points can be found.

2.3.1.2 Detection of planes

The principle of the Hough transform can be extended to 3D for the detection of planes in point clouds [Vosselman, 1999; Maas and Vosselman, 1999]. A plane in a 3D space can be described by

$$d = X \cos \alpha \cos \beta + Y \sin \alpha \cos \beta + Z \sin \beta \qquad (2.11)$$

Every point (X, Y, Z) in a point cloud defines a surface in the 3D parameter space spanned by the plane parameters α, β and d. Points in the point cloud that are coplanar, will correspond to surfaces in the parameter space that intersect at a point. The α, β and d coordinates of this intersection point yield the parameters of the plane in the point cloud. When discretising the parameter space with constant increments of α and β, the bin areas on the Gaussian sphere of all normal directions will vary. The selection of the best plane will then be biased towards the bins with the larger sizes. To obtain an unbiased selection, the number of points in a bin can be divided by the size of the bin on the sphere spanned by α and β, or the parameter space can be discretised with increments of α that depend on the value of β [Rabbani, 2006].

Before transforming the points to the parameter space first reducing all coordinates to the centre of the point cloud is recommended. By adopting this local coordinate system the range of possible values of the plane parameter d is minimised. This results in a smaller parameter space and thus reduced memory requirements.

After the selection of a maximum in the parameter space, it needs to be verified whether the points that chose this maximum are adjacent to each other. To perform this connected component analysis, a data structure such as a triangulation, octree or k-D tree is required.

Once the points that belong to a planar surface have been selected, a least-squares plane fitting is often used to improve the accuracy of the plane parameters. In Equations (2.4) to (2.6) it was derived that the normals of the best-fitting plane (in a least-squares sense) can be obtained by a singular value decomposition of a weighted variance matrix. The eigenvector corresponding to the smallest eigenvalues specifies the normal direction of the plane. It can also be shown that the centre of gravity of the set of points is located on the best-fitting plane [Weingarten *et al.*, 2004]. Hence, the least-squares fitting of a plane to a point cloud can be obtained directly from the eigenvector of the smallest eigenvalue in combination with the centre of gravity. This direct least-squares solution does not require iterations with linearised observation equations.

The procedure for detecting multiple planes in a 3D point set is equivalent to detecting multiple lines in a 2D point set. After removing the contributions of the points of an extracted plane from the parameter space, the location of the next maximum in the parameter space is sought and the adjacency of the points voting for this maximum is evaluated. This process continues until no further planes with a minimum number of points can be determined.

While the Hough transform is known to reliably detect planes in a point cloud, not all maxima in the parameter space necessarily correspond to planar surfaces in the object space. In particular, when many planes or other shapes are present in the point cloud, the maximum in the parameter space may be caused by an arbitrary set of coplanar points belonging to various surfaces. An artificial example of a profile of a roof landscape is shown in Figure 2.19. In this case, the Hough transform does not extract one of the roof faces but suggests a plane intersecting several roof faces. Because of the discretisation of the Hough space, a bin in the Hough space corresponds to a zone (volume) around a plane in the object space. This zone represents the set of all planes in object space that fall within one bin of the Hough space. All points in this zone will therefore vote for the same plane in the Hough space. In the example in Figure 2.19, this zone is shown as the grey area. Because the number of points in this area is larger than the number of points on any of the roof faces, the plane intersecting the roof faces obtains the highest count in the parameter space. By checking the adjacency of the points voting for the "best" plane, it will, however, be found that this hypothesis should not be accepted. The detection of a plane then proceeds with the evaluation of points voting for the second maximum in the Hough space.

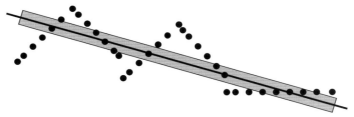

Figure 2.19 Point cloud with an incorrect hypothesis of a Hough transform (solid line). Points in the grey area voted for the hypothesised line.

2.3.1.3 Detection of cylinders

Cylinders are often encountered in industrial landscapes. A cylinder is described by five parameters. Although a 5D parameter space could be defined, the number of bins in such a space make the detection of cylinders very time and memory consuming and unreliable. To reduce the dimension of the parameter space, the cylinder detection can be split into two parts: the detection of the cylinder axis direction (two parameters) and the detection of a circle in a plane (three parameters). For this procedure the availability of normal vectors is required [Vosselman *et al.*, 2004; Rabbani, 2006]

In the first step, the normal vectors are plotted on a Gaussian sphere. Because all normal vectors on the surface of a cylinder point to the cylinder axis, the Gaussian sphere will show maxima on a large circle. The normal vector of that circle is the direction of the cylinder axis. Figure 2.20 shows an industrial scene with a few cylinders. The Gaussian half-sphere of this scene shows maxima on several large circles that correspond to the different axis directions of the cylinders. By extracting large circles with high counts along a larger part of the circle, hypotheses for cylinder axis directions are generated.

All points that belong to a selected large circle are now projected onto a plane perpendicular to the hypothesised cylinder axis. In this plane, the points of a cylinder should be located on a circle. The detection of a circle in a 2D space is a well-known variant of the Hough transformation for line detection [Kimme *et al.*, 1975]. The 3D parameter space consists of the two coordinates of the circle centre and the circle radius. Each point is mapped to a cone in the parameter space, i.e. for each possible value of the radius, each point is mapped to a circle in the parameter space. The circle centre is known to lie on this circle. If use is made of the normal vector information, the number of bins that need to be incremented can again be reduced. In this case each point can be mapped to two lines in the parameter space (assuming that the sign of the normal vector is unknown), i.e. for each radius and a known normal vector, there are only two possible locations of the circle centre.

2.3.2 The random sample consensus algorithm

The random sample consensus (RANSAC) algorithm was developed as a general approach to robust model fitting in computer vision [Fischler and Bolles, 1981]. Its

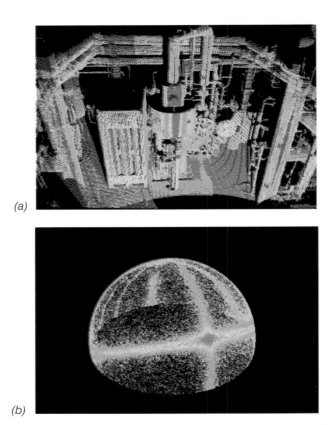

(a)

(b)

Figure 2.20 (a) Industrial scene with points coded by the angle between the surface normal direction and the viewing direction; and (b) Gaussian half-sphere with circles corresponding to the dominant cylinder axis directions. (Vosselman *et al.*, 2004; courtesy the authors)

many applications range from correspondence search and epipolar geometry estimation to detection of simple shapes. Its particular strengths are its conceptual simplicity and resilience to large numbers of gross outliers in the data. To illustrate, a shape fitting to a point cloud is used as an example (see Figure 2.21).

The RANSAC paradigm is based on the notion of minimal sets. A minimal set is the smallest number of data samples required to uniquely define a model. In the example used this is the minimal number of points required to uniquely define a given type of geometric primitive, e.g. three for a plane and four for a sphere. Now, given a set P of data points and a model that requires at least n points to uniquely determine its parameters, i.e. $n = 3$ in the case of planes, RANSAC randomly chooses a predetermined number t of minimal subsets $s_i \in \{S \subset P, |S| = n\}$ of cardinality

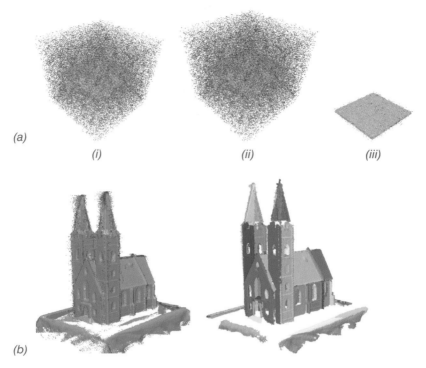

(a)

(i)   (ii)   (iii)

(b)

Figure 2.21 (a) Fitting of a plane to noisy data containing 20% outliers: (i) the data points; (ii) the consensus set of $M*$ in the data points depicted in red; and (iii) the extracted consensus set. (b) A scan of a church. Lots of noise and outliers are contained near the top of the towers. Nonetheless planes are extracted reliably. [Schnabel *et al.*, 2007b]

n in P. For each such subset s_i the uniquely defined model M_i is instantiated and the consensus set $C_i \subset P$, sometimes also referred to as a score, is determined. The consensus set is comprised of all points in P that are within a certain error tolerance ε of the model M_i, i.e. $C_i = \{\mathbf{p} \in P, d\,(M_i, \mathbf{p}) < \varepsilon\}$ where $d(M_i, \mathbf{p})$ denotes the deviation of $\mathbf{p}$ from the model M_i. The consensus set is the critical part of RANSAC that is responsible for the robustness of the algorithm. In effect, the consensus set performs outlier detection: it partitions the data into inliers, which are the points contained in the set, and outliers which are all other points. In the case of fitting simple shapes, C_i is typically comprised of all points that are within a distance ε to the shape defined by M_i. After all t minimal subsets and their consensus sets have been established, the model $M^* \in \{M_i\}$ with the largest consensus set, i.e. the most inliers, is chosen as the best-fitting result. Optionally, M^* can be further refined by least-squares fitting to its consensus set C^*.

There are two critical parameters in the algorithm above: the number t of minimal subsets that are processed and the error tolerance ε. Given a desired success probability z, an appropriate value for t is easily derived as follows

$$(1-a)^t = 1 - z \qquad (2.12)$$

$$t = \frac{\log(1-z)}{\log(1-a)} \qquad (2.13)$$

where a is the probability that a chosen minimal set is "good", i.e. is comprised only of points lying on the correct model. If, for instance, a point cloud contains 20% outliers and we are trying to fit a plane to the data then $a = (1 - 0.2)^n = 0.8^3 = 0.512$. In this case, in order to achieve a success probability of 99%, only seven subsets have to be drawn and tested. Unfortunately, real data, besides containing outliers, typically are also corrupted by a certain amount of noise, so that the above computations do not directly apply since the effect of the noise was not considered in the probability for good minimal sets. The parameter ε in the RANSAC algorithm is used to handle noisy situations. In theory ε should depend on the error in the data points and the error in the models which is caused by using noisy samples for instantiation. In practice it is generally not possible to analytically capture the effect of noise in the minimal set on the instantiated model, so that error tolerances have to be determined empirically. If the noise level of the data is known (e.g. from scanner specifications) a reasonable value for ε usually is a small multiple of the scanning precision. The choice of ε can have a significant impact on the result delivered by the RANSAC algorithm. If ε is too small then the correct solution can have a small consensus set because not all correct points are included due to noise. On the other hand, if ε is large the consensus set of an incorrect solution may be large because it contains many outliers.

A slightly modified version of the RANSAC algorithm, dubbed MSAC (M-estimator sample consensus), which was introduced by Torr and Zisserman (1998), alleviates the problem of large ε. Since RANSAC returns the model M^* with the largest consensus set, it can be considered as finding the model that satisfies

$$M^* = \underbrace{\arg\min}_{M} \sum_{\mathbf{p} \in P} \rho\left(M, \mathbf{p}\right) \qquad (2.14)$$

where

$$\rho\left(M, \mathbf{p}\right) = \begin{cases} 0 & d\left(M, \mathbf{p}\right) < \varepsilon \\ 1 & d\left(M, \mathbf{p}\right) < \varepsilon \end{cases} \qquad (2.15)$$

MSAC replaces ρ with the following measure

$$\tilde{\rho}\left(M, \mathbf{p}\right) = \begin{cases} d\left(M, \mathbf{p}\right)^2 & d\left(M, \mathbf{p}\right) < \varepsilon \\ \varepsilon^2 & d\left(M, \mathbf{p}\right) \geq \varepsilon \end{cases} \tag{2.16}$$

With this cost function robustness is retained as outliers are still given a constant penalty. But now, inliers are scored based on their deviation from the model. This means that if ε is large, the instantiated models can still be distinguished based on their fitting quality.

The problem of fitting shapes to a point cloud is now further considered. Usually a single point cloud contains an unknown number of simple shapes, which all contain a different number of points, e.g. a scan of a house contains planes for the walls, doors, windows and roof, cylinders for the gutters and so on. Using RANSAC, what is the probability of detecting one of these shapes in the data? Consider a point cloud P of size N and a shape ψ consisting of k points. If for simplicity it is assumed that any n points of the shape will lead to an appropriate estimate of the shape, then the probability of detecting ψ in a single pass is the probability of drawing a minimal set from among the k points of ψ

$$P(\psi) = \binom{k}{n} \bigg/ \binom{N}{n} \approx \left(\frac{k}{N}\right)^n \tag{2.17}$$

The probability of a successful detection $P(\psi, s)$ after s candidates have been drawn equals the complementary of s consecutive failures

$$P(\psi, s) = 1 - (1 - P(\psi))^s \tag{2.18}$$

Again, solving for s gives the number of candidates t required to detect shapes of size k with a probability $P(\psi, t) = p_t$

$$t \geq \frac{\ln\left(1 - p_t\right)}{\ln\left(1 - P(\psi)\right)} \tag{2.19}$$

For small $P(\psi)$ the logarithm in the denominator can be approximated by its Taylor series $\ln(1 - P(\psi)) = -P(\psi) + O(P(\psi)^2)$ so that

$$t \approx \frac{-\ln\left(1 - p_t\right)}{P(\psi)} \tag{2.20}$$

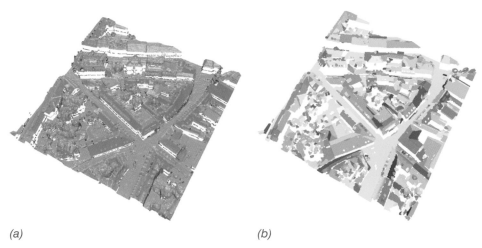

(a)   *(b)*

Figure 2.22 A RANSAC-based segmentation of an urban scan using planar shapes.

Thus, given the complexity F of evaluating the cost function, the asymptotic complexity of the shape detection is $O(t \cdot F) = O\left(\dfrac{1}{P\psi}F\right)$.

See Figure 2.22 for a segmentation result using the RANSAC method [Schnabel *et al.*, 2007b] on an urban scene reconstructed from aerial images. Despite the noise and reconstruction errors, planes are reliably detected on the streets and the roofs but unfortunately also in some of the vegetation. These false detections have to be handled by further classification.

2.3.3 Surface growing

Surface growing can be regarded as an extension to three dimensions of the well-known region growing algorithm. Region growing is used in image processing to group adjacent pixels into segments that fulfil some homogeneity criterion, such as similarity in grey value [Ballard and Brown, 1982]. For processing point clouds surface growing can be formulated as a method of grouping nearby points based on some homogeneity criterion, such as planarity or smoothness of the surface defined by the grouped points [Hoffman and Jain, 1987; Besl and Jain, 1988; Sapidis and Besl, 1995; Vosselman *et al.*, 2004; Vieira and Shimada, 2005]. Surface growing, like region growing, consists of two steps: seed detection and growing.

In the seed detection step a small set of nearby points is sought that forms a planar surface. All points within some radius around an arbitrarily selected point are analysed with a Hough transform, RANSAC search or robust plane fitting to determine whether some percentage of the point set fits to a plane. If this is not the case, another point is chosen arbitrarily and the neighbourhood of this point is analysed. Once a set of coplanar points has been found, this set constitutes the seed surface that will be expanded in the growing phase. All points of the seed surface obtain a label with the surface number.

In the growing phase all points of the seed surface are put onto a stack. Points on the stack are processed one by one. For each of these points, the neighbouring points are determined using a data structure such as a k-D tree. If a neighbouring point has not yet been assigned to a surface (i.e. has no label), it is tested to determine if the point can be used to expand the surface. If the point is within some distance of the plane fitted to the surface points, the surface label is assigned to the point and the point is put onto the stack. In this way all points on the stack are processed and the surface is grown until no more neighbouring points fit to the surface plane.

When a point has been added to the surface, the parameters of the plane can be re-estimated to improve the accuracy. Doing so after the addition of every point, however, leads to an unnecessarily large computational effort. In particular, when a surface already contains many points, the addition of one point near the plane will hardly affect the plane parameters. It is therefore more efficient to only re-estimate the plane parameters once the surface has grown by a certain percentage (20–50%).

As an alternative to growing planar surfaces, the growing phase can also be used to extract smooth, but not necessarily planar, surfaces. In this case a point is taken from the stack and a plane (or higher-order surface) is fitted to the points in the neighbourhood of the stack point that are already assigned to the surface. Hence, the surface parameters are computed locally and do not make use of all points assigned to the surface. The distance of neighbouring unlabelled points to this local surface is then used as a criterion for expanding the surface with these points. The distance threshold for accepting unlabelled points determines the smoothness of the extracted surfaces. The lower the threshold, the smoother the extracted surface will become.

In addition to checking the distance of a neighbouring point to a surface, it may sometimes be useful to also compare the direction of an estimated local surface normal vector around the neighbouring point with the normal vector of the growing plane. Near surface edges, normal vectors may, however, be difficult to estimate.

The above-described procedure of accepting unlabelled points within a specified distance of the growing surface may lead to non-optimal segmentations around edges where two surfaces meet. An example is shown in Figure 2.23(a). The order in which the roof surfaces were grown was from left to right. When growing the roof face on the left side, the surface also accepts some points of the right part of the gable roof as these points are still within the specified distance tolerance. When growing the right part of the gable roof, points of the horizontal part are accepted for the same reason. The amount of incorrect assignments depends on the distance tolerance. The larger the

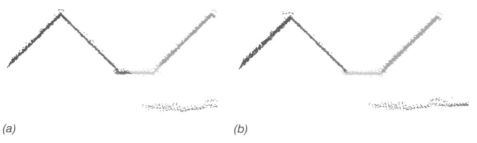

(a)   (b)

Figure 2.23 (a) A segmentation considering only unlabelled points during growing; and (b) a segmentation reconsidering assignments of already labelled points.

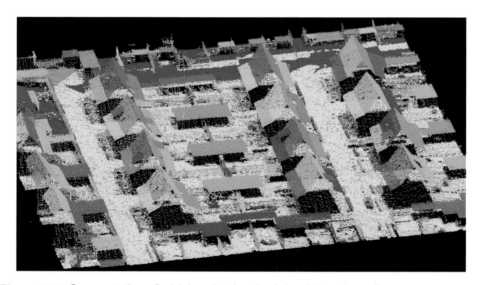

Figure 2.24 Segmentation of a high point density dataset into planar faces.

tolerance, the greedier the surface growing will become. To avoid this behaviour it may therefore be better to test not only unlabelled points but also those that are already assigned to another surface. For the labelled points it needs to be decided which of the surfaces fits the best. Although this increases the processing time, results around edges between surfaces will become more accurate (Figure 2.23b).

Figure 2.24 shows the results of a surface growing algorithm applied to a dataset with 20 points/m². All larger roof faces are clearly identified as well as all dormer windows (of 3 × 3 m in size). Several garden fences and building walls are also extracted as planar segments. As the streets in the north part of the scene are not in the same plane as other points at street level, additional segments were generated in this part.

2.3.4 Scan line segmentation

Scan line segmentation has become popular as a method for segmentation of range images in computer vision [Jiang and Bunke 1994; Hoover *et al.*, 1996]. This approach initially processes the scan lines (rows of a range image) one by one. Each scan line is split into straight 3D line segments. Once all scan lines have been segmented, a surface growing process is used to form the surface segments by merging the detected straight 3D line segments of neighbouring scan lines.

Jiang and Bunke (1994) selected seed surfaces for the growing phase from triples of line segments that satisfied three conditions: all line segments should have a minimum length, the segments should overlap for some percentages, and all pairs of neighbouring points on two line segments should be within a specified distance [Hoover *et al.*, 1996]. A line segment on a neighbouring scan line is added in the surface growing phase. Its end points are close to the plane fitted to the current surface points.

Although designed for range images, scan line segmentation can, of course, also be applied to airborne or terrestrial laser scanner data. The scan lines are already defined by the scanning mirrors of a laser scanner. Alternatively, a dataset can be split into thin slices and all points within a slice considered as points on a scan line [Sithole and Vosselman, 2003; Sithole, 2005].

2.4 Data compression

The huge amount of data acquired with 3D scanners or vision-based techniques is often hard to handle due to its sheer size. Therefore representations that allow compact storage and quick transfer of these data are often required for an efficient workflow when processing the acquired information. A common technique for point cloud compression is to use a hierarchical representation. Early approaches such as QSplat [Rusinkiewitz and Levoy, 2000] were mainly motivated as level-of-detail hierarchies for accelerated rendering but achieved moderate compression as a by-product as well. Here a progressive compression technique for point clouds based on octrees that gives high compression rates and can also be used as a level-of-detail representation will be discussed. Indeed Botsch *et al.* (2002) have shown that high-performance point splatting is well facilitated using a compressed octree data structure.

If an octree is fine enough, i.e. contains many levels, the octree structure allows implicit representation of the points in the cloud as the centres of its leaves. The depth of the tree controls the fidelity of this representation. Indeed, the depth specifies the quantisation of the points with respect to the bounding cube. If, for instance, the octree is constructed with a depth of 16, then the centres of the leaves give a representation of the original point cloud that is quantised to 16 bits in each dimension. The hierarchical structure of the octree representation can now be exploited to derive a compact representation of the implicitly contained quantised points.

The coding of the octree proceeds in a top-down and breadth-first fashion. The root cell can always be assumed to be non-empty. Then, for each cell it is only necessary to encode which child cells are non-empty. The decoder can then faithfully recover the octree by following the same traversal rules as the encoder, while always constructing those child cells that have been specified as occupied by the encoder. Note that no further information has to be encoded for empty cells. Therefore the compression of the octree only depends on the way the non-empty child cells are encoded.

In the case of an octree 8 bits (a single byte) are needed to specify the occupancy of the children for any given cell. A compacted representation is already obtained if one byte per cell is indeed written to disk. If entropy coding, e.g. arithmetic coding, is used, the compression rate can be further increased. Moreover, it is usually possible to predict the occupied child cells from occupied child cells in the neighbourhood and in conjunction with entropy coding even higher compression rates can be achieved.

To analyse the compression achievable with such an octree representation the assumption is made that the point cloud samples the surface densely with respect to the cell size of the leaves, i.e. the quantisation precision. Let L_i denote the occupied cells on level i of the tree. Then L_i can be reconstructed from the cells on the next coarser level L_{i-1} and approximately $\frac{1}{4}|L_i|$ byte codes, if it is assumed that the surface is a two-manifold. Thus, the number of byte codes needed to encode an octree of depth s is approximately

$$\sum_{i=1}^{s-1} \frac{1}{4^i}|L_s| \le \frac{1}{3}|L_s| \tag{2.21}$$

In effect this means that less than 1/3 bytes are required per quantised point even without any further optimisations such as predictive entropy coding. In practice, however, point clouds are rarely sampled densely enough to fulfil the above assumptions so that achievable compression rates are somewhat lower.

The compression introduced by the octree coding is best understood if each leaf in the tree is thought of as being represented by the string that contains the unique path from the root to that leaf. The set of all these strings defines all leaves and thereby implicitly all the points. The octree encoding now effectively performs prefix compression on these strings. Many paths from the root to the leaves share a common prefix and by means of the hierarchical encoding each prefix is indeed encoded only once (Figure 2.25).

On top of the compaction achieved by the common prefix coding, entropy coding can further increase the compression rate by exploiting the fact that byte codes containing four set bits are more likely to appear than other configurations.

The highest compression rates are obtained if predictive entropy coding is employed [Schnabel and Klein, 2006]. Since the octree is traversed in breadth-first order, the

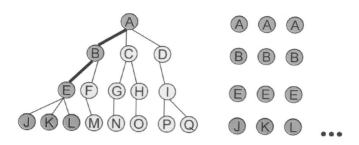

Figure 2.25 The common prefix reduction inherent to the octree coding. The paths from the root to the leaves J, K and L share the common prefix A, B, and E. This section of the path is encoded only once for J, K and L.

Figure 2.26 Progressive evolution of the point cloud during decoding of the octree. The decoder can be stopped at any time and the centres of the leaves of the tree that has been decoded so far give a coarse approximation of the original point cloud. The images were rendered using splatting.

centres of the cells on the traversal front at all times provide a coarse approximation Q of the complete point cloud P (Figures 2.26 and 2.27).

During traversal, Q is refined progressively as additional child cells are visited (or created in the case of the decoder). Indeed the decoder can stop processing at any time and return Q as a preliminary result of the decompression, e.g. in streaming applications. Moreover, as Q is an approximation of the original point cloud and is accessible to both encoder and decoder, it can be used to predict the child cell configurations of the cell subdivisions.

The prediction is based on an approximation of the surface induced by Q. For a leaf cell C that is to be subdivided, a planar approximation of the true underlying surface in C is obtained. The approximation is obtained as the least-squares plane fit to the k nearest neighbours in Q of the cell centre of C. The intersection of this plane with the

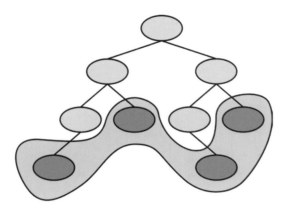

Figure 2.27 The traversal front Q.

child cells and the distance of the child cell centres to the plane can then be used for prediction of the non-empty child cells.

A major deficit of the octree representation with respect to real-time rendering is that it does not provide random access to the nodes at the different levels. For instance it is frequently the case during rendering that only a certain part of the object is actually visible, e.g. the front of a building, and it would therefore be advantageous to decompress and render only the necessary points. Since the octree compression approach needs a breadth-first traversal of the tree to achieve high compression, it is not possible to selectively refine the tree only in the required parts. Therefore several alternative approaches for point cloud compression have been proposed that allow real-time rendering exploiting the parallel computing capabilities of modern graphics hardware. To this end a method based on encoding runs in a uniform grid of trapezo rhombic dodecahedra was proposed by Krüger *et al.* (2005). For each cell in a run the next cell in the run is specified as one of the cell's neighbours in a bit-code. As the method is based on a uniform grid it does not inherently support level-of-detail. Instead several independent grids at different resolutions are created and depending on the viewer's distance to the points a grid with sufficient resolution is chosen for rendering.

Another method focusing on real-time decompression has been proposed [Schnabel *et al.*, 2007a]. Here the point cloud is decomposed into patches that are each approximated by a primitive shape, i.e. a plane, sphere, cylinder, cone or torus (see Section 2.3). The fine-scale geometry is then encoded as height-fields over these patches. These height-fields are compressed progressively using vector quantisation. Vector quantisation has the advantage that decompression can be trivially parallelised and random access is supported. Therefore decompression can easily be performed on the GPU during rendering. For scenes such as buildings, cities or other manmade environments where the simple shapes are predominant, the overall compression rate can be very high since the geometry is already closely approximated by shape primitives.

References

Ballard, D.H. and Brown, C.M., 1982. *Computer Vision*, Prentice-Hall, Inc., Englewood Cliffs, NJ, USA.

Besl, P. and Jain, R., 1988. Segmentation through variable-order surface fitting. *Transactions on Pattern Analysis and Machine Intelligence*, 10(2), 167–192.

Botsch, M., Wiratanaya, A. and Kobbelt, L., 2002. Efficient high quality rendering of point sampled geometry. *Proceedings of the 13th Eurographics Workshop on Rendering (EGRW '02)*, 26–28 June 2002, Pisa, Italy, 53–64.

Fischler, A. and Bolles, C., 1981. Random sample consensus: a paradigm for model fitting with applications to image analysis and automated cartography. *Communications of the ACM*, 24(6), 381–395.

Fugro-Inpark, 2007. FLI-MAP 400. http://www.flimap.nl/ (accessed 17 March 2009).

Guennebaud G., Barthe L. and Paulin M., 2006. Splat/mesh blending, perspective rasterization and transparency for point-based rendering. *Proceedings of the IEEE/Eurographics/ACM Symposium on Point-Based Graphics '06*, 29–30 July 2006, Boston, MA, USA, 49–58.

Gross, M. and Pfister, H. (Eds.), 2007. *Point-Based Graphics*, Morgan Kaufmann Series in Computer Graphics, Elsevier, Amsterdam, The Netherlands.

Guthe, M., Borodin, P., Balázs, Á. and Klein, R., 2004. Real-time appearance preserving out-of-core rendering with shadows. Keller, A. and Jensen, H.W. (Eds.), *Rendering Techniques 2004, Proceedings of Eurographics Symposium on Rendering*, 21–23 June 2004, Norrköping, Sweden, 69–79.

Hoffman, R.L. and Jain, A.K., 1987. Segmentation and classification of range images. *IEEE Transactions on Pattern Analysis and Machine Intelligence*, 9(5), 608–620.

Hoover, A., Jean-Baptiste, G., Jiang, X., Flynn, P.J., Bunke, H., Goldgof, D.B., Bowyer, K., Eggert, D.W., Fitzgibbon, A. and Fisher, R.B., 1996. An experimental comparison of range image segmentation algorithms. *IEEE Transactions on Pattern Analysis and Machine Intelligence*, 18(7), 673–689.

Hough, P.V.C., 1962. US Patent 3 066 954, Method and means for recognizing complex patterns.

Jiang, X.Y. and Bunke, H., 1994. Fast segmentation of range images into planar regions by scan line grouping. *Machine Vision and Applications*, 7(2), 115–122.

Kimme, C., Ballard, D.H. and Sklansky, J., 1975. Finding circles by an array of accumulators. *Communications of the ACM*, 18(2), 120–122.

Krüger, J., Schneider, J. and Westermann, R., 2005. Duodecim – a structure for point scan compression and rendering. *Proceedings of the Symposium on Point-Based Graphics 2005*, 21–22 June 2005, New York, NY, USA, 99–146.

Maas, H.-G. and Vosselman, G., 1999. Two algorithms for extracting building models from raw laser altimetry data. *ISPRS Journal of Photogrammetry and Remote Sensing*, 54(2–3), 153–163.

Okabe, A., Boots, B. and Sugihara, K., 1992. *Spatial Tessellations: Concepts and Applications of Voronoi Diagrams*, Wiley, New York, NY, USA.

Rabbani, T., 2006. Automatic Reconstruction of Industrial Installations Using Point Clouds and Images. Ph.D. thesis. Publications on Geodesy 62, Netherlands Geodetic Commission, 154 p., http://www.ncg.knaw.nl/Publicaties/Geodesy/pdf/62Rabbani.pdf (accessed 17 March 2009).

Rusinkiewicz, S. and Levoy, M., 2000. QSplat: a multiresolution point rendering system for large meshes. *Proceedings of the 27th Annual Conference on Computer Graphics and Interactive Techniques*, 23–28 July 2000, New Orleans, LA, USA, ACM Press/Addison-Wesley Publishing Co., New York, NY, USA.

Samet, H., 1990. *Applications of Spatial Data Structures: Computer Graphics, Image Processing, and GIS*, Addison-Wesley, Reading, MA, USA.

Sapidis, N.S. and Besl, P.J., 1995. Direct construction of polynomial surfaces from dense range images through region growing. *ACM Transactions on Graphics*, 14(2), 171–200.

Schnabel, R. and Klein, R., 2006. Octree-based point-cloud compression. Botsch, M. and Chen, B. (Eds.), *Proceedings of Symposium on Point-Based Graphics 2006*, 29–30 July 2006, Boston, MA, USA.

Schnabel, R., Möser, S. and Klein, R., 2007a. A parallelly decodeable compression scheme for efficient point-cloud rendering. *Proceedings of Symposium on Point-Based Graphics 2007*, 2–3 September 2007, Prague, Czech Republic, 214–226.

Schnabel, R., Wahl, R. and Klein, R., 2007b. Efficient RANSAC for point-cloud shape detection. *Computer Graphics Forum*, 26(2), 214–226.

Sedgewick, R., 2002. *Algorithms in C++*, Addison-Wesley, Reading, MA, USA.

Shewchuk, J.R., 1996. Triangle: Engineering a 2D quality mesh generator and delaunay triangulator. *Applied Computational Geometry: Towards Geometric Engineering*, Lin, M.C. and Manocha, D. (Eds.), *Lecture Notes in Computer Science 1148*, Springer-Verlag, Berlin, Germany, 203–222. The Triangle software is available at http://www.cs.cmu.edu/~quake/triangle.html (accessed 17 March 2009).

Shirley, P., 2005. *Introduction to Computer Graphics*, 2nd edn., A. K. Peters, Ltd, Wellesley, MA, USA.

Sithole, G. and Vosselman, G., 2003. Automatic structure detection in a point cloud of an urban landscape. *Proceedings of the 2nd GRSS/ISPRS Joint Workshop on Remote Sensing and Data Fusion over Urban Areas, URBAN2003*, 22–23 May 2003, Berlin, Germany, 67–71.

Sithole, G., 2005. Segmentation and Classification of Airborne Laser Scanner Data. Ph.D. thesis. Publications on Geodesy 59, Netherlands Geodetic Commission, 184 p., http://www.ncg.knaw.nl/Publicaties/Geodesy/pdf/59Sithole.pdf (accessed 17 March 17 2009).

Torr, P. and Zisserman, A., 1998. Robust computation and parametrization of multiple view relations. *Proceedings of the Sixth International Conference on Computer Vision (ICCV'98)*, 4–7 January 1998, Bombay, India, IEEE Computer Society, Washington, DC, USA, 727.

Vieira, M. and Shimada, K., 2005. Surface mesh segmentation and smooth surface extraction through region growing. *Computer Aided Geometric Design*, 22(8), 771–792.

Vosselman, G., 1999. Building reconstruction using planar faces in very high density height data. *International Archives of Photogrammetry and Remote Sensing*, 32(Part 3/2W5), 87–92.

Vosselman, G., Gorte, B.G.H., Sithole, G. and Rabbani, T., 2004. Recognising structure in laser scanner point clouds. *International Archives of Photogrammetry, Remote Sensing and Spatial Information Sciences*, 46(Part 8/W2), 33–38.

Wahl, R., Guthe, M. and Klein, R., 2005. Identifying planes in point-clouds for efficient hybrid rendering. *Proceedings of the 13th Pacific Conference on Computer Graphics and Applications*, 12–14 October 2005, Macao, China, on CD-ROM.

Weingarten, J.W., Gruener, G. and Siegwart, R., 2004. Probabilistic plane fitting in 3D and an application to robotic mapping. *Proceedings of the 2004 IEEE International Conference on Robotics and Automation, ICRA 2004*, 26 April to 1 May, New Orleans, LA, USA, 927–932.

Zwicker, M., Pfister, H., van Baar, J. and Gross, M., 2001. Surface splatting in computer gaphics. *Proceedings SIGGRAPH 2001*, 12–17 August, 2001, Los Angeles, CA, USA, ACM Press, New York, NY, USA, 371–378.

Registration and Calibration | 3

Derek Lichti and Jan Skaloud

This chapter focuses on two key factors that influence data quality for both terrestrial and airborne laser scanning: registration and calibration. Generally speaking both are essential processes that should be performed prior to the application of point cloud processing algorithms, though registration may follow the application of some initial processing. The task of registration is concerned with the coordinate system in which the point cloud is represented. It is performed in order to transform acquired data from one coordinate system, which may be an internal, instrument-defined system, to another, such as a local mapping frame. Calibration entails identification, modelling and estimation of systematic error sources inherent to laser scanning systems in order to maximise data accuracy.

This chapter is structured as follows. First, the geometric models for terrestrial and airborne laser scanning observations, the coordinate systems in which they are parameterised and the transformations between systems are described. Second, the error sources that act as perturbations to the geometric observation models and their analytical models are reviewed. The third section is devoted to registration methods. The fourth section of this chapter is devoted to calibration methods with particular emphasis on system self-calibration for both terrestrial and airborne laser scanners. A short summary concludes this chapter.

3.1 Geometric models

This section presents the basic geometric models for laser scanner point observations. It commences with a discussion of the relevant rotation matrix parameterisations that are used throughout the chapter. This is followed by the observation equations and the transformation model for terrestrial laser scanning data. The various reference frames involved in airborne laser scanning are then described, culminating in the final positioning equation.

3.1.1 Rotations

Among the many means available to parameterise 3D rotation, three are of particular relevance in laser scanning. The first is a sequence of three rotations. For the elementary rotation of an arbitrary Cartesian coordinate frame about the x_j-axis ($j = 1, 2, 3$) by an angle θ, the transformation matrices representing rotations about each of the coordinate axes are given by

$$\mathbf{R}_1(\theta) = \begin{pmatrix} 1 & 0 & 0 \\ 0 & \cos\theta & \sin\theta \\ 0 & -\sin\theta & \cos\theta \end{pmatrix} \tag{3.1}$$

$$\mathbf{R}_2(\theta) = \begin{pmatrix} \cos\theta & 0 & -\sin\theta \\ 0 & 1 & 0 \\ \sin\theta & 0 & \cos\theta \end{pmatrix} \tag{3.2}$$

$$\mathbf{R}_3(\theta) = \begin{pmatrix} \cos\theta & \sin\theta & 0 \\ -\sin\theta & \cos\theta & 0 \\ 0 & 0 & 1 \end{pmatrix} \tag{3.3}$$

In the case of a right-handed frame, the matrices of Equations (3.1) to (3.3) describe a clockwise rotation as viewed from the origin towards the positive x_j-axis. The rotation from a p-frame to a q-frame can be performed by three sequential rotations about the x_i^p-axes ($i = 1, 2, 3$), an Euler angle sequence, which yields the rotation matrix $\mathbf{R}_p^q = \mathbf{R}_3(\kappa)\mathbf{R}_2(\varphi)\mathbf{R}_1(\omega)$. Many possible rotation sequences exist but only those relevant to laser scanning are introduced where appropriate in this chapter.

The second rotation representation comprises one rotation angle, θ, about an axis, $\mathbf{n}$, given by a vector

$$\mathbf{n} = \begin{pmatrix} n_1 & n_2 & n_3 \end{pmatrix}^T \tag{3.4}$$

which can be shown to be the unit eigenvector corresponding to the real, unit eigenvalue of the rotation matrix $\mathbf{R}$. The analytical form of the rotation matrix derived from this parameterisation is

$$\mathbf{R} = \begin{pmatrix} \cos\theta + n_1^2(1-\cos\theta) & n_1 n_2(1-\cos\theta) + n_3\sin\theta & n_1 n_3(1-\cos\theta) - n_2\sin\theta \\ n_1 n_2(1-\cos\theta) - n_3\sin\theta & \cos\theta + n_2^2(1-\cos\theta) & n_2 n_3(1-\cos\theta) + n_1\sin\theta \\ n_1 n_3(1-\cos\theta) + n_2\sin\theta & n_2 n_3(1-\cos\theta) - n_1\sin\theta & \cos\theta + n_3^2(1-\cos\theta) \end{pmatrix} \tag{3.5}$$

The third rotation representation uses four algebraic parameters as the unit quaternion

$$\mathring{q} = q_0 + q_1 \mathbf{i} + q_2 \mathbf{j} + q_3 \mathbf{k}$$
$$= q_0 + \mathbf{q} \tag{3.6}$$

where

$$q_0^2 + q_1^2 + q_2^2 + q_3^2 = 1 \tag{3.7}$$

A quaternion can be considered as a vector plus a scalar or a complex number with three imaginary components. The corresponding rotation matrix is given by

$$\mathbf{R} = \begin{pmatrix} q_0^2 + q_1^2 - q_2^2 - q_3^2 & 2(q_1 q_2 + q_0 q_3) & 2(q_1 q_3 - q_0 q_2) \\ 2(q_1 q_2 - q_0 q_3) & q_0^2 - q_1^2 + q_2^2 - q_3^2 & 2(q_2 q_3 + q_0 q_1) \\ 2(q_1 q_3 + q_0 q_2) & 2(q_2 q_3 - q_0 q_1) & q_0^2 - q_1^2 - q_2^2 + q_3^2 \end{pmatrix} \tag{3.8}$$

The unit quaternion can be expressed in terms of the angle and axis parameters as follows

$$\mathring{q} = \cos\left(\frac{\theta}{2}\right) + \sin\left(\frac{\theta}{2}\right) n_1 \mathbf{i} + \sin\left(\frac{\theta}{2}\right) n_2 \mathbf{j} + \sin\left(\frac{\theta}{2}\right) n_3 \mathbf{k} \tag{3.9}$$

The angle–axis and quaternion representations find application in point cloud registration and, in particular, in inertial data processing as they are free from the singularity problem known as gimbal lock [Jekeli, 2001].

3.1.2 The geometry of terrestrial laser scanning

As depicted in Figure 3.1, a point scanned by terrestrial laser scanning is modelled here in terms of spherical coordinate observations and the rigid body transformation that links the scanner and object spaces. The basic observation equations of range, horizontal direction and elevation angle are given by

$$\rho_{ij} = \sqrt{x_{ij}^2 + y_{ij}^2 + z_{ij}^2} \tag{3.10}$$

$$\theta_{ij} = \arctan\left(\frac{y_{ij}}{x_{ij}}\right) \tag{3.11}$$

$$\alpha_{ij} = \arctan\left(\frac{z_{ij}}{\sqrt{x_{ij}^2 + y_{ij}^2}}\right) \tag{3.12}$$

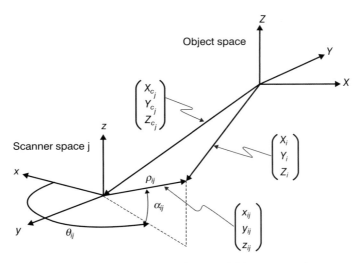

Figure 3.1 Observation equation geometry of terrestrial laser scanning.

where (x, y, z) are the coordinates of object point i in scan j (sensor frame s_j), which are related to object space through the following rigid body transformation

$$\begin{pmatrix} x_{ij} \\ y_{ij} \\ z_{ij} \end{pmatrix} = \mathbf{R}_o^{s_j} \left(\begin{pmatrix} X_i \\ Y_i \\ Z_i \end{pmatrix} - \begin{pmatrix} X_{c_j} \\ Y_{c_j} \\ Z_{c_j} \end{pmatrix} \right) \qquad (3.13)$$

where (X_c, Y_c, Z_c) are the object space coordinates of scanner position j and $\mathbf{R}_o^{s_j}$ is the rotation matrix from object space (o) to the sensor frame (s_j), which is parameterised here in terms of the sequential angles, ω, φ and κ

$$\mathbf{R}_o^{s_j} = \mathbf{R}_3 \left(\kappa_o^{s_j} \right) \mathbf{R}_2 \left(\varphi_o^{s_j} \right) \mathbf{R}_1 \left(\omega_o^{s_j} \right) \qquad (3.14)$$

The scanner position and rotation parameters comprise the elements of exterior orientation.

3.1.3 The geometry of airborne laser scanning

While the nature of most terrestrial laser scanning assignments allows operation within a local coordinate system, the character of airborne laser scanning requires the employment of a "global frame" as well as several intermediate frames. Table 3.1 provides an overview of the reference frames used together with their respective abbreviations (frame IDs). All frames are defined to be right-handed Cartesian systems. The relations between the laser scanner and the navigation sensors are schematically depicted in Figure 3.2 and will be defined in the following subsections.

Table 3.1 Overview of the reference frames

ID	Description
s	Sensor frame, defined by the principal axes of the laser scanner
b	Body frame, realised by the triad of accelerometers within an inertial navigation system (INS)
l	Tangent frame of the global Earth ellipsoid ("local-level frame"), spanned by the ellipsoidal north-, east- and down-axes (NED)
e	Earth-centred, Earth-fixed (ECEF) frame, realised by a variant of the International Terrestrial Reference Frame (ITRF)
m	Mapping frame spanned by the grid east-, north-, and up-axes (ENU), usually established by the national projection (denoted as the *p*-frame) and the national datum (denoted as the *n*-frame)

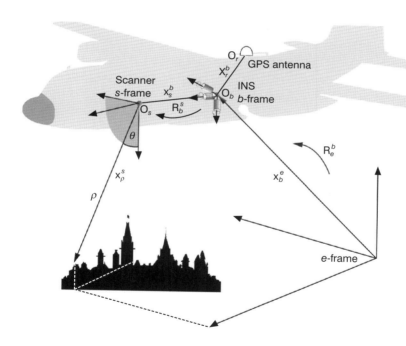

Figure 3.2 Observation geometry of airborne laser scanning.

3.1.3.1 Sensor frame – s

Airborne lasers scanners are typically line-based (2D) scanners; other architectures are described in Chapter 1. Hence, obtaining the 3D model of the scene requires moving the scanner across it. A target in the scanner frame can be expressed by means of the range ρ that follows from the run-time measurement of the laser pulse and the encoding angle θ of the scanner that reflects the current orientation of the laser beam with respect to this frame. The actual decomposition depends on the definition of θ with respect to the scanner axes, which are assumed to form a right-handed Cartesian frame. One possibility is described in Equation (3.15).

$$\mathbf{x}^s = \rho \cdot \begin{pmatrix} 0 \\ \sin\theta \\ \cos\theta \end{pmatrix} \tag{3.15}$$

3.1.3.2 Earth-centred, Earth-fixed frame – e

As the satellite orbits used in determining the path of the laser scanner carrier are related to some Earth-centred, Earth-fixed (ECEF) frame, the calculated trajectory is defined with respect to such a frame. Following Seeber (1993), this frame is defined as follows. The origin is the geocentre, the x_1^e-axis, points towards the Greenwich meridian, the x_3^e-axis is a mean direction of the rotation axis of the Earth, and the x_2^e-axis completes the system to a 3D right-handed Cartesian system. A geocentric ellipsoid of revolution is usually associated with the ECEF frame, which together with geophysical parameters defines a World datum (e.g. WGS-84). As shown in Figure 3.3, a point in space in such a system can be represented either by its Cartesian coordinates x_1^e, x_2^e, x_3^e or by its ellipsoidal coordinates, i.e. latitude (φ), longitude (λ) and height (h), where the reference meridian is that of Greenwich and the reference parallel is the equator. The latter is often used in GPS/INS software as the output of the carrier trajectory.

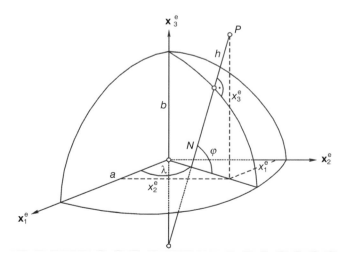

Figure 3.3 Cartesian and ellipsoidal coordinates.

The relation between the Cartesian and ellipsoidal coordinates is expressed in Equation (3.16).

$$\mathbf{X}^e = \begin{pmatrix} x_1^e \\ x_2^e \\ x_3^e \end{pmatrix} = \begin{pmatrix} (N+h)\cos\varphi\cos\lambda \\ (N+h)\cos\varphi\sin\lambda \\ \left(\dfrac{b^2}{a^2}N+h\right)\sin\varphi \end{pmatrix} \tag{3.16}$$

where N is the radius of curvature in the prime vertical [Seeber, 1993] and a, b are the semi-major and semi-minor axes of the ellipsoid.

3.1.3.3 Local-level frame – l

The local-level frame usually serves as a reference for the orientation output in GPS/INS processing. Its origin is arbitrary and defined by a point on the Earth's surface (topocentre). The x_1^l-axis points to the north, the x_2^l- axis points to the east and the x_3^l-axis points to the ellipsoidal normal (down) or to the ellipsoidal zenith (up). The down direction defines the *l*-NED frame (north-east-down), whereas the up direction identifies the *l*-ENU frame (east-north-up). Note that both frames are right handed, but the former is often preferred in inertial/aircraft navigation. The rotation matrix $\mathbf{R}_l^e$ for the coordinate transformation from the *l*- to the *e*- frame is

$$\mathbf{R}_{l_{NED}}^e = \begin{pmatrix} -\sin\varphi\cos\lambda & -\sin\lambda & -\cos\varphi\cos\lambda \\ -\sin\varphi\sin\lambda & \cos\lambda & -\cos\varphi\sin\lambda \\ \cos\varphi & 0 & -\sin\varphi \end{pmatrix} \tag{3.17}$$

3.1.3.4 Body frame – b

The body frame is a right-handed 3D Cartesian frame related to the system carrier (e.g. an aircraft or a helicopter) and is used to determine the relative orientation or attitude of the carrier with respect to a local-level frame. The origin of the body frame is situated at a specific point within the carrier, which in the case of airborne laser scanning is conveniently chosen to be at the navigation centre of the inertial navigation system (INS) or, alternatively, at the centre of the laser scanner. The axes of an INS either coincide with or are easily related to the principal rotation axes of the carrier, e.g. the x_1^b-axis points forward along the fuselage, the x_2^b-axis points right, and the x_3^b-axis points down. The parameters used to describe the 3D attitude in navigation are denoted as roll (r), pitch (p) and yaw (y). The convention defining the sequence of the rotation may vary, but most GPS/INS systems adopt the aerospace definition [Kuipers, 1998] that relates the reference local-level frame l_{NED} to the body frame as

$$\mathbf{R}^b_{l_{NED}} = \mathbf{R}_3(y)\mathbf{R}_2(p)\mathbf{R}_1(r). \tag{3.18}$$

The relationship between an arbitrary vector $\mathbf{x}^s$ in the scanner frame and the same vector expressed in the b-frame is given by

$$\mathbf{x}^b = \mathbf{x}^b_s + \mathbf{R}^b_s \cdot \mathbf{x}^s, \tag{3.19}$$

where $\mathbf{x}^b_s = \mathbf{R}^b_s \mathbf{x}^s_b$ denotes the origin of the s-frame (i.e. the centre of projection or "firing point" of the scanner) in the b-frame, also known as the lever-arm. The rotation matrix $\mathbf{R}^b_s$ in Equation (3.19) is called the bore-sight and represents the misalignment between the s- and b-frames. It is usually parameterised by three Euler angles ω, φ, κ. The magnitude of the bore-sight and the lever-arm needs to be determined by calibration.

3.1.3.5 Airborne laser scanner observation equation

The observation equation for airborne laser scanning follows from the geometrical relation depicted in Figure 3.2. First, the individual vectors are added as $\mathbf{x}^e = \mathbf{x}^e_b + \mathbf{R}^e_b(\mathbf{x}^b_s + \mathbf{R}^b_s \mathbf{x}^s)$. This relation is then expanded using Equations (3.15) to (3.19). The observation equation in the e-frame is obtained after rearrangements and reads

$$\mathbf{x}^e_p(t) = \mathbf{x}^e_b(t) + \mathbf{R}^e_l\big(\varphi(t),\lambda(t)\big)\mathbf{R}^l_b\big(r(t),p(t),y(t)\big)\mathbf{R}^b_s\big(\omega,\varphi,\kappa\big)\left(\mathbf{x}^s_b + \rho(t)\begin{pmatrix}0\\ \sin\theta(t)\\ \cos\theta(t)\end{pmatrix}\right), \tag{3.20}$$

where all the components were defined previously. The symbol (t) indicates quantities that vary with time. The trajectory-related parameters of position $\mathbf{x}^e_b(t)$ and attitude $\mathbf{R}^l_b$, parameterised by the three Euler angles $r(t)$, $p(t)$, $y(t)$, the range $\rho(t)$ and the encoder angle $\theta(t)$ are considered as the measurements. The lever-arm $\mathbf{x}^b_s$ and the bore-sight $\mathbf{R}^b_s$ are considered as constants previously obtained from calibration, while the rotation $\mathbf{R}^e_l(t)$ is a function of geographic position on the reference ellipsoid. If the scanner is mounted on a stabilised platform then the projection of the lever-arm axis is no longer constant and will vary in time as a function of the gimbals' orientation. In such a case, it is important to record the gimbals' pickoff per axis and synchronise them with the rest of data. Denoting the gimbals' orientation $\mathbf{R}^s_u$ as a user frame that is coincident with the scanner frame for nil stabilisation ($\mathbf{u} = 0$), the definite lever-arm is evaluated as $\mathbf{x}^b_s(t) = \mathbf{R}^s_u(t)\mathbf{x}^b_{s,u} = 0$.

3.1.3.6 Mapping frame – m

Although the coordinates of a laser footprint are usually generated via Equation (3.20), the final point cloud is often needed in some other datum and projection labelled as the "mapping frame", habitually representing a national coordinate system. The problems related to the registration of the laser point cloud directly in such a frame are described in Section 3.3.2.

3.1.3.7 Trajectory determination by GPS/INS

In airborne laser scanning the trajectory of the laser carrier is determined by integrating satellite positioning with inertial navigation [Schwarz *et al.*, 1993]. Loosely coupled integration is frequently adopted, which means that the satellite data are processed first, yielding a profile of positions and velocities that is subsequently smoothed by the inertial data. Currently, the integration technique, almost exclusively, employs Kalman filtering [Skaloud and Schwartz, 2000; Colomina, 2007]. The filter setup contains error states related to the trajectory (i.e. position, velocity and attitude) and to those linked to the inertial sensors (e.g. gyro drift, accelerometer bias). The evolution of these states is driven via a model fed by inertial measurements and controlled by satellite positioning acting as external observations. The complementary character of both datasets limits the effect of systematic errors in the inertial measurements while smoothing the short-term effects in satellite positioning (e.g. cycle slips, satellite drop-offs). The resulting trajectory is optimal in terms of position and attitude accuracy and the high data rate of inertial measurements assures that it spans the whole spectrum of aircraft motion. It can therefore be used for the georeferencing of the laser scanners operating at any frequency. The whole processing, including point cloud generation, can be also achieved in real-time [Schaer *et al.*, 2008].

Airborne laser scanners are usually sold with integrated GPS/INS instruments and corresponding software. However, off-the-shelf scanners are also available and can be retrofitted with GPS/INS devices. Moreover, larger service providers sometimes combine equipment of different origins to either optimise its selection with respect to the mission or simply to "cover up" for an inoperative apparatus. Irrespective of the reason, the integration mechanism poses a few potential pitfalls, especially for less experienced users or in the case where a processing sequence involves several users. These are mainly related to the definition of the instrument axes, the determination and application of the lever-arms towards the GPS antenna and scanner origin ($\mathbf{x}_r^b$, $\mathbf{x}_s^b$), respectively, and bore-sight ($\mathbf{R}_s^b$), as well as to the non-standardised output of the trajectory (e.g. NED vs. ENU, sign and sequence of Euler angles). Finally, hardware problems related to synchronisation offsets or out-of-spec inertial sensors are not uncommon and their detection and identification remain difficult.

3.2 Systematic error sources and models

It is well known that systematic errors exist in the instrumentation (e.g. total stations, cameras, etc.) used to collect measurements. Systematic errors in laser scanners exist due to imperfections in instrument manufacture and assembly. For example, the vertical, trunnion and collimation axes of a terrestrial laser scanner in perfect adjustment are assumed to be mutually orthogonal and to intersect at a common point. In general these conditions are not met in real systems, so error modelling and calibration are necessary. In the airborne case, for example, the axes of the laser scanner sensor frame and the inertial measurement unit (IMU) body frame cannot be perfectly aligned, which necessitates bore-sight angle modelling and calibration. This section details the systematic error sources and models for both terrestrial and airborne laser scanning systems.

3.2.1 Systematic errors and models of terrestrial laser scanning

A common approach for dealing with systematic errors is to treat them as perturbations – ideally additive corrections – to the observation equations and perform some form of calibration procedure to estimate the model coefficients, called the additional parameters. This approach has been successfully applied to terrestrial laser scanning systems [Gielsdorf *et al.*, 2004; Lichti and Franke, 2005; Lichti and Licht, 2006; Lichti, 2007; Reshetyuk, 2006; Rietdorf *et al.*, 2003; Schneider and Schwalbe, 2008]. Correction models for terrestrial laser scanning have emerged from two sources: the photogrammetric and surveying literature, owing to some similarities in instrument construction [e.g. Adams, 1999; Cooper, 1982; Rüeger, 1990] and experimentation. Errors that fall into the former category comprise the set of physical terms since their physical basis can be readily identified. The latter are called the empirical terms since they may lack physical interpretation but can be shown to exist as significant, systematic effects in repeated trials. General models for systematic errors of terrestrial laser scanners collected from the relevant literature and investigations are proposed in the following sections. These are followed by discussion of other systematic effects that cannot be modelled with a repeatable functional model and are thus termed artefacts.

3.2.1.1 Range

The range correction model is given by

$$\Delta\rho = A_0 + A_1\rho + A_2 \sin \alpha + \sum_{k=1}^{n}\left(A_{2k+1} \sin\left(\frac{2\pi k\rho}{U_1}\right) + A_{2k+2} \cos\left(\frac{2\pi k\rho}{U_1}\right)\right) + ET \quad (3.21)$$

The first two additional parameters are well known from electronic distance measurement (EDM) instrument error modelling. A_0 is the additive constant or zero error and A_1 is the scale error, which may or may not be estimable in a self-calibration procedure depending on network design (see Section 3.4). The third coefficient, A_2, models the vertical offset of the laser axis from the trunnion axis. Though empirically identified by Lichti and Franke (2005) in a Faro scanner, it is included here as a physically interpretable error source. Its effect on the estimated observational residuals and the superimposed model trend are shown in Figure 3.4. The sinusoidal terms in the summation represent the cyclic error terms that model range errors caused by internal optical or electrical interference. U_1 is the shortest unit length, which is equal to one-half of the shortest modulating wavelength. One can generally expect the first- and second-order terms to be the most significant, though Salo *et al.* (2008) report higher-order terms for their Faro terrestrial laser scanner. Multitone systems having more than one modulating wavelength may require additional terms with unit lengths of U_2, U_3, etc. These are, of course, only relevant for rangefinders that operate using the phase-difference measurement principle; they are not applicable for pulsed rangefinders. The notation ET represents empirical terms. For example, Lichti (2007) reports sinusoidal terms as a function of horizontal direction having a period of 90° for a Faro 880 terrestrial laser scanner.

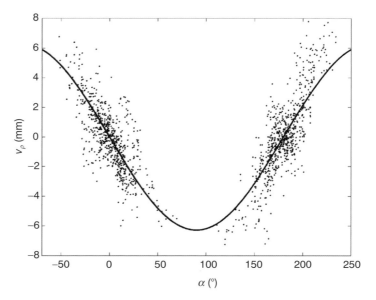

Figure 3.4 The effect of the vertical offset of the laser axis relative to the trunnion axis in Faro 880 terrestrial laser scanner data. The dots indicate the range observation residuals, v_ρ, and the solid line denotes the estimated model trend. (From Lichti and Franke, 2005)

3.2.1.2 Horizontal direction

The general correction model for errors in the horizontal direction is given by

$$\Delta\theta = B_1\theta + B_2\sin\theta + B_3\cos\theta + B_4\sin 2\theta + B_5\cos 2\theta + B_6\sec\alpha + B_7\tan\alpha + B_8\rho^{-1}$$

$$+ \sum_{k=1}^{n}\left(B_{2k+7}\sin(k\alpha) + B_{2k+8}\cos(k\alpha)\right) + ET \tag{3.22}$$

The first term, B_1, is the horizontal encoder (circle) scale error. B_2 and B_3 model the two components of the horizontal circle eccentricity, which is included in the error model of Schneider and Schwalbe (2008), and B_4 and B_5 model the non-orthogonality of the plane containing the horizontal encoder and the vertical axis, as shown in Figure 3.5. The non-orthogonality of the collimation axis (line of sight) and the trunnion (horizontal) axis, referred to hereafter as the collimation axis error, is given by B_6 and the non-orthogonality of the trunnion axis and the vertical axis, called the trunnion axis error, is given by B_7. Their combined effect and corresponding models are shown in Figure 3.6. If the zenith angle parameterisation is chosen instead of the elevation angle, then these errors' functions change to cosecant and cotangent, respectively. B_8 is the eccentricity of the collimation axis relative to the vertical axis whose

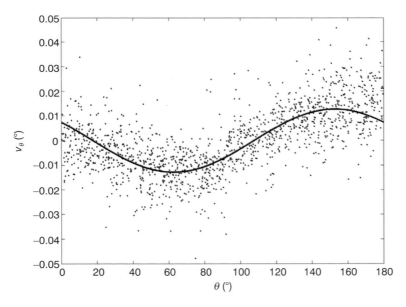

Figure 3.5 The effect of the non-orthogonality of the plane containing the horizontal encoder and the vertical axis in Faro 880 terrestrial laser scanner data. The dots indicate the horizontal direction observation residuals v_θ, and the solid line denotes the estimated model trend. (From Lichti and Franke, 2005)

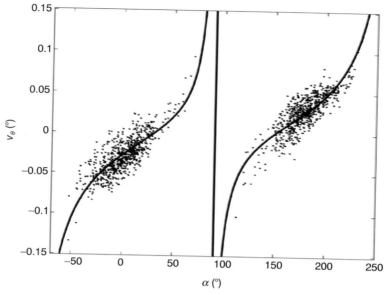

Figure 3.6 The combined effect of collimation and trunnion axis errors in Faro 880 terrestrial laser scanner data. The dots indicate the horizontal direction observation residuals v_θ, and the solid line denotes the combined estimated model trend. (From Lichti and Franke, 2005)

effect is most significant at short ranges. The summation represents the terms of a Fourier series used to model trunnion axis wobble, as reported by Harvey and Rüeger (1992) for theodolites and found experimentally in a terrestrial laser scanner by Lichti (2007). Care must be taken in selecting the terms of the series to be used since some may be highly correlated to other additional parameters in the model. ET again represents the empirical terms. Note that there is no index error, B_0, since it is functionally equivalent to the tertiary rotation angle of the instrument exterior orientation (i.e. κ). Its inclusion in a self-calibration process would render the solution singular as these parameters are perfectly correlated.

3.2.1.3 Elevation angle

The general correction model for errors in elevation angle is given by

$$\Delta\alpha = C_0 + C_1\alpha + C_2\sin\alpha + C_3\sin 2\alpha + C_4\cos 2\alpha + C_5\rho^{-1} + C_6\sin 3\theta + C_7\cos 3\theta + ET \quad (3.23)$$

The C_0 term is the well-known vertical circle index error [e.g. Rüeger, 2003] and C_1 is the scale error term. The vertical circle eccentricity is modelled with a single term, C_2; its effect is shown in Figure 3.7. The corresponding cosine term is nearly perfectly

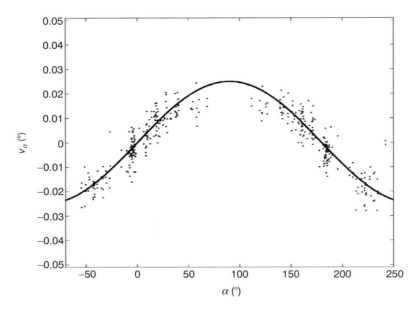

Figure 3.7 The effect of vertical circle eccentricity in Faro 880 terrestrial laser scanner data. The dots indicate the elevation angle observation residuals v_α, and the solid line denotes the estimated model trend.

correlated to exterior orientation elements and, therefore, omitted [Lichti, 2007]. The C_3 and C_4 coefficients model the non-orthogonality of the plane containing the vertical circle and the trunnion axis and C_5 is the eccentricity of the collimation axis relative to the trunnion axis that has been observed in reflector-less total station instruments, which share many salient properties with terrestrial laser scanners [Lichti and Lampard, 2008]. The last two physical terms, C_6 and C_7, model the mechanical wobble of the vertical axis caused by mass imbalance in the instrument that has been reported by Kersten *et al.* (2005) and Lichti (2007). Though identified empirically, these terms are included in the model since this is a physically interpretable error source. The period of 120° is believed to be linked to the angular separation between the support posts of the instrument in the tribrach and that of the tribrach foot screws. ET represents the empirical terms that may be necessary.

Blind application of all model terms to all scanners is of course not recommended. Not all terms may be relevant to a particular system due to its construction, whilst others may not be significant due to the quality of its assembly and/or manufacturer calibration. In addition, some additional parameters may only be weakly estimable or not estimable at all due to the scanner's narrow field of view (FOV). Camera-style scanners feature narrow horizontal and vertical FOVs and rotating prism scanners have a narrow vertical FOV. A discussion of significance assessment of additional parameters is given in Section 3.4.1.

Additional terms may also be required. For example, rotating polygonal mirror scanners (e.g. Riegl models) are known to have a sinusoidally varying rangefinder error due to the offset of the mirror surfaces from the rotation axis [Riegl, 2001]. The error model might be more complex for systems that acquire data through two separate windows with independent galvanometers. An example of such a system is the Leica Scan Station, which captures the upper and lower portions of a near-spherical point cloud separately in such a way that the two segments overlap. In summary, a thorough understanding of the instrument's construction and operation, as well as a process of model identification using data from a strong, highly redundant self-calibration network, are recommended in order to determine the most appropriate set of coefficients.

3.2.1.4 Data artefacts

Other systematic effects, called artefacts here, may exist in laser scanner data due to a variety of reasons that are very much dependent on the type of instrument used and the characteristics of reflecting surfaces in the scene. They are realised as outlier points within a point cloud and their magnitude varies from one situation to the next. Their detection and subsequent removal can require a great deal of tedious manual editing. Some of the common errors are listed in the following subsections.

Due to the fact that the laser beamwidth of scanning systems is non-zero, there is an inherent uncertainty in the angular location of the point to which a range measurement is made. The apparent location of the range observation is along the centreline of the emitted beam but the actual location cannot be predicted since it could lie anywhere within the projected beam footprint. The effect is angular displacement of features, whose shape appears distorted. The magnitude depends on the sample scene phase (i.e. the angular position of the scene relative to the measurement sampling lattice) and, therefore, cannot be reliably predicted without the assistance of geometric form fitting, as shown in Figure 3.8. The scanned plumb lines themselves are not visible in the resulting point clouds of Figures 3.8(a) and 3.8(b); best-fit, vertical lines have been superimposed to indicate their location. It should be stated, though, that the significance of this effect is diminished for systems with narrow beams.

When an incident laser beam illuminates two surfaces that are spatially adjacent in the range direction, the responses from both surfaces are integrated together. A pulsed laser scanner without full-waveform recording capability cannot discriminate between the returns from two surfaces separated by less than half the pulse length, so point measurements appear in the range discontinuity region between the two surfaces. An example of this effect, known as mixed-pixels [Hebert and Krotkov, 1992], is shown for Riegl LMS-Z210 data in Figure 3.9.

Detector saturation errors occur when scanning highly reflective surfaces at close range. A range bias (walk) results when the returned energy exceeds the dynamic range of the detector [Amann *et al.*, 2001]. It appears as a trail of points in front of or behind

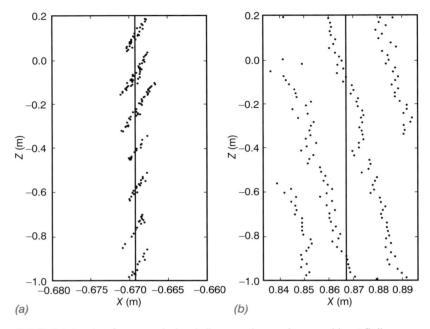

(a)   (b)

Figure 3.8 Pointclouds of scanned plumb lines and superimposed best-fit lines illustrating angular displacement due to the beamwidth: (a) Leica HDS 2500 data; and (b) Riegl LMS-Z210 data. (After Lichti and Jamtsho, 2006)

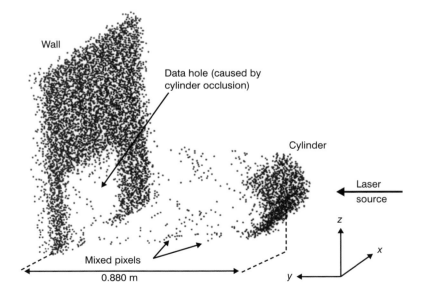

Figure 3.9 Mixed-pixels effect in Riegl LMS-Z210 scanner data of a cylinder and wall separated by less than one-half of the pulse length. (After Lichti *et al.*, 2005)

the reflector along the sight line. Its magnitude is difficult to quantify and it is more easily identified for extreme cases (e.g. retro-reflectors such as traffic signs).

Closely related to the saturation problem is blooming, where returns are received from the features surrounding a target such as a retro-reflector making it appear larger in size. The highly reflective material facilitates reflection of energy from the periphery of the incident laser beam, which may not normally occur with diffuse surfaces. The result is an increase in the width of a feature in the angular domain.

Saturation resulting from the measurement of a 10 mm diameter retro-reflective target affixed to a 20 mm wide bar is illustrated for Leica HDS 2500 scanner data in Figure 3.10(a). This side view of the point cloud shows that several range measurements are biased, with the largest being 0.135 m. Figure 3.10(b) shows the blooming effect in the front view of the same data. Here the apparent diameter of the target, as indicated by the circular pattern in the point cloud around it, has inflated by a factor of four due to the blooming effect.

Pesci and Teza (2008) report on the performance of an Optech ILRIS-3D terrestrial laser scanner with retro-reflective targets. They found severe range biases for distances up to 300 m as well as halo effects (i.e. blooming). They also state that generalised modelling of the effects is not possible. They conclude that retro-reflective targets should not be used except for scanning over very long distances and at normal incidence.

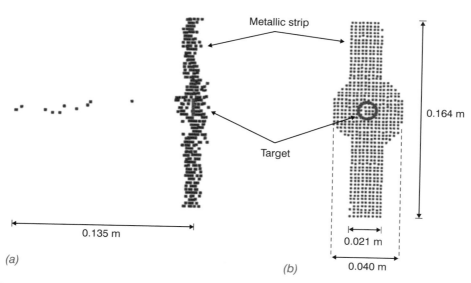

Figure 3.10 (a) Saturation and (b) blooming effects in Leica HDS 2500 data of a retro-reflective target. (After Lichti *et al.*, 2005)

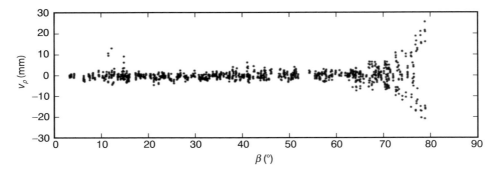

Figure 3.11 Self-calibration range residuals as a function of incidence angle, β, in Faro 880 terrestrial laser scanner data. (After Lichti, 2007; courtesy ISPRS)

Specularly-reflecting surfaces within a scanned scene can also cause multipath, in which range measurements are made from backscattered laser energy that has been reflected by more than one surface. It is particularly prevalent when scanning highly reflective surfaces and at close range or in the proximity of corners [Runne *et al.*, 2001].

The finite beamwidth of a laser can also give rise to range bias when the beam illuminates a surface at non-normal incidence. Baltsavias (1999) proposes a model for range error due to incidence angle in the context of airborne laser scanning. Figure 3.11 shows the effect in terms of range residuals from the self-calibration of a Faro 880 system. It can be seen that with only a few exceptions between 10° and 15°, the large outliers occur at incidence angles greater than about 65° or 70°.

3.2.2 Errors and models for airborne laser scanning

The airborne laser scanning position vector, $\mathbf{l}_p$, of observed point p is a function of the estimated GPS/INS trajectory, parameterised by the GPS/INS position $\mathbf{x}_b^e = (x_b^e \, y_b^e \, z_b^e)^T$ and orientation $(r \, p \, y)^T$ vectors, respectively, the observed laser scanner range, ρ, and the encoding angle, θ, of the laser beam, i.e. $\mathbf{l}_p = (x_b^e \, y_b^e \, z_b^e \, r \, p \, y \, \rho \, \theta)^T$. Note that p denotes both pitch and position. The formulation of the error budget needs to consider not only the accuracy of the individual observations but also the imperfections within the system assembly. Hence, considering the individual terms of Equation (3.20) and rearranging yields the following general error model

$$x_p^e = x_b^e + \Delta\hat{x}_b^e + \mathbf{R}_l^e \mathbf{R}_{b'}^l \left(\hat{r}\hat{p}\hat{y}\right) \mathbf{R}_b^{b'} \left(\Delta r \Delta p \Delta y\right) \mathbf{R}_s^b \left(\Delta\omega \, \Delta\varphi \, \Delta\kappa\right) \mathbf{T}_s^{s'} \left(\hat{\omega} \, \hat{\varphi} \, \hat{\kappa}\right)$$

$$\times \begin{pmatrix} -\left(\hat{\rho}+\Delta\rho\right)\sin\left(\Delta\eta\right) \\ \left(\hat{\rho}+\Delta\rho\right)\sin\left(\hat{\theta}+\Delta\theta\right)\cos\left(\Delta\eta\right) + \hat{x}_b^s + \Delta x_b^s \\ \left(\hat{\rho}+\Delta\rho\right)\cos\left(\hat{\theta}+\Delta\theta\right)\cos\left(\Delta\eta\right) \end{pmatrix} \qquad (3.24)$$

where

Δx_b^e are the GPS/INS positioning errors;

$R_b^{b'}(\Delta r \Delta p \Delta y)$ are the GPS/INS orientation errors;

$R_s^{b'}(\Delta \omega \Delta \varphi \Delta \kappa)$ is the bore-sight orientation error (see Equation 3.25);

$T_s^{s'}(\hat{\omega}\hat{\varphi}\hat{\kappa})$ is the known part of the installation matrix (see Equation 3.25);

Δx_b^s is the lever-arm error between the scanner and the IMU expressed in the scanner frame;

$\Delta \rho$ is the overall ranging error;

$\Delta \theta$ is the overall encoder/scanner error; and

$\Delta \eta$ is the misalignment between the angular encoder and the scanning plane.

The misalignment matrix R_s^b is defined as a product of two matrices

$$R_s^b = R_{s'}^b T_s^{s'},\tag{3.25}$$

where $T_{s'}^s$ denotes an installation matrix (usually composed of 90° or 180° rotations only), and $R_{s'}^b$ is a residual bore-sight matrix of small angles. Thus, the latter may be approximated by Equation (3.26).

$$R_{s'}^b \approx \begin{pmatrix} 1 & -\Delta \kappa & \Delta \varphi \\ \Delta \kappa & 1 & -\Delta \omega \\ -\Delta \varphi & \Delta \omega & 1 \end{pmatrix} = I + dR_{s'}^b,\tag{3.26}$$

where $dR_{s'}^b$ is a skew-symmetric matrix. Such definition is advantageous in the process of calibration, while the use of $T_s^{s'}$ allows generalising the definition of the scanner-axes independently of system and installation. In the following subsections the possible structure and magnitude of the individual error terms will be studied as well as their relative contribution to the total error budget.

3.2.2.1 Trajectory positioning errors Δx_b^e

The performance of kinematic GPS/INS positioning is mainly governed by two types of factors. The first is related to the satellite configuration and signal observability (i.e. signal blockage, number of satellites in view, baseline to satellite geometry, number of baselines, etc.), while the second stems from the accuracy of the observables. In precise *relative* positioning, the accurate but ambiguous phase measurements are of prime interest. When the carrier-phase ambiguities can be correctly determined, the residual error sources affecting the baseline accuracy are relatively small, for receiver separations of up to 100 km. Table 3.2 indicates that for a baseline length less than 10 km,

Table 3.2 Residual effects in kinematic *relative* positioning for baseline length ≈ 10 km [Tiberius, 1998]

	Troposphere	Ionosphere	Multipath	Noise	Orbits
Error (m)	10^{-2}–10^{-3}	10^{-2}–10^{-3}	10^{-2}	10^{-3}	10^{-2}

a relative accuracy of 10–20 ppm (i.e. 1–2 cm) can be considered as an accuracy limit for a good satellite constellation. However, if the ambiguities are not correctly determined as integers, the relative positioning accuracy is typically at the level of 100 ppm for baselines of this length. For longer baselines, precise post-mission satellite orbits need to be used and atmospheric delays should be modelled and estimated [Kim *et al.*, 2004]. The estimation of these quantities is often provided by regional or nationwide reference networks. The use of multiple or network–receiver configurations is a modern means of retaining the positioning errors within the aforementioned limit [Mostafa, 2007]. The tropospheric delays due to height differences should be modelled in all cases [Kleusberg and Teunissen, 1998].

As previously explained, the inertial data provide smoothing of some short-term error phenomena in the GPS output. Their ability to maintain the expected precision in *relative* positioning after a complete outage of GPS signal (rare, but nevertheless possible, in airborne surveys but frequent in terrestrial surveys) is limited in time. For a tactical-grade INS system (used in most airborne laser scanners) this interval is only 10–20 s while for navigation-grade instruments this extends to 1–2 minutes. This also means that in cases of satellite drop-offs or weak satellite geometry, the GPS/INS positioning accuracy will settle at the level of actual GPS accuracy.

Precise point positioning (PPP) is a positioning concept that uses data from a single GPS receiver and precise satellite orbit and clock information generated by the International GPS Service (IGS). This technique is reported to achieve decimetre or sub-decimetre accuracy without the need for GPS reference station data [Gao and Wojciechowski, 2004]. The drawback with PPP is usually long algorithm initialisation and periodic instability as well as the need for an on-line access to IGS-derived products which are not immediately available post-mission. Nevertheless, there is a significant potential for this already commercially available technique for mapping applications with relaxed accuracy requirements or those executed over large, remote areas.

3.2.2.2 Trajectory orientation errors $R_b^{b'}(\Delta r \Delta p \Delta y)$

The orientation error budget of a typical GPS/INS system used in airborne laser scanners is summarised in Table 3.3, where most of the indicated numbers have been confirmed experimentally. Tactical-grade INSs are used in most off-the-shelf systems, thanks to their smaller size and, especially, their better cost–performance ratio.

Table 3.3 Inertial attitude determination performance with GPS aiding

Time	Navigation grade (~ 0.01°/h)		Tactical grade (~ 0.1–3°/h)	
	$r/\omega - p/\varphi$ (°)	y/κ (°)	$r/\omega - p/\varphi$ (°)	y/κ (°)
1 s	0.0008–0.0014	0.0008–0.002	0.001–0.02	0.001–0.05
1–3 min	0.0014–0.003	0.004–0.005	0.005–0.04	0.008–0.1
Longer time	same as over 1–3 min but manoeuvre dependent			

The orientation errors are classified per individual INS axis and time because there is no such thing as a perfect instrument and, as strong as it is, the GPS/INS integration cannot completely remove all the errors. In other words, the data integration handled by a GPS/INS Kalman filter/smoother cancels only the non-overlapping part of the error spectra of the respective sensors. Thus the process of error cancellation by filtering/smoothing is not complete and the amount of residual errors varies with the type of the instrument and the dynamics of an aircraft. For example, the GPS data cannot compensate high frequency orientation errors due to platform vibration. For that reason, denoising inertial data [Skaloud *et al.*, 1999] prior to processing, or using a high sampling rate (0.4-2 kHz) has proven in some cases to be indispensable for attitude determination in a vibrating environment. Another significant portion of the residual orientation errors is due to the quality of the in-flight alignment. Usually, the data-integrating filter/smoother keeps refining the inertial platform error models during the whole flight. The strength of this process is in its ability to decorrelate the misalignment errors from other error sources when sufficient flight dynamics are encountered. Its weakness lies in the algorithm correlation of these models with changes of the accelerometer errors and the anomalous gravity field, both of which act as wrongly sensed accelerations that are "eliminated" by readjusting the previously aligned platform.

3.2.2.3 Scanner component errors $\Delta\rho$, $\Delta\theta$

The modelling of ranging errors is considerably complex as it depends on the number of physical parameters and the specific design of the rangefinder electronic. For pulse systems, the absolute range accuracy depends on how accurately the time difference between the emission and consecutive returning of the laser pulse can be measured [Baltsavias, 1999]. Currently, temporal differences corresponding to range differences of a few centimetres can be measured, provided conditions are optimal. These values are relatively homogeneous across systems of different types (Table 3.4). Sub-centimetre resolution is currently possible only with continuous-wave systems, which are, however, rarely used in practice. The range measurements are routinely corrected for atmospheric refraction and signal intensity [Morin, 2002].

Table 3.4 Ranging and angular accuracy specification provided by manufacturers (source: riegl.com, toposys.de, optech.ca, leica-geosystems.com)

Sensor type	Range Error (m)	Angular resolution (°)	Beam divergence (mrad)	Angular uncertainty (°)	Total angular error (°)
Mid-range					
LMS Q-240	0.02	0.005	2.7	0.0387	0.0390
Long-range					
LMS Q-560	0.02	0.001	0.5	0.0072	0.0073
Falcon III	0.02	0.002	0.7	0.0100	0.0102
ALTM (3100, Gemini)	0.02	0.001	0.15–0.3	0.00215–0.0043	0.0024–0.0044
ALS50-II/60	0.02	0.001	0.15	0.00215	0.0024

The errors related to the scanning mechanism are closely linked to the system design and measurement method. This can generally be modelled as

$$\Delta \theta(t) = \Delta \theta_0 + \theta(t) \cdot s_\theta \tag{3.27}$$

where $\Delta \theta_0$ is the zero-offset bias and s_θ is the scale factor [Katzenbeisser, 2003]. The latter is important to consider in oscillating scanning mechanisms, where the mirror is subject to high accelerations causing torsions that may lead to shifts in the mirror position. A difference in angular position between the scan mirror and the encoder then causes a misregistration of the observed distance. The encoder errors are usually present only at the noise level that corresponds to the minimum resolution (Table 3.4), although effects related to mirror imperfection have been reported [Latypov, 2005]. The typical angular and range measurement accuracies for various altitude classes of lidar sensors are listed in Table 3.4.

The second, and very important, angular uncertainty of the laser beam is related to its divergence. The divergence of the laser beam gives rise to uncertainty in location of the actual point of range measurement. As discussed in Section 3.2.1.4, the instrument records the apparent position of the point along the emitted beam centreline. The actual location is, however, uncertain and could be anywhere within the beam footprint. This effect can be modelled under the assumptions that the beam cross-section is circular and that the probability governing the angular position of the range measurement is uniform within the cross-section [Lichti *et al.*, 2005]. The standard deviation of beamwidth uncertainty under these conditions is one-quarter of the diameter. Note that there is no assumption about the distribution of power within the model, which is more likely to be Gaussian [e.g. Glennie, 2007]. It is, however, easy to show that for a

Gaussian beam with circular cross-section having a diameter of 4σ, which corresponds to the standard e^{-2} diameter definition [e.g. Kamerman, 1993], the angular positional uncertainty is also equal to one-quarter of the beamwidth, i.e. 1σ. The amplitude and the centre (i.e. the expectation) of the signal power distribution are further modulated by the incidence angle, that is the relative alignment between the beam direction and the normal of the reflecting surface, as shown in Figure 3.12. In this respect, the behaviour of airborne laser scanning is similar to that of terrestrial laser scanning (see Figure 3.11) as further discussed in Section 3.2.2.7.

3.2.2.4 Lever-arm error Δx_b^s

The system assembly errors are related either to uncertainties in the position or in the alignment between sensors. The former is denoted as a lever-arm error. Since the magnitude of the lever-arm x_b^s is small (i.e. at the decimetre level) and determined once by the system manufacturer, its practical influence on Δx_b^s is negligible. On the other hand, the lever-arm between the body frame and GPS antenna, x_r^b (Figure 3.2), is considerably larger (i.e. at the metre level) and varies per installation. Therefore, the accuracy of GPS/INS lever-arm determination has a predominant influence on Δx_b^s.

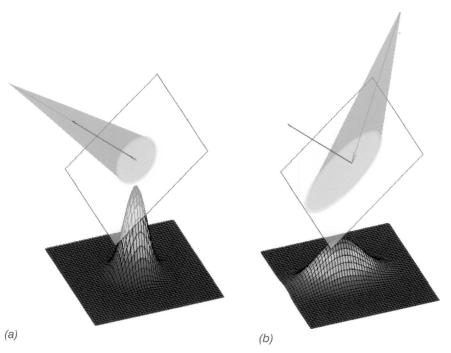

(a)   (b)

Figure 3.12 Influence of the incidence angle on relative power distribution: (a) incidence angle = 0°; and (b) incidence angle = 60°. (After Schaer *et al.*, 2007; courtesy the authors)

3.2.2.5 Bore-sight error $R_s^{s'}(\Delta\omega\Delta\varphi\Delta\kappa)$

The bore-sight error corresponds to the uncertainty of the relative orientation between the IMU body frame and the laser scanning plane. As will be shown later, the bore-sight represents one of the critical error components of the error budget. Its magnitude depends on the system design and can reach the degree level. The practical influence of the bore-sight calibration on the mapping accuracy is demonstrated in Figure 3.13, which shows a cross-section of a building recovered from eight flight lines (two clover-leaf patterns flown at two different heights). As can be seen from the profile, the discrepancies due to roll errors (Figure 3.13a) are clearly visible in both the inclined and horizontal planes. The errors due to pitch (Figure 3.13b) are not apparent in horizontal planes, however they have an opposite effect on inclined planes. Finally, the errors in yaw/heading (Figure 3.13c) cannot be discerned from the profile, which is why cross-section analyses are not suitable for recovery of this component of the bore-sight.

3.2.2.6 Scanner assembly error $\Delta\eta$

Although the resolution of modern scanner-angle encoders is as fine as 0.001° (Table 3.4), the encoder is not mounted exactly perpendicular to the mirror's rotation angle

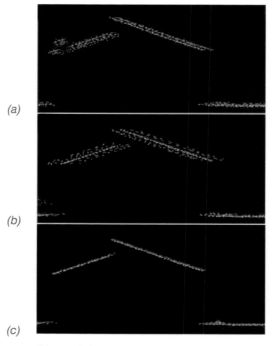

(a)

(b)

(c)

Figure 3.13 Influence of bore-sight error on a cross-section plotted separately for (a) roll, (b) pitch, and (c) yaw angles from eight flight lines of different directions and heights. (After Skaloud and Lichti, 2006; courtesy ISPRS)

[Latypov, 2005]. The projection of this misalignment $\Delta\eta$ causes the spacing between encoder bins to appear as changing with the scan angle. This effect is sometimes referred to as "pitch-slope error" [Leica Geosystems, 2005] and needs to be calibrated with the challenge of separating it from other error sources, namely the bore-sight.

3.2.2.7 Target accuracy error budget

The summation of the individual error terms and their projection by Equation (3.20) determines the ultimate accuracy of the target. The theoretical target accuracy can be predicted by variance propagation. This type of analysis is presented in Glennie (2007) for different systems and mission scenarios. There it is shown that when considering the relatively small performance gap between current airborne laser scanners (Table 3.4), the target error is mainly governed by angular uncertainty, which is in turn amplified by the flying height above the terrain. Such "amplification" is almost linear within the considered ranges and the upper limit of the error budget can be deduced from Table 3.5. The same source also presents the relative distribution of the individual error components, which are synthesised in Figures 3.14 to 3.16 for a fixed-wing aircraft, a helicopter and a kinematic terrestrial laser scanning system, respectively [Glennie, 2007]. It can be seen from these figures that horizontal errors for fixed-wing laser scanners are dominated by attitude errors; the combined IMU and bore-sight error make up from 60% to 75% of the overall horizontal error, depending on flight elevation. The vertical errors also show significant contributions by these two factors. Here, the attitude errors contribute 25% to over 50% of the error as a function of altitude. On the other hand, the error breakdown of a helicopter system is significantly different from that of a fixed-wing system; compare Figure 3.14 with Figure 3.15. The dominant error for the helicopter system is clearly the scanner angle error, which is principally a result of the large beam divergence of the laser. It is shown in Glennie (2007) that the error budget for a helicopter is more uniform when a laser beam of lower divergence is used (0.5 mrad vs. 2.7 mrad). Finally, the situation changes again for a terrestrial-based

Table 3.5 Target accuracy in airborne laser scanning for a fixed-wing aircraft [after Glennie, 2007]

	Vertical error per altitude (m)		Horizontal error per altitude (m)	
Altitude (m)	500	3000	500	3000
Tactical IMU	0.07	0.20	0.30	1.25
Navigation IMU	0.06	0.15	0.20	0.60

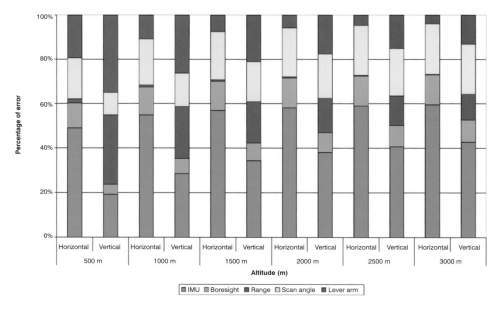

Figure 3.14 Subsystem error contribution (without GPS) to target error for a fixed-wing system. (After Glennie, 2007)

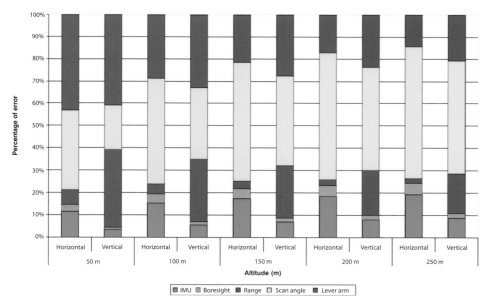

Figure 3.15 Subsystem error contribution (without GPS) to target error for a helicopter system (LMS-Q240). (After Glennie, 2007)

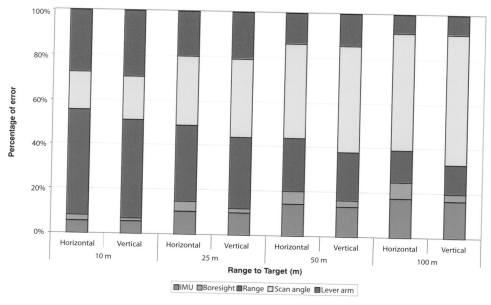

Figure 3.16 Subsystem error contribution (without GPS) to target error for a terrestrial vehicle system (LMS-Q240). (After Glennie, 2007)

mobile scanning system, where range errors and scanning angles are the predominant error contributors (Figure 3.16).

3.2.2.8 Scanning geometry

It has been shown that the scanning geometry has a large impact on the error budget [Schaer *et al.*, 2007]. Figure 3.17 illustrates the distribution of errors modelled by error propagation without considering the scanning geometry. The errors thus vary proportionally with the variation of the range on a steep slope from 500 to 1500 m. In general, the accuracy evolution follows a very homogeneous pattern with no sudden changes. The situation changes considerably when the effect of the incidence angle on target accuracy is considered (Figure 3.18). The high incidence angles on this particular slope with rock faces cause strong accuracy degradation mainly in the vertical component. Overall, the scanning geometry has a substantial influence on the target accuracy and should be considered when deriving precise digital terrain models. Furthermore, mission planning can be improved when considering this factor.

3.3 Registration

The point cloud produced by a terrestrial or an airborne laser scanner is very often required in a reference system different to the one in which the data were acquired.

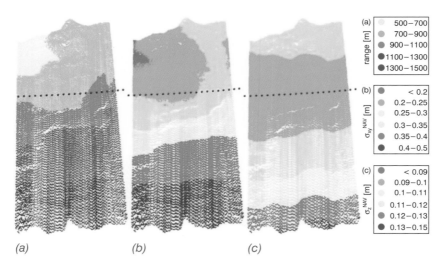

(a)   *(b)*   *(c)*

Figure 3.17 Predicted mapping accuracies without geometry analysis: (a) range length; (b) estimated horizontal accuracy (1σ); and (c) estimated vertical accuracy (1σ). The blue dots represent the vertical projection of the flight path. (After Schaer *et al.*, 2007; courtesy the authors)

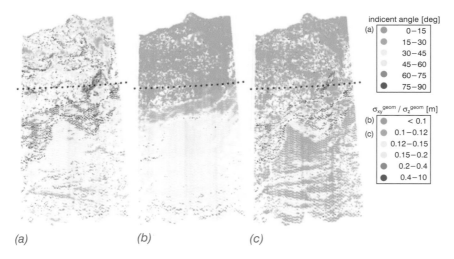

(a)   *(b)*   *(c)*

Figure 3.18 Impact of scanning geometry on target accuracy: (a) computed laser incidence angles (in degrees); (b) predicted horizontal geometric accuracy (1σ); and (c) predicted vertical geometric accuracy (1σ). (After Schaer *et al.*, 2007; courtesy the authors)

Terrestrial scanners measure the positions of points with reference to their internally defined sensor frame. Large and/or complex objects cannot typically be adequately sampled from only one instrument location, so multiple setups are required to obtain complete coverage. These individual datasets must be transformed or registered from their respective internal systems into an externally defined system, i.e. object space. Object space may be defined by a national mapping coordinate system, a local-level system or the internal (sensor) system of one of the scans, the so-called master scan. Whatever the case, object space is generally a right-handed Cartesian system. In the airborne case the point cloud generated according to Equation (3.20) is expressed in an Earth-centred Earth-fixed frame. More often than not, the final coordinates must be transformed into a mapping frame. In this section the various methods for registration of both terrestrial and airborne laser scanning data are discussed.

3.3.1 Registration of terrestrial laser scanning data

The registration process in the terrestrial laser scanner context typically involves two distinct steps: estimation of the registration parameters; and application of the transformation to the entire point cloud. Some or all of the parameters may be observed rather than estimated. Typically the transformation of choice is a 3D rigid body. If there is reason to believe a scale difference exists between the scanner and object spaces, then the 3D similarity transformation can be used.

Point-based registration procedures are perhaps the most obvious and they can be broadly classified into two categories: registration with point correspondences; and registration where correspondences do not exist. The former is rather straightforward but is treated here for completeness. A great deal of research effort has been and continues to be devoted to the latter, the iterative closest point (ICP) method, and its variants, since the first papers on the subject were published in the early 1990s [e.g. Besl and McKay, 1992; Chen and Medioni, 1992]. More recent research has concentrated on feature-based methods that use geometric primitives extracted from the point cloud scene to estimate the registration parameters.

3.3.1.1 Target-based registration

This subsection addresses the case of point-based registration for which exact point correspondences between coordinate systems exist. This is generally achieved with signalised targets placed within the scanner's fields of view. The targets are generally oversampled, and the highly redundant set of points is used to estimate the 3D centre coordinates of the target in the sensor frame. These coordinates are then used as observations for the parameter estimation. The targets may be planar or a geometric primitive such as a sphere.

The exact structure of the planar targets varies depending on the system manufacturer. Some are specially constructed for compatibility with surveying equipment or are fabricated with magnetic backing for easy mounting on metallic surfaces in industrial environments. Others are provided as digital templates that can be printed on a high-quality laser printer. The common feature they generally share is a high degree of contrast between target components to allow high-accuracy centre determination using some form of signal- or feature-based approach. Some examples of planar-target

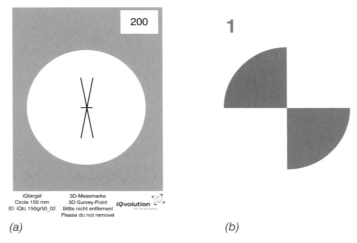

Figure 3.19 Planar target templates for terrestrial laser scanning supplied by system manufacturers: (a) Faro; and (b) Leica.

templates are given in Figure 3.19. Retro-reflective targets, favoured for photogrammetric measurement, are generally avoided due to the artefacts they can cause (see Section 3.2.1.4).

The most common geometric-primitive targets are spheres and hemispheres. The latter comprises both a planar target and a half-sphere whose centres coincide to allow easy target coordination by tacheometry. Estimation of the sphere parameters, three centre coordinates and radius is a straightforward application of least-squares adjustment. The solution is strengthened greatly if the target radius is known and enforced as a parameter constraint. High correlation can exist between the radius and the centre coordinate in the direction parallel to the vector between the scanner and target since the sphere (or hemisphere) cannot be completely sampled. In theory 50% of the surface can be seen from an instrument location, but in reality the coverage is much lower due to the high incidence angles of near-tangential observations, which can also cause data artefacts.

Ideally, the observation equations for the transformation parameter estimation are expressed in terms of the spherical coordinates. For their unique solution, six coordinate observations from three non-collinear points are required, though more points are desirable for redundancy. The least-squares solution by an iterative Gauss–Markov model is straightforward. The linearised observation equations for the spherical observation case are given by

$$
\begin{pmatrix} \mathbf{A}_{\rho e} \\ \mathbf{A}_{\theta e} \\ \mathbf{A}_{\alpha e} \end{pmatrix} \hat{\boldsymbol{\delta}}_e + \begin{pmatrix} \mathbf{w}_\rho \\ \mathbf{w}_\theta \\ \mathbf{w}_\alpha \end{pmatrix} = \begin{pmatrix} \hat{\mathbf{v}}_\rho \\ \hat{\mathbf{v}}_\theta \\ \hat{\mathbf{v}}_\alpha \end{pmatrix}
\tag{3.28}
$$

where $\mathbf{A}_{\rho e}$, $\mathbf{A}_{\theta e}$ and $\mathbf{A}_{\alpha e}$ are the (Jacobian) matrices of partial derivatives of the observations ρ, θ and α (Equations 3.10–3.12) taken with respect to the exterior orientation parameters e; $\mathbf{w}$ and $\mathbf{v}$ respectively represent the misclosure and residual vectors; δ_e is the vector of corrections to the exterior orientation parameters and $\wedge$ indicates an estimable quantity. Weights are generally assigned on a group basis so the weight matrix, $\mathbf{P}$, is given by

$$\mathbf{P} = \begin{pmatrix} \mathbf{P}_\rho & 0 & 0 \\ 0 & \mathbf{P}_\theta & 0 \\ 0 & 0 & \mathbf{P}_\alpha \end{pmatrix} \tag{3.29}$$

where $\mathbf{P}_\rho$, $\mathbf{P}_\theta$, and $\mathbf{P}_\alpha$ are the weight matrices for the range, direction and elevation angle measurements, respectively. The corresponding least-squares normal equations are therefore given by

$$\left(\mathbf{A}_{\rho e}^T \mathbf{P}_\rho \mathbf{A}_{\rho e} + \mathbf{A}_{\theta e}^T \mathbf{P}_\theta \mathbf{A}_{\theta e} + \mathbf{A}_{\alpha e}^T \mathbf{P}_\alpha \mathbf{A}_{\alpha e} \right) \hat{\delta}_e + \left(\mathbf{A}_{\rho e}^T \mathbf{P}_\rho \mathbf{w}_\rho + \mathbf{A}_{\theta e}^T \mathbf{P}_\theta \mathbf{w}_\theta + \mathbf{A}_{\alpha e}^T \mathbf{P}_\alpha \mathbf{w}_\alpha \right) = 0 \tag{3.30}$$

The methods for deriving this equation can be found in Mikhail (1976), for example.

In the more common case where the Cartesian coordinates, (x, y, z), are treated as the observables, the matrix form of the observation equations is

$$\begin{pmatrix} \mathbf{A}_{xe} \\ \mathbf{A}_{ye} \\ \mathbf{A}_{ze} \end{pmatrix} \hat{\delta}_e + \begin{pmatrix} \mathbf{w}_x \\ \mathbf{w}_y \\ \mathbf{w}_z \end{pmatrix} = \begin{pmatrix} \hat{\mathbf{v}}_x \\ \hat{\mathbf{v}}_y \\ \hat{\mathbf{v}}_z \end{pmatrix} \tag{3.31}$$

Special cases that allow either parameter elimination or parameter observation include: levelling of the instrument, in which case two of the scanner orientation angles are known (i.e. $\omega = \varphi = 0$); setup of the instrument over a known location, where X_c, Y_c and Z_c are known; and orientation of the instrument by means of observation to a known target, for which the κ angle is known.

3.3.1.2 Iterative closest point methods

Registration algorithms for which exact point-to-point correspondence between point clouds does not exist can be broadly termed iterative closest point (ICP) methods. The basic ICP algorithm is presented in this subsection and whilst many variants designed to improve it exist, no attempt is made to discuss all of them. However, some important variants and recent advances are reported. For details about the many variants, readers are referred to the listed references and the work of Rusinkiewicz and Levoy (2001) and Grün and Akca (2005).

Iterative closest point objective function

Consider two sets of points, $\mathbf{x}$ and $\mathbf{y}$, related by the rigid body transformation

$$\mathbf{y}_i = \mathbf{R}\mathbf{x}_i + \mathbf{y}_0 \tag{3.32}$$

where $\mathbf{R}$ is the rotation matrix and $\mathbf{y}_0$ is the translation vector. The basic premise of the ICP is that the slave point cloud or data shape is rigidly transformed to best fit the master point cloud or model shape such that the sum of squares of Euclidean distances between the nearest (transformed) slave point, $\mathbf{x}_i$, and the master point, $\mathbf{y}_i$, is minimised, i.e.

$$e^2 = \sum_i \left\| \mathbf{R}\mathbf{x}_i + \mathbf{y}_0 - \mathbf{y}_i \right\|^2 \rightarrow \min \tag{3.33}$$

where the summation is taken over all data points in the overlap area.

Closed-form orientation estimation

Prior to describing the ICP method in more detail, it is first necessary to briefly review Horn's (1987) method for estimating the transformation parameters since it is central to the ICP algorithm. Horn's method can also be used as an alternative to the traditional iterative least-squares approach described in the previous subsection.

Each set of points is first reduced to their respective centroids so that they share a common origin

$$\mathbf{x}_i' = \mathbf{x}_i - \mathbf{x}_m$$
$$\mathbf{y}_i' = \mathbf{y}_i - \mathbf{y}_m \tag{3.34}$$

where the centroids are given by

$$\mathbf{x}_m = \frac{1}{n_x} \sum_{i=1}^{n_x} \mathbf{x}_i$$
$$\mathbf{y}_m = \frac{1}{n_y} \sum_{i=1}^{n_y} \mathbf{y}_i \tag{3.35}$$

This allows independent estimation of the rotation and translation elements of the transformation. The rotation is estimated first. Following the centroid reduction, the objective function becomes

$$e^2 = \sum_i \left\| \mathbf{R}\mathbf{x}_i' - \mathbf{y}_i' \right\|^2 = \sum_i \left(\mathbf{x}_i'^T \mathbf{x}_i' + \mathbf{y}_i'^T \mathbf{y}_i' - \mathbf{y}_i'^T \mathbf{R}\mathbf{x}_i' \right) \rightarrow \min \tag{3.36}$$

Since the first two terms are sums of squares of the point coordinates, e^2 is minimised when the third term is maximised. To do so, the rotation is parameterised in terms of a unit quaternion, $\mathring{q} = q_0 + q_1\mathbf{i} + q_2\mathbf{j} + q_3\mathbf{k}$, and the third term is manipulated into the form

$$\sum_i \mathbf{y}_i'^T \mathbf{R} \mathbf{x}_i' = \mathring{q}^T \mathbf{N} \mathring{q} \tag{3.37}$$

where

$$\mathbf{N} = \begin{pmatrix} s_{xx} + s_{yy} + s_{zz} & s_{yz} - s_{zy} & s_{zx} - s_{xz} & s_{xy} - s_{yx} \\ & s_{xx} - s_{yy} - s_{zz} & s_{xy} + s_{yx} & s_{zx} + s_{xz} \\ & & -s_{xx} + s_{yy} - s_{zz} & s_{yz} + s_{zy} \\ \text{sym.} & & & -s_{xx} - s_{yy} + s_{zz} \end{pmatrix} \tag{3.38}$$

whose entries are functions of the elements of the cross-covariance matrix of point coordinates, $\mathbf{S}$

$$\mathbf{S} = \sum_{i=1} \mathbf{x}_i' \mathbf{y}_i'^T = \begin{pmatrix} s_{xx} & s_{xy} & s_{xz} \\ s_{yx} & s_{yy} & s_{yz} \\ s_{zx} & s_{zy} & s_{zz} \end{pmatrix} \tag{3.39}$$

The unit quaternion that maximises $\mathring{q}^T \mathbf{N} \mathring{q}$ is the eigenvector corresponding to the largest positive eigenvalue of $\mathbf{N}$. The coordinate centroids and estimated rotation matrix are then substituted into Equation (3.32) to calculate the translation vector

$$\mathbf{y}_0 = \mathbf{y}_m - \mathbf{R} \mathbf{x}_m \tag{3.40}$$

For greater detail on the method, see Horn (1987). Its elegance is that it is a closed-form, least-squares solution. The drawback is that weights cannot be introduced.

Iterative closest point algorithm description

Besl and McKay (1992) proposed the ICP algorithm for shape registration. Though nowadays it is most commonly used to register two or more point clouds, they present a general framework for the registration of points, line segments and triangulated surfaces as well as implicit and parametric curves and surfaces. Here the focus is on the registration of two point clouds. Besl and McKay (1992) also parameterise the rotation in terms of the unit quaternion and define the translation vector, $\mathbf{y}_0$, as

$$\mathbf{y}_0 = (q_4 \quad q_5 \quad q_6)^T \tag{3.41}$$

Good *a priori* alignment of the two point clouds is required. This is often done manually by selecting a small set of corresponding points and using them for estimating the initial transformation parameters. (Recent research has tried to eliminate this step with the goal of completely automating the algorithm.) The estimated transformation is then applied to the slave point cloud. The basic ICP algorithm then proceeds as follows [Besl and McKay, 1992]:

1. Approximate values for the transformation parameters are all set to zero, except for q_0, which is set to 1.

2. Nearest neighbour points are located either by k-D tree or other methods (see Chapter 2).

3. The transformation parameters are determined by Horn's method and applied to the slave point cloud.

4. The RMS registration error (i.e. e in Equation 3.33) is calculated and compared with a convergence criterion. If this test fails, then the process is repeated from step 1.

A high number of iterations is typically required and the method is prone to outliers. However, research efforts since the early 1990s have improved on these points.

Other iterative closest point approaches
Chen and Medioni (1992) take a slightly different approach in their formulation. Among the differences are the rotation parameterisation in terms of three independent angles, the use of homogeneous coordinates and, most significantly, how the distance measure is reckoned. Their objective function is to minimise the sum of squares of Euclidean distances between slave points and the master point cloud *surface*. The surface of the master point cloud is locally approximated by a tangent plane and the distance is taken along the surface normal from the slave point – so-called normal shooting. The principal advantage in avoiding the use of nearest points is more rapid convergence. The difference between these two approaches is schematically depicted in Figure 3.20.

Williams and Bennamoun (2001) present an ICP method for simultaneous registration of multiple point clouds. Theirs is a global approach that is immune to the problem of systematic error propagation. This phenomenon occurs when point clouds are added sequentially in pair-wise registration. Grün and Akca (2005) outline their least-squares 3D surface matching algorithm, which allows one or more slave surfaces to be matched to a 3D master (template) surface. The objective function of minimum sum of squares of distances is the same as in other methods, but the parametric least-squares formulation allows incorporation of observational weights, rigorous error propagation, outlier detection and shaping parameters to model deformation between point clouds. Good initial alignment is required, but convergence is more rapid than with the standard ICP method.

Rusinkiewicz and Levoy (2001) present some of the many variants to the ICP. They classify these according to the following stages of the ICP algorithm: point selection, point matching, weighting, outlier rejection, the error metric used, and minimisation

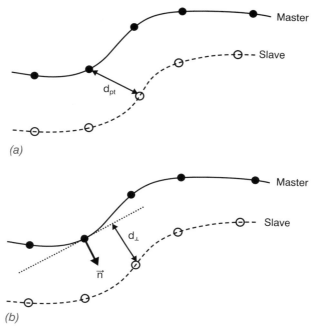

Figure 3.20 ICP distance reckoning methods: (a) point–point distance; and (b) point–plane distance.

of the error metric. Several researchers have reported their more recent efforts to automatically determine the initial alignment required for ICP-type registration methods. Bae and Lichti (2008) determine the approximate alignment using features extracted from the 3D point cloud. Extraction and subsequent matching are performed using approximate curvature measures computed from the eigenvalue decomposition of each point's local neighbourhood covariance matrix. Beinat *et al.* (2006) use more rigorously computed Gaussian curvature. Barnea and Filin (2008) reduce the dimension of the registration problem by extracting features or key points from the 2D range image. They use a modified corner detector that searches for large gradients in the range image and distances between key points to improve the identification of corresponding points. Kang *et al.* (2007) also derive interest points from the 2D intensity image and match points across intensity images. The range data are used to remove false matches by comparing invariant distance measures calculated from the 3D coordinates of candidate match points. Boström *et al.* (2008) introduce a weighting scheme that is a function of point precision and incidence angle and use cut-off criteria to exclude points.

In addition to the RMS registration error, e, the success of ICP registration can be effectively demonstrated through colour maps indicating the distance between a point

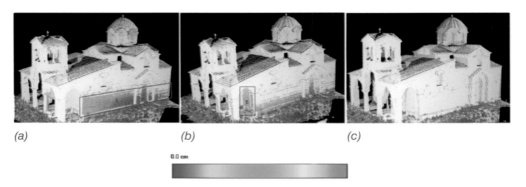

(a)   (b)   (c)

0.0 cm

Figure 3.21 Colour maps of the distance between a point and its local surface after registration by (a) target-based resection; (b) Leica Cyclone ICP; and (c) the ICP method proposed by Bae and Lichti (2008). (After Bae and Lichti, 2008; courtesy ISPRS)

and its local, best-fit planar surface. Figure 3.21 is an example of this type of presentation in which registration results from three methods are shown.

3.3.1.3 Feature-based registration

Rabbani *et al.* (2007) present a registration method that uses simple geometric primitives (planes, spheres, cylinders and tori) extracted from point clouds of industrial scenes rather than points. They propose two approaches:

1. An indirect method in which the point clouds are segmented to identify the features. The parameters of the features are then estimated by geometric form fitting. The registration parameters are estimated by minimising the sum of squared differences between feature model parameters.

2. A direct method that is more rigorous but converges more slowly. The parameter estimates from the indirect method are used as approximate values. The objective function in this case is to minimise the sum of squares of orthogonal distances between the points and their respective geometric features. Thus, the processes of registration and modelling, traditionally performed independently, are combined into one.

Though the formulation of the method is presented in terms of the registration of two point clouds, extension to global, multiple point cloud registration is straightforward. One of the real advantages is that, unlike the ICP-based approaches, little or no overlap between point clouds is needed. The only requirement is that portions of common features must appear in the two (or more) point clouds.

Brenner *et al.* (2008) present two algorithms. The first is fundamentally feature based as it uses planar patches segmented from the point cloud. The transformation parameters are estimated after establishing correspondences by a priority search. The second uses the normal distribution transform but presumes the scanner is upright.

3.3.2 Registration of airborne laser scanning data

Although the coordinates of the laser footprint are usually generated via Equation (3.20), the final point cloud is often needed in some other datum and projection labelled as the mapping *m*-frame (Table 3.1), which habitually represents a national coordinate system. A general schema of the possible subsequent transformations is shown in Figure 3.22, which reflects the fact that a national datum differs from that of the GPS orbits and some form of mapping projection is applied.

The transformation of 3D point coordinates from the *e*-frame to the national *n*-frame is generally achieved as

$$\mathbf{x}_{PC}^{n} = \mathbf{x}_{e}^{n} + \mu^{n}\mathbf{R}_{e}^{n}\mathbf{x}_{PC}^{e} \tag{3.42}$$

where $\mathbf{x}_{e}^{n}$ is the origin of the *e*-frame given in the *n*-frame, μ^{n} is the common scale factor of the national datum, $\mathbf{R}_{e}^{n}$ is the datum rotation matrix and $\mathbf{x}_{PC}^{e}$ is the observed position vector. An additional transformation may be required when the system of national coordinates is based on a triangulation network where local distortions with respect to the global reference frame exist. Such effects are usually planimetric and may be compensated by, for example, position-dependent affine transformations. These procedures cannot be generalised and are usually provided by the state mapping authority. A national altimetry system, on the other hand, requires transformation from the ellipsoid to the geoid (*g*-frame) or some other height-reference surface based on physical parameters. Finally, some form of mapping projection is applied. This is a common process depending on projection definition and is therefore not detailed here [Snyder and Bugayevskiy, 1995]. Depending on the implementation, laser point cloud generation software may skip the passage from *e* to *n* and *g*, respectively, but offer the output in a global (e.g. UTM) or other projection.

3.3.2.1 Direct registration in national coordinates

An alternative, and perhaps more efficient, way of obtaining the laser point cloud coordinates in the *m*-frame is to transform the GPS/INS trajectory to that frame and then calculate the laser point cloud coordinates (dashed sequence in Figure 3.22). The theoretical benefit is in a reduction of the computational effort by a factor of 100 or more [Legat, 2006]. Nevertheless, this approach also requires the transformation of aircraft

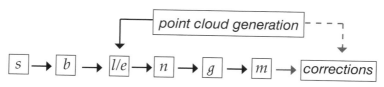

Figure 3.22 Transformation of laser point cloud to mapping *m*-frame.

attitude in parallel. Following Skaloud and Legat (2008), the transformation of the attitude to a projection p-frame (e.g. $m = p$) can be obtained as

$$\mathbf{R}_p^b = \mathbf{R}_l^b \mathbf{R}_3\left(-\gamma_{\mathrm{PC}}\right)\mathbf{R}_{\mathrm{ENU}}^{\mathrm{NED}} \tag{3.43}$$

where the third matrix on the right-hand side involves exchanging the first and second axes and reflecting the third; the second term is a standard rotation matrix about the third axis of the (modified) p-frame with the (negative) local convergence of meridians, γ_{PC}, computed at the scanner perspective centre (PC); the first term represents the attitude matrix $\mathbf{R}_l^b$ on the national ellipsoid obtained as matrix product

$$\mathbf{R}_l^b = \mathbf{R}_l^b \mathbf{R}_l^l \mathbf{R}_n^e \mathbf{R}_l^n \tag{3.44}$$

where $\mathbf{R}_l^n$ is defined by the perspective centre coordinates on the national ellipsoid; $\mathbf{R}_n^e$ is the transpose of the datum rotation matrix; $\mathbf{R}_l^l$ is defined by the coordinates on the global ellipsoid, and $\mathbf{R}_l^b$ is the observed attitude matrix. Finally, individual corrections need to be applied to every terrain point to compensate for the effects of the Earth's curvature and the mapping length distortion (Figure 3.23). Following Legat (2006), the distortion can be effectively reduced from several decimetres (under adverse conditions) to the sub-centimetre and, therefore, negligible level. The 3D laser vector is first transformed from the b-frame to the p-frame by $\Delta\mathbf{x}^p=\mathbf{R}_b^p\Delta\mathbf{x}^b$, where $\Delta\mathbf{x}^p$ involves the uncorrected position differences Δx^p, Δy^p, $\Delta\bar{h}$ of point A from the perspective centre. Without going into the details and by denoting the national height of the perspective

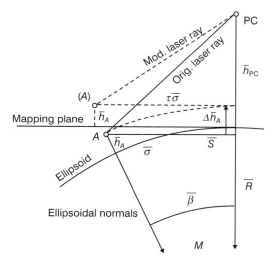

Figure 3.23 Coordinate corrections due to the Earth's curvature and the length distortion of the map projection. (After Legat, 2006; courtesy ISPRS)

centre as $\overline{h}_{PC}$ and the preliminary correction to the point A as $\overline{h}_{A_{pre}} = \overline{h}_{PC} + \Delta\overline{h}$, the correction for the horizontal coordinates is obtained as

$$\Delta x^p_{corr} = \Delta x^p \tau \left(1 - \overline{h}_{A_{pre}} \left| \overline{R} \right|\right)$$
$$\Delta y^p_{corr} = \Delta y^p \tau \left(1 - \overline{h}_{A_{pre}} \left| \overline{R} \right|\right) \qquad (3.45)$$

where τ denotes the local scale factor and $\overline{R}$ is a mean radius of curvature of the national ellipsoid. The correction of the vertical component is then

$$\overline{h}_{A_{corr}} = \overline{h}_{A_{pre}} + \sqrt{\left(\Delta x^p\right)^2 + \left(\Delta y^p\right)^2} \Big/ \left(2\overline{R}\right) \qquad (3.46)$$

with all terms as defined previously.

3.3.2.2 Strip adjustment

A popular approach to correct for unmodelled systematic errors, primarily from the navigation system components, in airborne laser scanning data draws on the principles of photogrammetric block adjustment using the Gauss–Markov model. It relies on strong block geometry in the form of a series of overlapping strips and cross-strips. The process of improving the correspondences between adjunct strips can be performed with or without self-calibration [e.g. SCOP++; IFP, 2004]. As the self-calibration is treated separately (Section 3.4.2) the objective considered here is to estimate global shift and rotation between individual strips. The strips must have sufficient side-lap to allow identification and measurement of both tie and control features, which may be discrete points, intensity values, or planar patches [e.g. Filin and Vosselman, 2004; Pfeifer *et al.*, 2005]. The objective is to estimate the parameters of a correction function for each strip [Pfeifer *et al.*, 2005], i.e.

$$\mathbf{x'}^e_p(s) = \mathbf{x}^e_p(s) + f(s) \qquad (3.47)$$

where $\mathbf{x}^e_p(s)$ is an observed point in strip s, $\mathbf{x'}^e_p(s)$ is the corrected point and $f(s)$ is the correction function for strip s. The error modelling is most commonly data-driven in which the estimable parameter set comprises low-order polynomial coefficients and the objective is to minimise a sum of squares of discrepancies in the coordinates of observed points. The usual implementation includes strip-wise translation and rotation (also known as tilt) parameters, often with corrections applied only in the height dimension. The set of unknowns also includes tie point coordinates or plane parameters or a combination of both. The observables comprise the laser scanning points and control point coordinates necessary for absolute translation and rotation parameter estimation. Though easy to implement and quite practical to use [e.g. Latypov, 2002], the shortcoming of this approach is that the polynomials are blindly applied without any attempt to identify the physical error source. The preferred approach is sensor driven, in which the error modelling is done with reference to the navigation system observables [e.g. Friess, 2006].

3.4 System calibration

In Section 3.2 the various systematic error sources and their analytical models for both terrestrial and airborne laser scanners were described in detail. Calibration is concerned with estimation of the parameters of these models. This section reviews the techniques for calibration of terrestrial and airborne lasers scanners with particular emphasis on system self-calibration.

3.4.1 Calibration of terrestrial laser scanners

The calibration of terrestrial laser scanners can be approached from at least two perspectives. One is individual component calibration, while the other is complete system self-calibration. Whilst the former is intuitive and can in principle make use of existing procedures and infrastructure, its direct application to terrestrial laser scanning can pose some practical problems. The latter is more ideal, in the sense that all system errors are estimated simultaneously, but requires strong network geometry.

Outdoor calibration baselines for electronic distance meters (EDM) are somewhat ubiquitous and are therefore natural facilities to use for terrestrial rangefinder calibration. They would seem to be ideal for estimating the rangefinder offset and the scale error since they are calibrated to a high accuracy. Problems with their use lie, however, in the inability of some scanners to be centred over a known point, though current hardware developments are overcoming this obstacle, and the baselines are typically designed for the unit lengths of EDM equipment used for surveying [e.g. 10 m; Rüeger, 1990]. Since the finest unit lengths of phase-difference terrestrial scanners are typically shorter than this (e.g. 1.2 m in the case of the Faro 880), cyclic error estimation from EDM baseline observations is, in general, not possible. Indoor baselines are adaptable and therefore overcome this problem, but they are less common.

Since terrestrial laser scanners do not have a telescope, they cannot be manually pointed to a distant point on two faces in order to determine the collimation axis, trunnion axis and vertical circle index errors using standard surveying techniques that are employed for theodolites and total stations. Instead, terrestrial laser scanners sample a scene in uniform angular increments. The errors can however be estimated through a highly redundant, network-based self-calibration procedure.

The basis of the network method is observation of a large number of targets in point clouds captured from several locations under different orientations. A least-squares adjustment of all observations in all scans is performed to simultaneously estimate the object space coordinates of the target points, the exterior orientation elements of each scanner frame and all additional parameters. The basic observation equations (Equations 3.10–3.12) are therefore augmented with Equations (3.21)–(3.23), i.e.

$$\rho_{ij} = \sqrt{x_{ij}^2 + y_{ij}^2 + z_{ij}^2} + \Delta\rho \tag{3.48}$$

$$\theta_{ij} = \arctan\left(\frac{y_{ij}}{x_{ij}}\right) + \Delta\theta \tag{3.49}$$

$$\alpha_{ij} = \arctan\left(\frac{z_{ij}}{\sqrt{x_{ij}^2 + y_{ij}^2}}\right) + \Delta\alpha \tag{3.50}$$

The matrix form of the linearised observation equations then becomes

$$\begin{pmatrix} \mathbf{A}_{\rho e} & \mathbf{A}_{\rho a} & \mathbf{A}_{\rho o} \\ \mathbf{A}_{\theta e} & \mathbf{A}_{\theta a} & \mathbf{A}_{\theta o} \\ \mathbf{A}_{\alpha e} & \mathbf{A}_{\alpha a} & \mathbf{A}_{\alpha o} \end{pmatrix} \begin{pmatrix} \hat{\delta}_e \\ \hat{\delta}_a \\ \hat{\delta}_o \end{pmatrix} + \begin{pmatrix} \mathbf{w}_\rho \\ \mathbf{w}_\theta \\ \mathbf{w}_\alpha \end{pmatrix} = \begin{pmatrix} \hat{\mathbf{v}}_\rho \\ \hat{\mathbf{v}}_\theta \\ \hat{\mathbf{v}}_\alpha \end{pmatrix} \tag{3.51}$$

where $\mathbf{A}_{xy}$ represents the Jacobian matrix of observation group x (ρ, θ and α) taken with respect to parameter set y (exterior orientation parameters, e; additional parameters, a; object point coordinates, o); δ_y represents the vector of corrections to the approximate parameter values for set y; and $\mathbf{w}_x$ and $\mathbf{v}_x$ are the misclosure vector (calculated minus observed values) and estimated residuals for observation group x, respectively. This system may be augmented by other independent observations or constraints if available. Examples might include inclinometer levelling angles, spatial distances between targets (required for scale error estimation) or position constraints if multiple scans are captured with the instrument optically centred over the same planimetric location.

The preferred datum definition is by inner constraints. The functional form of the model for constraints imposed on object points takes the form

$$\begin{pmatrix} \mathbf{0} & \mathbf{0} & \mathbf{G}_o^T \end{pmatrix} \begin{pmatrix} \hat{\delta}_e \\ \hat{\delta}_a \\ \hat{\delta}_o \end{pmatrix} = \mathbf{0} \tag{3.52}$$

where the form of $\mathbf{G}_o^T$ is described in Kuang (1996), for example. Though not implemented here, inner constraints can also be imposed on the exterior orientation parameters. See Dermanis (1994) for their analytical form for different rotation parameterisations.

Using the Lagrange multiplier method, the least-squares normal equations corresponding to Equations (3.51) and (3.52) are given by

$$
\begin{pmatrix}
\begin{pmatrix} \mathbf{A}_{\rho e}^T \mathbf{P}_\rho \mathbf{A}_{\rho e} \\ +\mathbf{A}_{\theta e}^T \mathbf{P}_\theta \mathbf{A}_{\theta e} \\ +\mathbf{A}_{\alpha e}^T \mathbf{P}_\alpha \mathbf{A}_{\alpha e} \end{pmatrix} & \begin{pmatrix} \mathbf{A}_{\rho e}^T \mathbf{P}_\rho \mathbf{A}_{\rho a} \\ +\mathbf{A}_{\theta e}^T \mathbf{P}_\theta \mathbf{A}_{\theta a} \\ +\mathbf{A}_{\alpha e}^T \mathbf{P}_\alpha \mathbf{A}_{\alpha a} \end{pmatrix} & \begin{pmatrix} \mathbf{A}_{\rho e}^T \mathbf{P}_\rho \mathbf{A}_{\rho o} \\ +\mathbf{A}_{\theta e}^T \mathbf{P}_\theta \mathbf{A}_{\theta o} \\ +\mathbf{A}_{\alpha e}^T \mathbf{P}_\alpha \mathbf{A}_{\alpha o} \end{pmatrix} & \mathbf{0} \\
 & \begin{pmatrix} \mathbf{A}_{\rho a}^T \mathbf{P}_\rho \mathbf{A}_{\rho a} \\ +\mathbf{A}_{\theta a}^T \mathbf{P}_\theta \mathbf{A}_{\theta a} \\ +\mathbf{A}_{\alpha a}^T \mathbf{P}_\alpha \mathbf{A}_{\alpha a} \end{pmatrix} & \begin{pmatrix} \mathbf{A}_{\rho a}^T \mathbf{P}_\rho \mathbf{A}_{\rho o} \\ +\mathbf{A}_{\theta a}^T \mathbf{P}_\theta \mathbf{A}_{\theta o} \\ +\mathbf{A}_{\alpha a}^T \mathbf{P}_\alpha \mathbf{A}_{\alpha o} \end{pmatrix} & \mathbf{0} \\
 & & \begin{pmatrix} \mathbf{A}_{\rho o}^T \mathbf{P}_\rho \mathbf{A}_{\rho o} \\ +\mathbf{A}_{\theta o}^T \mathbf{P}_\theta \mathbf{A}_{\theta o} \\ +\mathbf{A}_{\alpha o}^T \mathbf{P}_\alpha \mathbf{A}_{\alpha o} \end{pmatrix} & \mathbf{G}_o \\
\text{sym.} & & & \mathbf{0}
\end{pmatrix}
\begin{pmatrix} \hat{\delta}_e \\ \hat{\delta}_a \\ \hat{\delta}_o \\ \hat{\mathbf{k}}_c \end{pmatrix}
=
\begin{pmatrix}
\begin{pmatrix} \mathbf{A}_{\rho e}^T \mathbf{P}_\rho \mathbf{w}_\rho \\ +\mathbf{A}_{\theta e}^T \mathbf{P}_\theta \mathbf{w}_\theta \\ +\mathbf{A}_{\alpha e}^T \mathbf{P}_\alpha \mathbf{w}_\alpha \end{pmatrix} \\
\begin{pmatrix} \mathbf{A}_{\rho a}^T \mathbf{P}_\rho \mathbf{w}_\rho \\ +\mathbf{A}_{\theta a}^T \mathbf{P}_\theta \mathbf{w}_\theta \\ +\mathbf{A}_{\alpha a}^T \mathbf{P}_\alpha \mathbf{w}_\alpha \end{pmatrix} \\
\begin{pmatrix} \mathbf{A}_{\rho o}^T \mathbf{P}_\rho \mathbf{w}_\rho \\ +\mathbf{A}_{\theta o}^T \mathbf{P}_\theta \mathbf{w}_\theta \\ +\mathbf{A}_{\alpha o}^T \mathbf{P}_\alpha \mathbf{w}_\alpha \end{pmatrix} \\
\mathbf{0}
\end{pmatrix}
=
\begin{pmatrix} \mathbf{0} \\ \mathbf{0} \\ \mathbf{0} \\ \mathbf{0} \end{pmatrix}
\tag{3.53}
$$

where $\mathbf{k}_c$ is the vector of Lagrange multipliers.

Several important aspects of the network design are worthy of discussion since one of the goals in self-calibration is to minimise the correlation between exterior orientation elements and the additional parameters. First, observations both in front of and behind the instrument are needed to solve for the collimation axis error, and a large elevation angle range is needed for estimation of the trunnion axis error. The latter can be achieved by placing targets on the floor and ceiling. At least two locations and a variety of ranges are needed for rangefinder additive constant determination. A large variety of ranges is also needed to estimate cyclic errors. From sampling theory, $U_1/2$ is the required minimum sampling interval for recovery of the shortest-wavelength (U_1) cyclic error, though a much finer increment is desired. It is also recommended to capture orthogonal scans from the same nominal location, a practice borrowed from photogrammetric camera calibration, to further decorrelate the exterior orientation parameters and additional parameters. As shown in Lichti (2007), inclusion of inclinometer observations in the solution can have a strong, positive effect on the solution by completely decorrelating the "levelling" angles ω and φ from some of the additional parameters, notably A_2 and C_0, as well as the height component of the scanner position,

Z_c. High redundancy, and therefore a large number of targets, is desired so that the trends in the residual plots uncovered during the exploratory data analyses – which is advised in order to determine the most appropriate error model – can be safely hypothesised to be due to unmodelled systematic errors. The effectiveness of this approach is best exemplified by the clear trends that can be seen in Figures. 3.4 to 3.7.

One is often confronted with the problem of how to decide exactly which systematic error model parameters are needed. No definitive solution exists on how to evaluate the significance of additional parameters since they are frequently correlated with each other and with other model variables. Statistical testing procedures that examine the quadratic form of additional parameters under the null hypothesis that their expectation is zero have long been used and are convenient. Nevertheless, they suffer from several drawbacks that have motivated recent efforts focusing on simulation methods that examine the influence of additional parameters in object space [Habib and Morgan, 2005; Habib *et al.*, 2006; Lichti, 2008].

Self-calibration of terrestrial laser scanners can be seen as evolving into an extension of the feature-based registration approach of Rabbani *et al.* (2007) where the point-on-feature conditions are augmented with additional parameters. The advantages of this approach include greater redundancy, no need to place signalised targets and possible on-site calibration, which is particularly attractive if the instrument calibration is temporally unstable [e.g. Lichti, 2007]. An obvious drawback could be that the geometric fidelity of features may not be known, e.g. the flatness of a plane, the circularity of a cylindrical pipe in cross-section. Gielsdorf *et al.* (2004), Rietdorf *et al.* (2003) and more recently Bae and Lichti (2007) have reported on a calibration method for terrestrial laser scanners using planar features. Skaloud and Lichti (2006) report such a procedure for calibration of airborne laser scanners, which is reviewed in the next section.

3.4.2 Calibration of airborne laser scanners

In most airborne laser scanners the lengths of the lever-arms between the rangefinder, the IMU and the GPS sensors can be determined separately by close-range surveying, although there are difficulties related to the realisation of the IMU body frame. Independent rangefinder calibration can also be performed by close-range surveying [e.g. Vaughn *et al.*, 1996]. The orientation errors related to the assembly (e.g. boresight angles, encoder misalignment, mirror torsion) cannot, however, be determined through independent laboratory calibration and therefore an in-flight procedure must be adopted.

Several methods of in-flight calibration exist and they rely on the use of terrain gradients [e.g. Burman, 2000; Soininen and Burman, 2005], control points [e.g. Kager, 2004] or building profiles extracted from overlapping point clouds [e.g. Schenk, 2001]. Both semi-automated and manual procedures exist and some methods treat the parameter estimation problem on a strip-wise basis [Latypov and Zosse, 2002] rather than a global, truly sensor-driven approach (see Section 3.3.2.2). Many of these methods, while functional, are recognised as being suboptimal since they are labour intensive (i.e. they require manual procedures), non-rigorous and provide no statistical quality assurance measures [Morin and El-Sheimy, 2002]. Furthermore, some cannot reliably recover all three of the angular mounting parameters. A review of several of these methods is given by Morin (2002), who suggests a photogrammetric calibration approach using many signalised tie points. The drawback of this method is the need for interpolation of tie point coordinates in the point clouds, which limits the pointing accuracy and, hence, calibration accuracy.

The in-flight, surface-based self-calibration method discussed herein is a sensor-driven approach based on conditioning the georeferenced target points to lie on surfaces of known form, particularly planes [Friess, 2006; Skaloud and Lichti, 2006]. The parameters of these planes are estimated together with the calibration parameters. The three position and three orientation estimates from the GPS/IMU system and the range and encoder angle measurements from the lidar system comprise the eight observables per point. The observation equation for an object point i, $\mathbf{x}_{p_i}^e$, given by Equation (3.24), lying on plane j is given by

$$\left\langle \mathbf{s}_j, \begin{pmatrix} \mathbf{x}_{p_i}^e \\ 1 \end{pmatrix} \right\rangle = 0 \tag{3.54}$$

where the plane parameters are given by

$$\mathbf{s}_j = \begin{pmatrix} s_1 & s_2 & s_3 & s_4 \end{pmatrix}^T \tag{3.55}$$

where s_1, s_2 and s_3 are the direction cosines of the plane's normal vector and s_4 is the negative orthogonal distance between the plane and the coordinate system origin. Note that the direction cosines must satisfy the following unit length constraint

$$s_1^2 + s_2^2 + s_3^2 - 1 = 0 \tag{3.56}$$

The linearised systems of observation equations and weighted constraints of the plane normal vector parameters are respectively given by

$$\mathbf{A}_1 \hat{\boldsymbol{\delta}}_1 + \mathbf{A}_2 \hat{\boldsymbol{\delta}}_2 + \mathbf{B}\hat{\mathbf{v}} + \mathbf{w} = 0 \tag{3.57}$$

$$\mathbf{G}_c \hat{\boldsymbol{\delta}}_2 + \mathbf{w}_c = \hat{\mathbf{v}}_c \tag{3.58}$$

where $\mathbf{A}_1$ and $\mathbf{A}_2$, are the respective design matrices of partial derivatives of the function (Equation 3.54 with $\mathbf{r}_i$ substituted from Equation 3.24) taken with respect to (1) the calibration parameters and (2) the plane parameters; $\hat{\boldsymbol{\delta}}_1$ and $\hat{\boldsymbol{\delta}}_2$ are the corresponding correction vectors; $\mathbf{B}$ is the design matrix of partial derivatives of the same function taken with respect to the observations; $\mathbf{G}_c$ is the design matrix of derivatives of the unit length constraint taken with respect to the plane parameters; $\mathbf{v}$ and $\hat{\mathbf{v}}_c$ are the respective vectors of residuals; and $\mathbf{w}$ and $\mathbf{w}_c$ are the misclosure vectors.

The influence of correlations within the GPS/INS observations for the IMU–camera bore-sight determination is demonstrated by Skaloud and Schaer (2003) and applies also to airborne laser scanning. There are two types of correlation in the GPS/INS data. The first is reflected in the correlation between the six parameters of exterior orientation; the other is the time correlation between the successive estimates. As shown in Skaloud and Schaer (2003), the second is far more significant for optimal bore-sight estimation with respect to the camera and thus most likely also with respect to the laser system. Although similar influences can be expected for the IMU–lidar case, the present stochastic model assumes all observational errors (of the eight observations per point and of the constraint observations) to be zero-mean and uncorrelated with each other. These assumptions are suboptimal and should be partly compensated by increasing the time varying covariances of the individual GPS/INS observations. On the other hand, considering purely diagonal covariance matrices allows important simplifications in the model formulation when processing voluminous airborne laser scanning data. Considering this, the weight matrices for the observations, $\mathbf{P}$, and for the constraints, $\mathbf{P}_c$, are assumed to be diagonal. It is furthermore assumed that the constraints for all planes are to be weighted equally with a much smaller variance than those of the measurements. Again, following standard least-squares procedures, the resulting final form of the normal equations is given by

$$
\begin{pmatrix}
\mathbf{A}_1^T \left(\mathbf{B}\mathbf{P}^{-1}\mathbf{B}^T \right)^{-1} \mathbf{A}_1 & \mathbf{A}_1^T \left(\mathbf{B}\mathbf{P}^{-1}\mathbf{B}^T \right)^{-1} \mathbf{A}_2 \\
sym. & \mathbf{A}_2^T \left(\mathbf{B}\mathbf{P}^{-1}\mathbf{B}^T \right)^{-1} \mathbf{A}_2 + \mathbf{G}_c^T \mathbf{P}_c \mathbf{G}_c
\end{pmatrix}
\begin{pmatrix}
\hat{\delta}_1 \\
\hat{\delta}_2
\end{pmatrix}
$$
$$
+ \begin{pmatrix}
\mathbf{A}_1^T \left(\mathbf{B}\mathbf{P}^{-1}\mathbf{B}^T \right)^{-1} \mathbf{w} \\
\mathbf{A}_2^T \left(\mathbf{B}\mathbf{P}^{-1}\mathbf{B}^T \right)^{-1} \mathbf{w} + \mathbf{G}_c^T \mathbf{P}_c \mathbf{w}_c
\end{pmatrix}
= \begin{pmatrix} 0 \\ 0 \end{pmatrix}
\tag{3.59}
$$

The network design that is optimal for the recovery of the bore-sight angles (or other additional parameters) should include several inclined planes having different aspects. Roofs are a practical choice for the planes but the data must be carefully edited in order to remove points from chimneys and other features as well as points at the edge that may be biased due to the effects described in Section 3.2.1. Nevertheless, this procedure can be practically automated and the method is relevant for calibrating scanners of different types [Skaloud and Schaer, 2007]. The solution can be strengthened somewhat if the plane parameters are determined *a priori* [e.g. Filin, 2003], but this sacrifices the practical benefits of this approach. Data should be captured from flight lines in four different directions. Typically this can be achieved by flying the standard cloverleaf pattern. For the rangefinder offset, at least two flying heights are suggested to decorrelate it from the plane distance parameters, though errors in the GPS-derived height observations are the dominant factor precluding accurate in-flight recovery of the rangefinder offset.

The effectiveness of self-calibration can be seen in the residuals from the best fit planes. Figure 3.24 shows surface plots of the residuals as functions of position on the roof and the corresponding histograms for data with and without self-calibration. The plots are oriented in a plane-centric coordinate system for which the origin is at the roof centroid and the u-, v- and w-axes correspond to the principal axes from the planar point cloud covariance matrix eigenvalue decomposition. Correlation in the residuals is evident in the mottled texture of the out-of-plane w-axis and the large magnitude of the errors. The effects of unmodelled roll and pitch bore-sight angles

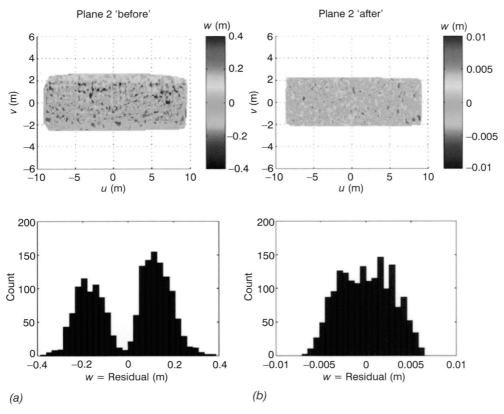

Figure 3.24 Surface plots of the plane-fit residuals with corresponding histograms: (a) before, and (b) after adjustment. (After Skaloud and Lichti, 2006; courtesy ISPRS)

are clearly visible as the bimodal structure in the histogram, which disappears after self-calibration.

As calibration remains an active field of research, the sophistication of this process is expected to evolve in the future. Although correct registration and system calibration act as prerequisites for further point cloud processing, most calibration approaches rely, to some extent, on correct point cloud preprocessing. Indeed, there is a feedback loop as the information extracted from the mismatched elements (e.g. patches, features, surface model gradients) obtained after classification is input to the calibration. For reasons of robustness and convergence it is important that the chosen characteristic of the elements is either intrinsic (at least locally) or that external knowledge can be imposed (e.g. form of the patches), as in the case of the presented method.

Summary

This chapter has focused on geometric positioning models, registration and system calibration for terrestrial and airborne laser scanning. The latter two are critical processes which, in general, should be completed prior to further point cloud processing. Registration is required in order for the data to be represented in a common coordinate system. Calibration is necessary to remove the effects of systematic errors and maximise point cloud accuracy.

References

Adams, M.D., 1999. *Sensor Modelling, Design and Data Processing for Autonomous Navigation*, World Scientific, Singapore.

Amann, M.C., Bosch, T., Lescure, M., Myllyla, R. and Rioux, M., 2001. Laser ranging: A critical review of usual techniques for distance measurement. *Optical Engineering*, 40(1), 10–19.

Bae, K.-H. and Lichti, D.D, 2007. On-site self-calibration using planar features for terrestrial laser scanners. *International Archives of the Photogrammetry, Remote Sensing and Spatial Information Sciences*, 36(Part 3/W52), 14–19.

Bae, K.-H. and Lichti, D.D., 2008. A method for automated registration of unorganised point clouds. *ISPRS Journal of Photogrammetry and Remote Sensing*, 63(1), 36–54.

Baltsavias, E. P., 1999. Airborne laser scanning: basic relations and formulas. *ISPRS Journal of Photogrammetry and Remote Sensing*, 54(2–3), 199–214.

Barnea, S. and Filin, S., 2008. Keypoint based autonomous registration of terrestrial laser point-clouds. *ISPRS Journal of Photogrammetry and Remote Sensing*, 63(1), 19–35.

Beinat, A., Crosilla, F. and Sepic, F., 2006. Automatic morphological prealignment and global hybrid registration of close range images. *International Archives of the Photogrammetry, Remote Sensing and Spatial Information Sciences*, 36(Part 5) (on CD-ROM).

Besl, P.J. and McKay, N.D., 1992. A method for registration of 3-D shapes. *IEEE Transactions on Pattern Analysis and Machine Intelligence*, 14(2), 239–256.

Boström, G., Gonçalves, J.G. and Sequeira, V., 2008. Controlled 3D data fusion using error-bounds. *ISPRS Journal of Photogrammetry and Remote Sensing*, 63(1), 55–67.

Brenner, C., Dold, C. and Ripperda, N., 2008. Coarse orientation of terrestrial laser scans in urban environments. *ISPRS Journal of Photogrammetry and Remote Sensing*, 63(1), 4–18.

Burman, H., 2000. Calibration and Orientation of Airborne Image and Laser Scanner Data Using GPS and INS, Ph.D. thesis, *Photogrammetry Reports No. 69*, Royal Institute of Technology, Stockholm.

Chen, Y. and Medioni, G., 1992. Object modelling by registration of multiple range images. *Image and Vision Computing*, 10(3), 145–155.

Colomina, I., 2007. From off-line to on-line geocoding: the evolution of sensor orientation, *Photogrammetric Week '07*, 3–7 September 2007, Stuttgart, Germany, Fritsch, D. (Ed.), Wichmann, Heidelberg, Germany, 173–183.

Cooper, M.A.R., 1982. *Modern Theodolites and Levels*, 2nd edn., Granada Publishing, London, UK.

Dermanis, A, 1994. The photogrammetric inner constraints. *ISPRS Journal of Photogrammetry and Remote Sensing*, 49(1), 25–39.

Filin, S., 2003. Recovery of systematic biases in laser altimetry data using natural surfaces. *Photogrammetric Engineering and Remote Sensing*, 69(11), 1235–1242.

Filin, S. and Vosselman, G., 2004. Adjustment of airborne laser altimetry strips. *International Archives of the Photogrammetry, Remote Sensing and Spatial Information Sciences*, 35(Part B3), 285–289.

Friess, P., 2006. Toward a rigorous methodology for airborne laser mapping, *Proceedings of Euro-COW*, 25–27 January 2006, Castelldefels, Spain, 7 (on CD-ROM).

Gao, Y. and Wojciechowski, A., 2004. High precision kinematic positioning using single dual frequency GPS receiver. *International Archives of the Photogrammetry, Remote Sensing and Spatial Information Sciences*, 35(Part B3), 845–850.

Gielsdorf, F., Rietdorf, A., and Gruendig, L., 2004. A Concept for the calibration of terrestrial laser scanners. *Proceedings of the FIG Working Week*, 22–27 May 2004, Athens, Greece (on CD-ROM).

Glennie, C.L., 2007. Rigorous 3D error analysis of kinematic scanning LIDAR systems. *Journal of Applied Geodesy*, 1(3), 147–157.

Grün, A. and Akca, D., 2005. Least squares 3D surface and curve matching. *ISPRS Journal of Photogrammetry and Remote Sensing*, 59(3), 151–174.

Habib, A. and Morgan, M., 2005. Stability analysis and geometric calibration of off-the-shelf digital cameras. *Photogrammetric Engineering and Remote Sensing*, 71(6), 733–741.

Habib, A., Pullivelli, A., Mitishita, E., Ghanman, M. and Kim, E.-M., 2006. Stability analysis of low-cost digital cameras for aerial mapping using different georeferencing techniques. *The Photogrammetric Record*, 21(113), 29–43.

Harvey, B.R. and Rüeger, J.M., 1992. Theodolite observations and least squares. *The Australian Surveyor*, 37(2), 120–128.

Hebert, M. and Krotkov, E., 1992. 3D measurements from imaging laser radars: how good are they? *Image and Vision Computing*, 10(3), 170–178.

Horn, B.K.P., 1987. Closed-form solution of absolute orientation using unit quaternions. *Journal of the Optical Society of America*, 4(4), 629–642.

Institute of Photogrammetry (IFP), 2004. SCOP++, Commercial software licensed by The Technical University Vienna, Austria.

Jekeli, C., 2001. *Inertial Navigation Systems with Geodetic Applications*, Walter de Gruyter, Berlin, Germany.

Kager, H., 2004. Discrepancies between overlapping laser scanning strips – simultaneous fitting of aerial laser scanner strips. *International Archives of the Photogrammetry, Remote Sensing and Spatial Information Sciences*, 35(Part B1), 555–560.

Kamerman, G. W., 1993. Laser radar. *Active Electro-Optical Systems*. vol. 6, Fox, C. S. (Ed.). In *The Infrared and Electro-Optical Systems Handbook*. Infrared Information Analysis Center, Ann Arbor, Michigan, USA, 1–76.

Kang, Z., Zlatanova, S. and Gorte, B. 2007. Automatic registration of terrestrial scanning data based on registered imagery. *Proceedings of the 2007 FIG Working Week*, 13–17 May 2007, Hong Kong SAR (on CD-ROM).

Katzenbeisser, R., 2003. About the calibration of lidar sensors. *International Archives of the Photogrammetry, Remote Sensing and Spatial Information Sciences*, 34(Part 3/W13), 59–64.

Kersten, T.P., Sternberg, H. and Mechelke, K., 2005. Investigations into the accuracy behaviour of the terrestrial laser scanning system Mensi GS100. In Grün., A. and Kahmen, H. (Eds.), *Proceedings of Optical 3-D Measurement Techniques VII*, 3–5 October, Vienna, Austria, vol. I, 122–131.

Kim, D., Bisnath, S., Langley, R.B. and Dare, P., 2004. Performance of long-baseline real-time kinematic applications by improving tropospheric delay modeling, *ION GNSS*, 21–24 September 2004, Long Beach, California, USA, 1414–1422.

Kleusberg, A. and Teunissen, P.J.G., 1998. *GPS for Geodesy*, Springer-Verlag, New York, NY, USA.

Kuang, S., 1996. *Geodetic Network Analysis and Optimal Design: Concepts and Applications*, Ann Arbor Press, Chelsea, MI, USA.

Kuipers, J.B., 1998. *Quaternions and Rotation Sequences*, Princeton University Press, Princeton, NJ, USA.

Latypov, D., 2002. Estimating relative lidar accuracy information from overlapping flight lines. *ISPRS Journal of Photogrammetry and Remote Sensing*, 56(4), 236–245.

Latypov, D., 2005. Effects of laser beam alignment tolerance on lidar accuracy. *ISPRS Journal of Photogrammetry and Remote Sensing*, 59(6), 361–368.

Latypov, D. and Zosse, E., 2002. LIDAR data quality control and system calibration using overlapping flight lines in commercial environment. *Proceedings of the ACSM/ASPRS Annual Conference*, 19–26 April 2002, Washington, DC (on CD-ROM).

Legat, K., 2006. Approximate direct georeferencing in national coordinates. *ISPRS Journal of Photogrammetry and Remote Sensing*, 60(4), 239–255.

Leica Geosystems, 2005. *Leica ALS50 Technical Notes*, Heerbrugg, Switzerland.

Lichti, D.D., 2007. Modelling, calibration and analysis of an AM-CW terrestrial laser scanner. *ISPRS Journal of Photogrammetry and Remote Sensing*, 61(5), 307–324.

Lichti, D.D., 2008. A method to test differences between additional parameter sets with a case study in terrestrial laser scanner self-calibration stability analysis. *ISPRS Journal of Photogrammetry and Remote Sensing*, 63(2), 169–180.

Lichti, D.D. and Franke, J., 2005. Self-calibration of the iQsun 880 laser scanner. In Grün., A. and Kahmen, H. (Eds.), *Proceedings of Optical 3-D Measurement Techniques VII*, 3–5 October 2005, Vienna, Austria, vol. I, 112–121.

Lichti, D.D., S.J. Gordon and T. Tipdecho, 2005. Error models and propagation in directly georeferenced terrestrial laser scanner networks, *ASCE Journal of Surveying Engineering*, 131(4), 135–142.

Lichti, D.D. and S. Jamtsho, 2006, Angular resolution of terrestrial laser scanners. *The Photogrammetric Record*, 21(114), 141–160.

Lichti, D.D. and Lampard, J., 2008. Reflectorless total station self-calibration. *Survey Review*, 40(309), 244–259.

Lichti, D.D. and Licht, M.G., 2006. Experiences with terrestrial laser scanner modelling and accuracy assessment. *International Archives of the Photogrammetry, Remote Sensing and Spatial Information Sciences*, 36(Part 5), 155–160.

Mikhail, E.M., 1976. *Observations and Least Squares*, IEP, New York, NY, USA.

Morin, K.W., 2002. Calibration of Airborne Laser Scanners. M.Sc. thesis, The University of Calgary.

Morin, K. and El-Sheimy, N., 2002. Post-mission adjustment of airborne laser scanning data, *Proceedings XXII FIG International Congress*, 19–26 April 2002, Washington, DC, USA (on CD-ROM).

Mostafa, M., 2007. New developments of inertial navigation systems at Applanix. In Fritsch, D. (Ed.), *Photogrammetric Week '07*, Stuttgart, Germany, Wichmann, Heidelberg, Germany, 77–87.

Pesci, A. and Teza, G., 2008. Terrestrial laser scanner and retro-reflective targets: an experiment for anomalous effects investigation. *International Journal of Remote Sensing*, 29(19), 5749–5765.

Pfeifer, N., Oude Elberink, S. and Filin, S., 2005. Automatic tie elements detection for laser scanner strip adjustment. *International Archives of the Photogrammetry, Remote Sensing and Spatial Information Sciences*, 36(Part3/W19), 174–179.

Rabbani, T., Dijkman, S., van den Heuvel, F. and Vosselman, G. 2007. An integrated approach for modelling and global registration of point clouds. *ISPRS Journal of Photogrammetry and Remote Sensing*, 61(6), 355–370.

Reshetyuk, Y., 2006. Calibration of terrestrial laser scanners for the purposes of geodetic engineering. *Proceedings of the 3rd IAG/12th FIG Symposium*, 22–24 May 2006, Baden, Austria (on CD-ROM).

Riegl, 2001. *LMS-Z210 laser mirror scanner technical documentation and users instructions*. Riegl, Horn, Austria.

Rietdorf. A., Gielsdorf, F. and Gruendig, L., 2003. Combination of hand measuring methods and scanning techniques for CAFM-data acquisition. *Proceedings of the FIG Working Week*, 13–17 April 2003, Paris, France (on CD-ROM).

Rüeger, J.M., 1990. *Electronic Distance Measurement: an Introduction*, 3rd edn., Springer-Verlag, Heidelberg, Germany.

Rüeger, J.M., 2003. *Electronic Surveying Instruments – A Review of Principles, Problems and Procedures*, University of New South Wales, Sydney, Australia.

Runne, H., Niemeier, W. and Kern, F., 2001. Application of laser scanners to determine the geometry of buildings. *Proceedings of Optical 3-D Measurement Techniques V*, 1–4 October 2001, Vienna, Austria, Wichmann, Karlsruhe, Germany, 41–48.

Rusinkiewicz, S. and Levoy, M., 2001. Efficient variant of the ICP algorithm. *Proceedings of 3-D Digital Imaging and Modelling (3DIM)*, 28 May to 1 June 2001, Quebec City, Canada, 145–152.

Salo, P., Jokinen, O. and Kukko, A., 2008. On the calibration of the distance measuring component of a terrestrial laser scanner. *International Archives of the Photogrammetry, Remote Sensing and Spatial Information Sciences*, 37(Part B5), 1067–1071.

Schaer, P., Skaloud, J., Landtwig, S. and Legat, K., 2007. Accuracy estimation for laser point cloud including scanning geometry. *International Archives of the Photogrammetry, Remote Sensing and Spatial Information Sciences*, 36(Part 5/C55), 279–353.

Schaer, P., Skaloud, J. and Tome, P., 2008. Towards in-flight quality assessment of airborne laser scanning. *International Archives of the Photogrammetry, Remote Sensing and Spatial Information Sciences*, 37(Part B5), 851–856.

Schenk, T., 2001. Modeling and analyzing systematic errors in airborne laser scanners. *Technical Notes in Photogrammetry*, 19, The Ohio State University, Columbus, OH, USA.

Schneider, D. and Schwalbe, E., 2008. Integrated processing of terrestrial laser scanner data and fisheye-camera image data. *International Archives of the Photogrammetry, Remote Sensing and Spatial Information Sciences*, 37(Part B5), 1037–1043.

Schwartz, K.P., Chapman, M.A, Cannon, M.E. and Gong, P., 1993. An integrated INS/GPS approach to the georeferencing of remotely sensed data. *Photogrammetry Engineering and Remote Sensing*, 59(11), 1167–1674.

Seeber, G., 1993. *Satellite Geodesy*, Walter de Gruyter, Berlin, Germany.

Skaloud J., Bruton, A.M. and Schwarz, K.P., 1999. Detection and filtering of short-term (1/f) noise in inertial sensors. *Navigation – Journal of The Institute of Navigation*, 46(2), 97–107.

Skaloud, J. and Schwartz, K.P., 2000. Accurate orientation for airborne mapping systems. *Photogrammetry Engineering and Remote Sensing*, 66(4), 393–401.

Skaloud, J. and Lichti, D.D., 2006. Rigorous approach to bore-sight self calibration in airborne laser scanning. *ISPRS Journal of Photogrammetry and Remote Sensing*, 61(1), 47–59.

Skaloud, J. and Schaer, P., 2003. Towards a more rigorous boresight calibration. *ISPRS International Workshop on Theory, Technology and Realities of Inertial/GPS/Sensor Orientation, Commission 1, WG I/5*, 22–23 September 2003, Castelldefels, Spain (on CD-ROM).

Skaloud, J. and Schaer, P., 2007. Towards automated lidar boresight self-calibration. *The 5th International Symposium on Mobile Mapping Technology*, 28–31 May 2007, Padua, Italy (on CD-ROM).

Skaloud, J. and Legat, K., 2008. Theory and reality of direct georeferencing in national coordinates. *ISPRS Journal of Photogrammetry and Remote Sensing*, 63(2), 272–282.

Snyder, J.P. and Bugayevskiy, L.M., 1995. *Map Projections: A Reference Manual*, Taylor and Francis, London, UK.

Soininen, A. and Burman, H., 2005. TerraMatch for MicroStation, Commercial software, Terrasolid Ltd, Finland.

Tiberius, C.C.J.M., 1998. *Recursive Data Processing for Kinematic GPS Surveying*, NCG-Nederlandse Commissie voor Geodesie, Delft, Nertherlands.

Vaughn, C.R., Bufton, J.L, Krabill, W.B. and Rabine, D., 1996. Georeferencing of airborne laser altimeter measurements. *International Journal of Remote Sensing*, 17(11), 2185–2200.

Williams, J. and Bennamoun, M., 2001. Simultaneous registration of multiple corresponding point sets. *Computer Vision and Image Understanding*, 81(1), 117–142.

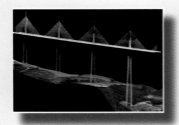

Extraction of Digital Terrain Models | 4

Christian Briese

This chapter focuses on the extraction of digital terrain models (DTMs) from laser scanning data. A DTM, sometimes also referred to as a digital elevation model (DEM), is a mathematical representation, i.e. a model, of the bare earth surface in digital form. The bare earth is here defined as the boundary surface between the solid ground and the atmosphere or objects attached to it (e.g. buildings, vegetation). Such a model is essential for many different application areas, e.g. flood risk management, infrastructure planning, environmental protection, and is nowadays a fundamental dataset within geographic information systems (GIS).

As well as the special focus on DTM generation based on airborne laser scanning data and some comments on the generation of a DTM from terrestrial laser scanning data, some further general remarks on DTM generation (data structures and interpolation techniques) can be found in this chapter. In addition to DTMs, the digital surface model (DSM, also often referred to as the digital canopy model, DCM), which describes the top surface that is visible from above, is often used in the area of topographic mapping (especially for visualisation purposes). The DSM is equivalent to the DTM in open areas, but in contrast to the DTM it describes the vegetation cover as well as the upper boundary surface of spatiotemporal fixed manmade objects such as buildings (see Figure 4.1). However, due to the difficulty of removing temporal objects on the terrain the DSM often also includes cars and other objects present during data acquisition. A graphical illustration that presents the difference between a DSM and DTM is provided in Figure 4.1. Within this chapter DSM determination from airborne laser scanning will only be briefly discussed. Further digital models, e.g. building models, will be covered in the subsequent chapters of the book.

In the past, analytical photogrammetry was the main measurement technique for gathering topographic information of large (from several square kilometres to country-wide) areas with a height accuracy in the decimetre or metre range. Based on manually observed surface features (points and lines on the bare earth surface) the DTM can be determined subsequently with the help of interpolation methods. In recent

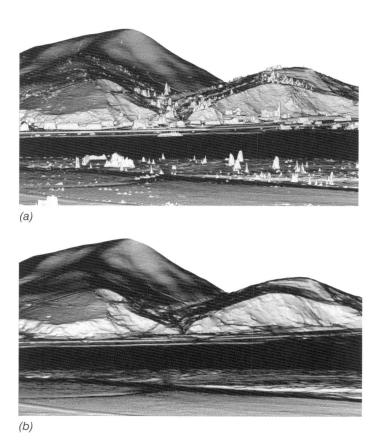

(a)

(b)

Figure 4.1 Perspective view of the same area: (a) Digital surface model (DSM, 1 m grid) determined from first echo airborne laser scanning data. The model describes the surface seen from above. (b) Digital terrain model (DTM, 1 m grid) determined from the classified terrain points. The DTM represents the bare earth surface. Data origin: Airborne laser scanning Klosterneuburg (Vermessung Schmid)

years, new and – compared to manual photogrammetric measurements – highly automated measurement techniques like airborne laser scanning, automated image matching, and interferometric synthetic aperture radar (InSAR) have become available. All these new techniques have in common that they allow a very dense sampling of the area of interest within a short time. However, compared to the "classical" manual data acquisition systems, where the interpretation and abstraction of the topography is performed interactively during the stage of data acquisition by human interpretation, the sampling of these automatic systems is not guided by a simultaneous interpretation of the scene. The interpretation is typically done separately in post-processing and can be adapted to the individual application. For DTM generation based on this automatically generated data, it is essential in this post-processing phase to extract the relevant bare earth data from the whole of the gathered information. This task is a classification of the automatically acquired topographic data into terrain and off-terrain information. In the airborne laser scanning community this process is typically referred to as filtering. Due to the huge amount of data that is generated by these automatic data acquisition techniques, there is a need to automate this process. Subsequently, based on the extracted terrain information, the DTM can be generated by appropriate interpolation methods.

The first section of this chapter discusses this classification or filtering step for the separation of terrain and off-terrain information based on laser scanning point clouds. Several filter algorithms have been developed for this task. The subsequent section presents a representative selection of some of these methods and provides literature references. Furthermore, the future potential of additional information provided by full-waveform airborne laser scanning systems for DTM generation is discussed. In order to give a high quality description of linear surface discontinuities and other linear features, the next section focuses on the determination of linear features based on airborne laser scanning data. In the final section of this chapter, general remarks on DTM generation are presented and some issues regarding DTM quality are discussed. For subsequent analyses it is often essential to provide the DTM in different levels of representation quality. Therefore, this section includes a discussion of the process of DTM data reduction.

4.1 Filtering of point clouds

As already mentioned in the introduction to this chapter, during laser scanning data acquisition no interpretation of the scene is performed. Therefore, in order to determine the relevant terrain information, filtering of the laser scanning data, i.e. the classification of the point cloud into terrain and off-terrain points, is essential. This section will focus on the filtering of airborne laser scanning data due to the fact that most of the research on the filtering of point cloud data was performed in this area. However, it has to be stressed that most of the main ideas are also relevant to the filtering of other point cloud data, e.g. terrestrial laser scanning data, multibeam echo sound data or point clouds obtained by automated image matching [Briese *et al.*, 2002; Bottelier *et al.*, 2005; Bauerhansl *et al.*, 2004]. However, within the process of DTM generation from these other data acquisition processes the individual data characteristics have to be considered, for example when working with terrestrial laser scanning data one has to take care of the typically strongly varying point density in the project coordinate frame.

Formally, the problem of filtering of point clouds for DTM generation can be stated as follows: for a given set of points $\mathbf{P} = \{\mathbf{p}_1(\mathbf{x}_1, c_1), \ldots, \mathbf{p}_n(\mathbf{x}_n, c_n)\}$, each point embedded in 3D space, $\mathbf{x}_i = (x_i, y_i, z_i)$, with its individual classification label c_i, finds a classifier function $f : \mathbf{x} \rightarrow c$, which maps each point $\mathbf{p}_i \in \mathbf{P}$ to its classification label $c_i \in \mathbf{C} = \{\text{terrain},$

off-terrain}. In this way the classification label c_i of each point $\mathbf{p}_i$ can be assigned to the attribute value "terrain" or "off-terrain".

Simultaneous use of all available point cloud data is important for the filtering, for example in the case of airborne laser scanning it is important to use the data from all airborne laser scanning strips, because data from another neighbouring strip can provide further valuable information for the filtering. In areas where two or more strips overlap, multiple view directions onto the surface are available that increase the probability of penetrating the vegetation canopy. Furthermore, use of the whole acquired point cloud increases redundancy and might allow shadow areas to be filled; these are areas not visible from one line of sight, for example due to objects such as houses above the terrain, of one airborne laser scanning strip or not visible from one terrestrial laser scanning station.

Therefore, as a precondition prior to the filtering of point cloud data, the transformation of all acquired echoes of all acquired airborne laser scanning strips, resp. terrestrial laser scanning stations, into one common coordinate frame has to be determined, i.e. all the data have to be georeferenced accordingly (see Chapter 3).

The description of a DTM based on airborne laser scanning as well as terrestrial laser scanning data will typically be based just on last echo points with the derived label "terrain", because the other echoes can be suggested to originate from objects above the bare earth. However, an exception might occur in areas where multipath reflections on mirror-like surfaces (typically in city areas with shop or car windows) can be found. This leads to so-called "long-range" measurements where the resulting position of the last echo may wrongly be recorded below the terrain surface. Another exception might occur on a terrain step edge where the first and last echo might result from the terrain surface. In such cases an echo prior to the last might be classified as originating from the bare earth.

As mentioned in Chapter 1 discrete echo airborne laser scanning systems typically provide, next to the ability to separate multiple echoes, radiometric information per echo. For this additional observation the term "intensity" measurement is widely used. This intensity value stored for each echo describes the strength of reflection of the illuminated surface elements contributing to one echo. However, it has to be considered that the echo intensity value stored by airborne laser scanning systems is typically not normalised to a certain standard range and standard sensor model [Wagner *et al.*, 2006; Höfle and Pfeifer, 2007]. Nowadays, this additional information per echo is only very rarely used in the process of filtering. Nearly all of the filtering algorithms that are discussed here are based only on geometric criteria. One exception is highlighted in Section 4.1.6. In that section the use of further information per echo determined based on full-waveform airborne laser scanning data will be discussed.

Some of the filter algorithms perform a rasterisation resp. gridding of the acquired points (a short discussion about rasterisation vs. gridding can be found in Section 4.3), which allows simple and fast neighbourhood operations in a raster data structure.

Filter algorithms operating on the raster data structure can be designed by analogy to methods developed for digital image processing. The advantage of these algorithms is that there are a lot of raster-based software packages available that can be used for the implementation of operations based on the raster data structure. For generation of the raster different operations can be used, taking the lowest, highest, median, mean, etc. point per raster. Furthermore, the filling of raster cells that contain no points has to be considered. All these issues about rasterisation operations are beyond the scope of this chapter. Section 2.1.1 discusses further issues and provides more details about this topic. In the context of this chapter it is only important to distinguish filter methods based on raster data from those operating on the point cloud, although one has to consider that a certain preprocessing has to be applied in order to rasterise the point cloud data. The disadvantage of the rastered data is that the rasterisation process is bought at the cost of a loss in precision [Axelsson, 1999] and may lead to undesired effects, for example gaps caused by occlusions, when the occlusion next to a building is represented by a tilted plane in the rasterised data (see Section 2.2).

Many different algorithms have been published for the filtering of airborne laser scanning data. An overview is provided in Pfeifer (2003), Sithole and Vosselman (2004), Sithole (2005) and Kobler *et al.* (2007). The simplest filter approaches apply only a local minimum determination (for example the lowest point within a certain raster cell is assigned the label "terrain"). However, these filters mostly do not lead to satisfactory results. The typically fixed cell size and the inherent assumption that the terrain is flat within the cell can produce systematic errors (for example in hilly terrain when the height of the lowest point within a cell is assigned to the cell centre) in the resulting DTM. However, before presenting a selection of advanced algorithms for filtering based on the previously mentioned overview papers, it has to be mentioned that sometimes within the practical workflow combinations of algorithms are also applied in order to increase the classification quality. For example, in the software package SCOP++ [Inpho, 2007] a prefiltering step, which tries to detect building regions (based on a raster data structure), is combined with a surface-based filtering method (hierarchical robust interpolation, see Pfeifer *et al.*, 2001 and Briese *et al.*, 2002) based on point cloud data (see Section 4.1.3). In this example a mixture of a raster-based and a point cloud based solution can be found. Such *a priori* building detection methods are often applied. These methods are not included here but will be discussed in the next chapter, which focuses on the detection and modelling of buildings based on laser scanning point clouds.

4.1.1 Morphological filtering

This group of filter algorithms is based on the concept of mathematical morphology [Haralick and Shapiro, 1992], which is a set-theoretical method of image analysis providing a quantitative description of geometrical structures based on a set of operators.

The two basic operators in the case of mathematical morphology are *erosion* and *dilation*. Erosion and dilation allow a simplification of the surface structure based on a certain structure element. With the help of a pair-wise combination of these two basic operations the so-called *closing* (erosion after dilation) and *opening* (dilation after erosion) operations can be performed, which can be used for the minimum and maximum determination within a certain structure element. Further details about mathematical morphology (applied on digital grey value images) can be found in Haralick *et al.* (1987) and Haralick and Shapiro (1992).

One of the first methods for filtering of laser point clouds was published by Lindenberger (1993) who used input data acquired with an airborne laser profiler. This approach for the classification is based on the previously mentioned concept of mathematical morphology. With the help of a structure element (window), which is moved point-wise along the profile, the operations opening (or closing) determine the upper (or lower) surface represented by the points within a profile (see Figure 4.2). The window size can be adapted and is defined in this approach by a certain number of points within the profile. For the erosion and dilation the height of the central element of the window is determined based on the lowest (or highest) point within the structure element. After this process the lower surface can be seen as an approximation of the DTM and can be used for the filtering of the points within the profile by a height tolerance band. In order to improve this initial classification based on mathematical morphology, Lindenberger (1993) proposes using an advanced modelling process of the terrain profile based on a robust time series analysis.

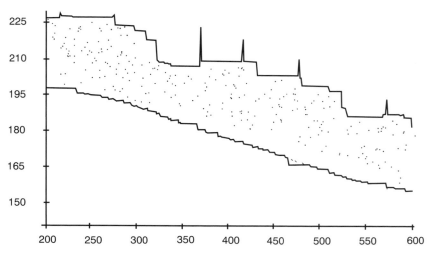

Figure 4.2 Resulting profiles (black lines) with acquired laser profile points (black dots) after an opening (lower profile) and closing (upper profile). The upper or lower profile (DSM or DTM, respectively) runs through the locally highest or lowest points, respectively. (Lindenberger, 1993)

A second algorithm based on mathematical morphology was published by Vosselman (2000). In contrast to the approach presented by Lindenberger (1993), Vosselman introduces a function that describes the maximum admissible height difference Δh within a structure element as a function of the distance d (see Figure 4.3a). The distance is calculated as the horizontal Euclidian distance between two points. Within the process of filtering, the structure element with an *a priori* defined 2D size (d_{max_x}, d_{max_y}) is positioned on each given point. Then, for the candidate point within the centre of the structure element the distance d_i and the height difference Δh_i to each point P_i within the area of the structure element is calculated. If one of the determined height differences Δh_i is higher than the admissible height difference at the

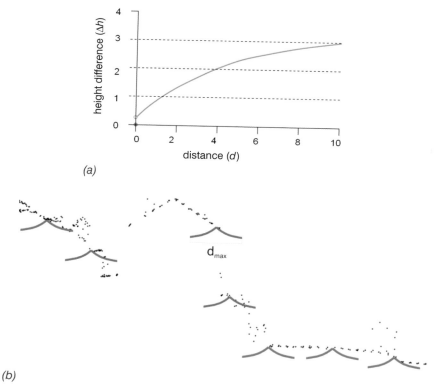

(a)

(b)

Figure 4.3 (a) Evaluation (filter) function which describes the admissible height difference Δh between two points as a function of the distance d (after Pfeifer, 2003; courtesy VGI). (b) Profile with airborne laser scanning points (black dots) that demonstrates the application of the filtering based on the evaluation function for a structure element of size d_{max} centred on each given point. Points where the evaluation function is displayed in green are classified as terrain points, while red evaluation functions indicate an off-terrain point classification.

given distance, the candidate point is assigned the classification label "off-terrain". Finally, all off-terrain points are labelled and the DTM can be reconstructed with the help of the classified terrain points. In practical applications a circle is used for the structure element and the predefined radius is typically set to 5 m. The evaluation with respect to the filter function itself can be determined based on the maximum terrain slope found in the area, from the height precision of the laser points or can be determined with the help of training areas. Based on training areas an optimal structure element which minimises commission and omission errors can be determined [Vosselman, 2000]. A profile showing the practical application of the filter method is presented in Figure 4.3.

An extension to the approach of Vosselman (2000) is presented by Sithole (2001). Instead of a fixed function that describes the admissible height differences, Sithole alters the function with respect to the local terrain slope, which leads to an adaptive slope depending filter. However, for the estimation of the terrain slope an *a priori* DTM is essential. Sithole (2001) proposes using a coarse DTM estimate with the help of a simple block minimum filter by taking the lowest point within a certain raster cell for the estimation of the local terrain slope. A similar extension to the adaptive slope based filter was proposed by Roggero (2001).

The previously mentioned algorithms are typically applied to point clouds, but mathematical morphology has its origin in the raster data structure. Applying these approaches to rasterised laser scanning data helps, due to the given topology, to speed up the process. Such a raster-based approach applied to airborne laser scanning data can be found in Kilian *et al.* (1996). A different approach that can be assigned to filters based on mathematical morphology and rasterised point clouds is the dual-rank algorithm [Eckstein and Munkelt, 1995] proposed by Lohmann *et al.* (2000). This approach replaces the height of the raster cell with the height of the k-lowest point within a certain window (erosion). Then the dilation is applied by selection of the height of the k-highest point within the window.

4.1.2 Progressive densification

The methods belonging to the progressive densification group follow a different filter strategy. These algorithms start with a small subset of the given point cloud (preclassified terrain points) and iteratively increase the amount of information used in order to classify the whole dataset step-by-step.

One representative approach of this group is the progressive triangular irregular network (TIN) densification introduced by Axelsson (2000). In this approach the first subset of the data is generated by a simple block minimum filter with a relatively big cell size. Subsequently, this first subset is triangulated in order to derive a coarse approximation of the bare earth. Then, for each of the triangles within the TIN an additional terrain point is added if certain criteria are fulfilled. This process is repeated

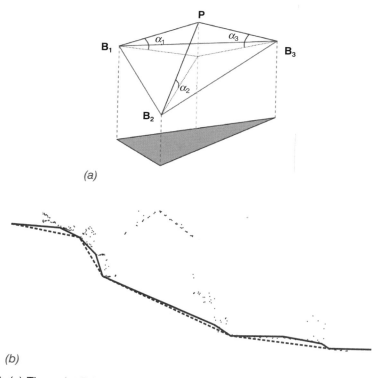

(a)

(b)

Figure 4.4 (a) The point **P** is tested using the angles α_1, α_2, α_3 with respect to the triangle defined by the points **B$_1$**, **B$_2$** and **B$_3$**, which are already classified as "terrain" points and currently form the local DTM (after Pfeifer, 2003; courtesy VGI). (b) Step-by-step process of DTM determination based on progressive TIN densification (initial DTM: dashed line; refined DTM: solid line)

iteratively until no further point can be added to the TIN or if a certain predefined point density is achieved, for example all edge lengths are smaller than a predefined threshold. Axelsson (2000) proposes using the offset of a point with respect to the triangle. For description of the offset the three angles α_1, α_2, α_3 between the triangle face and the line joining each of the three triangle nodes is calculated (see Figure 4.4). If a point can be found with angles below a predefined threshold, this point is classified as "terrain" and is added to the TIN.

A similar method is described by Hansen and Vögtle (1999). It is also based on a TIN, but in contrast to the approach introduced by Axelsson (2000), this approach uses a different solution for the determination of the first subset and, furthermore, they introduce a different densification criterion. They propose determining the first subset based on the lower part of the convex hull of the given points and as acceptance criteria they use the height difference Δh between each candidate point and the surface of the planar triangle.

A further similar approach is presented in Sohn and Dowman (2002). Within this approach the initial TIN is built by the lowest points in the four corners of the whole area. Then in the following "downward" step, the lowest point within each triangle is added to the TIN. This step is repeated until no point below the TIN can be added. In order to identify further points as "terrain" points Sohn and Dowman (2002) propose adding an upward densification step. Within this step they suggest adding in each triangle points which are in a certain buffer with respect to the surface defined be the triangle and which fulfil a certain minimum description length (MDL) criterion.

The methods in this group are characterised by the following: (i) they classify the given point set in a progressive manner; and (ii) the filtering is typically done together with a reconstruction of the DTM, which is in contrast to the methods presented in the section on mathematical morphology.

4.1.3 Surface-based filtering

Similar to the algorithms based on progressive densification, all the methods that can be assigned to the surface-based approaches use a surface reconstructed from the point cloud for the filtering of the given points. However, in contrast to the methods belonging to progressive densification groups, where the points assigned to the terrain class increase step-by-step, these methods typically start with the assumption that all given points belong to the terrain surface and then iteratively remove or reduce the influence of those points that do not fit the surface model by a step-by-step refinement of the surface description.

In the following, filtering based on robust interpolation [Kraus and Pfeifer, 1998] will be presented in detail. Robust interpolation integrates the filtering and DTM interpolation in one process. The aim of this algorithm is to determine an individual weight $p_i \in [0;1]$ for each irregularly distributed point $\mathbf{P}_i$ in such a way that the modelled surface represents the terrain. Finally, all given points can be classified into terrain and off-terrain points based on a height difference threshold value with respect to the final DTM. The process of robust interpolation consists of the following steps:

1. interpolation of the surface model considering individual weights for each point (at the beginning all points are equally weighted);

2. calculation of the filter values f_i (signed distance from the surface to the measured point (negative residuals)) for each point $\mathbf{P}_i$; and

3. computation of a new weight p_i for each point $\mathbf{P}_i$ according to its filter value f_i.

These steps are repeated until a stable situation is reached (the weight of each point does not change significantly and thus the run of the surface does not change from one iteration to the next) or a maximum number of iterations is reached. The results of this

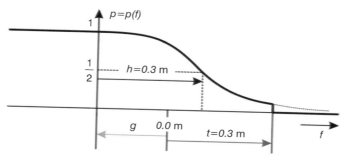

Figure 4.5 A typical weight function with asymmetric shape and shifted origin (by the value of *g*) for the elimination of off-terrain points above the terrain with robust interpolation. Points with a height *f* above the surface (*f* >> 0) are assigned a low weight, while points below the surface get a higher weight. In the approach presented by Kraus and Pfeifer (1998) the shift value *g* is determined automatically from the point cloud data. (Briese *et al.*, 2002; courtesy the authors)

process are a surface model and a classification of the points into the classes terrain and off-terrain.

The two most important entities of this algorithm are the functional model (step 1) and the weight model (step 3). For the functional model Kraus and Pfeifer (1998) use linear prediction [Kraus and Mikhail, 1972; Kraus, 2000][1] considering an individual weight (i.e. individual variance for the measurement error) for each point. The elimination of gross errors is controlled by a weight function. The input parameter of this function is the filter value f_i and its output is a (unitless) weight p_i. The weight function can be adapted to the distribution of the given points with respect to the DTM surface. In the case of a symmetrical distribution around the DTM it can be set to a bell curve (similar to the one used for robust error detection in bundle block adjustment) controlled by the half-width value (*h*) and tangent slope (*s*) at half-weight. However, for use with airborne laser scanning data, where the number of resulting points above the DTM are much higher than below, Kraus and Pfeifer (1998) propose using the bell curve in an asymmetric and shifted way (see Figure 4.5). The asymmetry means that the left and right branch are – in contrast to a symmetric bell curve – independent, and the shift means that the weight function is not centred at the zero point. With the help of two tolerance values (max *t–* and max *t+*) points with a certain distance to the computed surface can be excluded from the DTM determination process and are assigned the class label off-terrain. This process is applied patch-wise (slightly overlapping squares typically with 50 to 300 points/patch) for a certain project area. A practical example of this process is presented in Figure 4.6.

---

1 Linear prediction (also called linear least-squares interpolation) is equivalent to kriging, but the concept is extended by the ability to minimise random measurement errors.

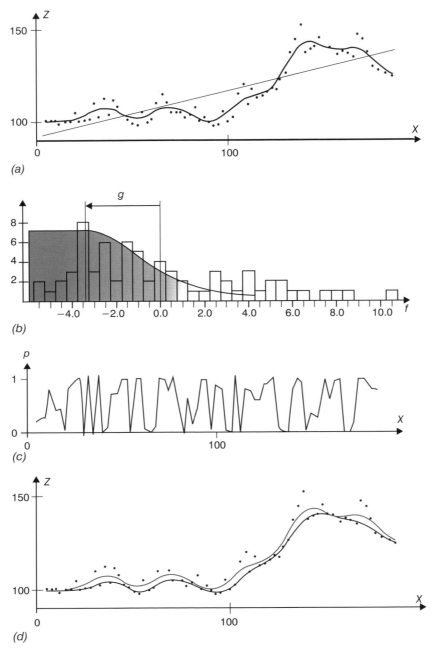

Figure 4.6 Robust interpolation demonstrated on a profile. (a) Averaging surface (black line) running in a medium layer between terrain and off-terrain points (black dots). The terrain points will lie below this intermediate surface and the off-terrain points above it (or on it). (b) Histogram of the filter values f_i overlaid with a shifted weight function (by the value of g; this value is estimated based on the filter values) which assigns a high weight to the terrain points (negative values) and a low weight to off-terrain points (points with a filter value bigger than $-g$). (c) Weights of the points over the profile ground plan positions. (d) The black line represents the resulting DTM surface after a few iterations, whereas the grey line indicates the surface from the first iteration. (Kraus and Pfeifer, 1998 with minor adaptations; courtesy the authors)

The robust interpolation algorithm relies on a "good" mixture of terrain and off-terrain points in order to iteratively eliminate the off-terrain points:

- The method works well for wooded areas, even if the penetration rate is about 50% or below. In such a case the off-terrain points are vegetation points and the number of required iterations is three or four. In the first iterations the canopy points are eliminated, and in the next iterations the understorey points are found.

- The method fails if the points are not mixed thoroughly, i.e. the vegetation points must not appear in clusters without a single terrain point. In particular, if the method is applied to city areas, the off-terrain points show this systematic behaviour (e.g. at a house roof).

In order to overcome this limitation in large building areas and in order to speed up the process Pfeifer *et al.* (2001) and Briese *et al.* (2002) embedded the robust interpolation technique into a hierarchical framework. For this, the original data are thinned out by data pyramids and the filtering is performed on different levels of resolution, starting with the coarsest level. The resulting coarse level DTM resulting from the coarse level data is used as the initial DTM for the next finer level, where a new DTM is again determined by robust interpolation. Then the DTM from the next finer level is determined by robust interpolation. The resulting DTM of the finest level consists of all measured terrain points. The process for two resolution levels is illustrated in Figure 4.7.

Another surface-based filtering method was introduced by Elmqvist *et al.* (2001). This approach is based on active shape models. The surface is determined by minimising an energy function that is guided on one hand by a certain surface stiffness (inner force) and on the other hand by the laser echoes (external force) that force the surface to run through them. The process starts with a horizontal surface below all laser echoes. Subsequently, the force of the points that influence the run of the surface is applied in an iterative manner. This leads to a deformation of the surface towards the points, but this deformation is restricted according to the defined surface stiffness. This stiffness prevents the surface from running through the points on the vegetation and on off-terrain objects (houses, cars, etc.) and leads to a surface that approximates the lower points. The final classification of the points is performed according to a tolerance band with respect to the final surface model.

4.1.4 Segment-based filtering

In contrast to all previously mentioned approaches where the classification entity is always one single point, the methods belonging to the segment-based approaches classify whole segments, i.e. a set of neighbouring points with similar properties. In this way these approaches try to overcome the problem of individual points being wrongly filtered, although the neighbouring points that belong to the same surface element

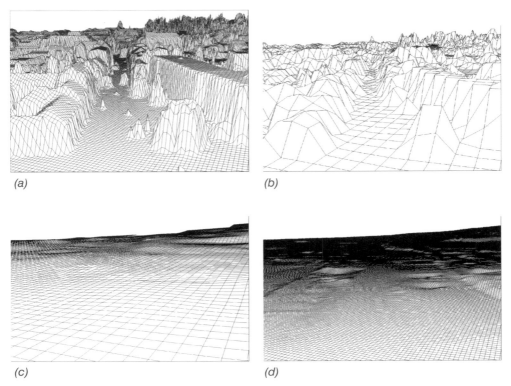

(a) *(b)*

(c) *(d)*

Figure 4.7 Different stages of DTM generation by hierarchic robust interpolation: (a) original DSM (1 m grid width); (b) initial coarse DSM; (c) coarse DTM after robust interpolation; and (d) final DTM obtained within the finest data level (1 m grid width). Data source: Airborne laser scanning Leithagebirge ("LiDAR-Supported Archaeological Prospection in Woodland", which is funded by the Austrian Science Fund (FWF P18674-G02))

are classified correctly (for example in the case of a surface discontinuity). Within the filtering workflow in the first step the individual segments have to be generated by aggregating points with similar properties. Afterwards these segments have to be classified. For both steps different algorithms can be chosen. Within this chapter two representative filter approaches are presented.

In general, the point cloud segmentation can be performed in object space or feature space. For segmentation in object space region growing techniques are typically used. Within this step-by-step process, neighbouring points are merged to form a segment as long as their properties are similar with respect to some threshold. For this similarity different criteria can be chosen, for example similar height value or normal vector. In contrast to object space segmentation, feature-based segmentation is performed by detecting clusters (separated accumulations of points with similar feature values) in the feature space. The same properties as in object-based segmentation can be used, but the grouping does, in contrast to segmentation in object space, only depend on the similarity in feature space. Further details about point cloud segmentation can be found in Section 2.3.

Sithole (2005) and Sithole and Vosselman (2005) presented one representative approach for the filtering of airborne laser scanning data based on segmented point clouds (see Figure 4.8). For the segmentation of the last echo point cloud they introduce an approach based on profile intersection (see Section 2.3.4). In the first step the point cloud is partitioned into a series of differently orientated profiles. Subsequently, the points within each profile are connected to line segments if certain criteria are fulfilled (within the filter approach different segmentation criteria are in use). Finally, the segmentation of the whole point cloud is derived by grouping line segments of different profiles. The subsequent classification of the segments is based on an analysis of the neighbourhood relation of the segments. Within the whole filter process several segmentation and classification steps using different criteria are applied. These aspects include a bridge, macro, micro, manmade and natural object detection.

A second representative approach for filtering based on segmented airborne laser scanning point cloud data was presented by Tovari and Pfeifer (2005). In contrast to Sithole and Vosselman (2005) they propose a segmentation of the point cloud based on region growing (see Section 2.3.3). Starting from one single point, points with similar characteristics are added to the segment in a step-by-step process. As grouping criteria Tovari and Pfeifer (2005) propose normal vector similarity and spatial proximity. After the process of segmentation the filtering is performed using a surface-based

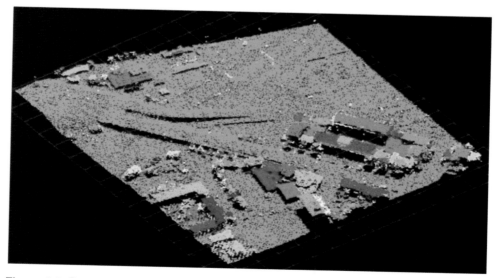

Figure 4.8 Segmentation example; all points that belong to one segment are displayed in the same colour. (Sithole and Vosselman, 2005; courtesy the authors)

approach. This process works similarly to the robust interpolation technique presented in Section 4.1.3., but instead of reweighting and classifying the individual points the segments are the entity within the robust estimation process.

Finally, it has to be mentioned that filtering based on segmentation can also be performed in a typically faster way (due to the given topology) based on a raster data structure. One representative approach was published by Jacobsen and Lohmann (2003). For the segmentation a raster-based region growing is used, while for the classification a set of different criteria per segment (shape) and between the neighboured segments is used.

4.1.5 Filter comparison

The practical performance of different algorithms was tested within an international filter test organised by the International Society of Photogrammetry and Remote Sensing (ISPRS, see Sithole and Vosselman, 2004). Within this test the filter results of eight different algorithms have been studied with airborne laser scanning datasets with different characteristics (different point density and landscape – rural vs. urban). A qualitative analysis as well as a quantitative analysis of the results was performed. While most of the filter algorithms worked well in areas with low complexity (gently sloped terrain, small buildings, sparse vegetation, and a high portion of terrain points) significantly different behaviour could be observed in more complex landscapes. In particular, complex city areas with large nested buildings and highly varying terrain levels were quite challenging for the automated filtering. The report concludes that progressive densification and surface-based filters mostly appear to yield better results. This might be caused by the usually bigger analysis areas which allow consideration of a broader context than other approaches.

It has to be mentioned that since the time of the study research in the area of filtering has been quite active (segmentation-based methods in particular have been introduced in the last few years). As well as further developments of new approaches, the combination of different algorithms within one filter workflow has been studied. However, for practical applications, a fully automated filter procedure that can be applied for all landscape areas has not yet been found.

Nowadays, the process of filtering and DTM generation based on airborne laser scanning data is already highly automated, but due to the complexity of the terrain surface manual checking and editing of the classification results is usually necessary. Most of the manual correction work is necessary in urban areas with complex terrain structure (different terrain levels, staircases, etc.) and near bridges, which are typically only partly handled correctly by current filter software (typical problem areas were identified in Sithole and Vosselman, 2004). The amount of manual work depends largely on the local surface characteristics. While nowadays typically no manual correction work is necessary for rural areas with small buildings and flat terrain, filtering of dense and complex city areas with different terrain levels cannot be carried out correctly with current automatic implementations of filtering procedures that rely only on the geometric relation of neighboured laser scanner points. In these complex areas additional data sources (e.g. building outlines) can help to reduce the manual correction work significantly. One reason is that all previously presented filter algorithms only analyse the geometric relationship between points in a certain local neighbourhood. In certain areas the bare earth surface cannot be characterised by geometric means only. In order to understand which surface parts belong to the terrain not only a geometric knowledge but also a local terrain context is required. With this aim, as well as geometric information provided by laser scanning sensors, radiometric information obtained by analysing the backscattered radiation of the laser scanning systems (intensity or full-waveform signal) or by other active or passive operating sensors might be helpful in order to filter complex landscapes.

Although it is difficult to provide a general statement, the influence of point density on the filtering process has to be mentioned. The influence of the point density is strongly dependent on the complexity of the topography. While for open terrain without steep slopes the point density is mainly important for guaranteeing a certain penetration rate in the presence of vegetation or other off-terrain objects, in complex environments, such as city areas with different surface levels or in mountainous regions with a very rough terrain surface characteristic, the point density is additionally important in guaranteeing a certain representation quality of the varying surface details. Sithole and Vosselman (2004) mention in their report, where they also studied the influence of the point density, that even at the highest resolutions (in this case it was about 0.67 points/m^2) the filters encounter difficulties in complex areas, which then masks the decaying performance of the filters at lower resolutions. In these environments the sampling process and the influence of the beam divergence [Briese, 2004b] have to be considered.

4.1.6 Potential of full-waveform information for advanced filtering

As mentioned in the introduction, the new generation of airborne laser scanning sensors, the so-called full-waveform (FWF) airborne laser scanning systems (see Chapter 1), are able to record the complete backscattered waveform. In contrast to the conventional discrete echo systems FWF airborne laser scanning sensors allow determination of the object echoes in post-processing and additionally allow further attributes per echo to be resolved, namely the echo width and echo amplitude [Wagner *et al.*, 2006]. Whereas the echo amplitude describes the backscattered signal strength, which depends on the range, atmosphere, the backscattering characteristics of the target, and the airborne laser scanning sensor, the echo width describes the range variation within one determined echo. If several close objects at slightly different ranges contribute to one echo, the echo width compared to the width of the emitted pulse will be increased.

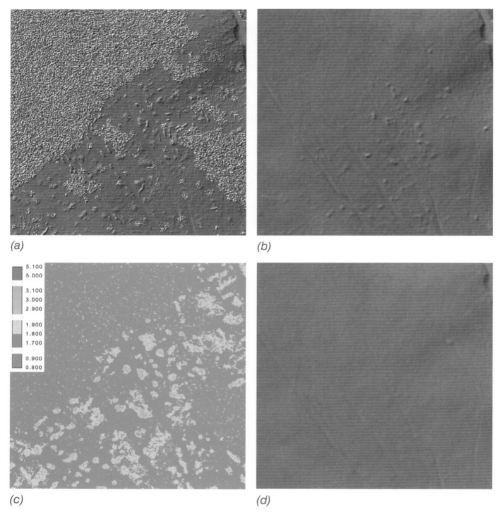

Figure 4.9 (a) Shading of a DSM based on airborne laser scanning data; (b) shading of a DTM after hierarchic robust interpolation without consideration of additional FWF information; (c) mapped echo widths determined from FWF airborne laser scanning data; and (d) shading of a DTM with pre-elimination of points with large echo width. Figure details are extracted from Doneus and Briese (2006)

Alternatively, if the range variation within the footprint is low, the echo width will be quite similar to the width of the emitted laser pulse. This additional information might be very useful in order to discriminate low vegetation from bare earth reflection and might help to reduce the problems of conventional airborne laser scanning systems in low vegetation areas [Pfeifer *et al.*, 2004].

The advanced capabilities of DTM generation of FWF airborne laser scanning systems were studied by Doneus and Briese (2006). The authors analysed the ability to use the echo width for the removal of low vegetation that can be eliminated only to a small extent during the filtering procedure. Figure 4.9 presents one sample dataset where the advanced capability of FWF data for DTM determination is demonstrated. One can recognise that the echo width in areas with low vegetation is – compared to open or wooded terrain with high trees – significantly increased. Therefore, a prefilter step that eliminates echoes with a significantly higher echo width was proposed and included in the DTM processing sequence. As can be seen in Figure 4.9(d) this prefilter step led to a significant improvement in the DTM with a better surface representation quality (especially in areas where the last echoes are a mixture of low vegetation and terrain reflection) compared to the DTM without this pre-elimination step (Figure 4.9b). Further investigation of the advanced filter capabilities of full-waveform information can be found in a number of publications [Wagner *et al.*, 2007; Doneus *et al.* 2008; Mücke, 2008]. Some first ideas for introducing this additional information as *a priori* weight for the individual points into filtering using robust interpolation (see Section 4.1.3) were presented by Mandlburger *et al.* (2007). Mücke (2008) investigated the precision of the determined echo widths with respect to the echo amplitude (see Figure 4.10). It could be demonstrated that echo width values of last echo points with low echo

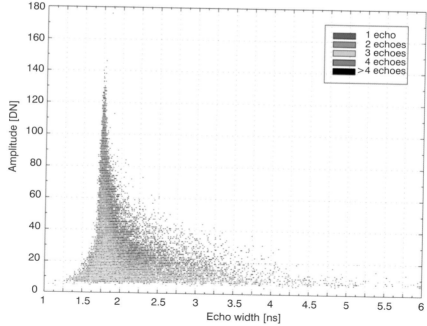

Figure 4.10 Echo amplitude of last echo points plotted against the respective echo width in a forested area. (Mücke, 2008; courtesy Werner Mücke)

amplitude are significantly noisier. Mücke (2008) therefore proposes considering an amplitude-dependent adaptation of the previously mentioned *a priori* weights.

Finally, it has to be mentioned that the use of the additional FWF information for advanced modelling tasks has not been studied in detail. However, the first examples demonstrate some of the potential of FWF airborne laser scanning data for advanced DTM generation from airborne laser scanning data.

4.2   Structure line determination

For a high quality representation of the bare earth, structure line information is essential in order to adequately describe linear features (e.g. a dam or a motorway) of the terrain surface. In the past these lines (breaklines, form lines, step edges) were typically determined by photogrammetric means by manually digitising a connected sequence of points. Due to the dense sensing capabilities of airborne laser scanning devices, the extraction of linear features is often of lower interest, because the dense models derived from an airborne laser scanning point cloud already present the structure line information implicitly. However, for a high accuracy representation of these linear features in the DTM, for example for hydrological applications, or for the task of data reduction (see Section 4.3.3), the explicit extraction of linear features based on airborne laser scanning data is essential. Methods for structure line extraction based on laser scanner data can be separated into raster and point cloud based approaches [Briese, 2004a, 2004b]. While the raster-based approaches typically use image-processing techniques such as first derivatives [Gomes-Pereira and Wicherson, 1999], gradient images [Sui, 2002], Laplacian operators [Gomes-Pereira and Janssen, 1999], the point cloud based approaches directly operate on the last echo airborne laser scanning data [Briese, 2004a, 2004b].

A summary of some of the previously mentioned raster-based approaches and an additional new method based on hypothesis testing (homogeneity measure: quadratic variation) can be found in Brügelmann (2000). In general these raster-based algorithms are applied to a previously generated (filtered) DTM. The results of these detection methods are pixels marked as edge pixels. In a further raster-to-vector conversion 2D breaklines can be generated. This conversion includes some smoothing in order to eliminate zigzag effects caused by the raster data structure. Finally, the height of the breakline is independently extracted from a slightly smoothed vegetation-free DTM at the planimetric position of the detected breakline. Approaches that share these basic aims can be found in a number of publications [Gomes-Pereira and Wicherson, 1999; Gomes-Pereira and Janssen, 1999; Rieger *et al.*, 1999; Sui, 2002]. A different semi-automated raster-based approach for 2D breakline modelling based on snakes, guided by locally determined main curvature values estimated by differential geometry, was introduced by Kerschner (2003). Borkowski (2004) presents two different 2.5D breakline modelling concepts. While one of these methods uses snakes, the second approach uses differential equations. Furthermore, Brzank *et al.* (2005) introduce an approach for the 2.5D detection and modelling of pair-wise structure lines in coastal areas. For the modelling the paper proposes using a hyperbolic tangent function.

Briese (2004a) introduces a different approach that uses the originally acquired point cloud. In contrast to the other approaches mentioned, this method does not rely on a previously filtered DTM and allows the simultaneous determination of all three coordinates of points along the lines within one process. This process can be executed prior to the filtering and the determined lines can be used subsequently as additional terrain information for filtering of the point cloud. Furthermore, the lines can be integrated into the final DTM data structure. In this approach, the lines are described with the help of continuously overlapping analytic surface patch pairs (see Figure 4.11). The patch pairs are determined by robust adjustment based on the points in the vicinity of the line. With the help of robust adjustment the influence of off-terrain points can be reduced (see Figure 4.11b) and with the help of an additional weight function the influence of points with a bigger distance to the 2D position of the line can be decreased. This modelling procedure relies on a given 2D approximation of the whole line. In order to reduce this effort Briese (2004a) introduces the concept of line growing. Starting from a line segment or just one given single 2D point, this procedure allows delineation of the whole line in a step-by-step procedure by growing in forward and backward directions. For modelling of the whole lines the growing procedure continues until the robust adjustment is successful or a certain break-off point (for example a too small intersection angle between the two analytical surfaces) is reached. Two results of this approach can be found in Figure 4.12. The advantage of this process

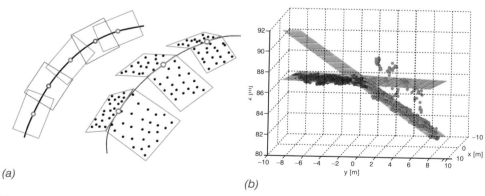

(a)

(b)

Figure 4.11 (a) Basic concept for the determination of breaklines with the help of overlapping patch pairs. (Kraus and Pfeifer, 2001; courtesy the authors). (b) Practical example of the local breakline estimation of one patch pair (10 m × 10 m) based on unclassified airborne laser scanning data (blue and green spheres). The local breakline is estimated as the intersection of the robustly estimated plane patch pair (Briese, 2004b)

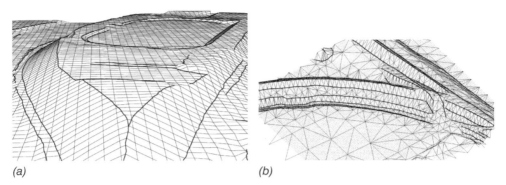

(a)   (b)

Figure 4.12 Structure lines estimated from airborne laser scanning data: (a) perspective view of a hybrid DTM; and (b) perspective view of a TIN with intermeshed structure lines. The hybrid DTM was generated in a German Federal Institute of Hydrology (Bundesanstalt fuer Gewaesserkunde, BfG) pilot project (Briese, 2004b)

is that as well as reduction of measurement noise in the z-direction, the planimetric accuracy of the line can be increased due to the redundant information provided by the number of points that contribute to the adjustment. Furthermore, the approach allows rigorous error propagation. Based on the law of the propagation of errors, the covariance matrix of one representative (intersection) point of the line can be determined and the covariance ellipsoid across the line direction can be estimated [Briese and Pfeifer, 2008]. This process for the estimation of breaklines can be extended to the different line types (form lines, step edges, boundary lines) and to 3D application with terrestrial laser scanning data [Briese and Pfeifer, 2008]. In Briese and Pfeifer (2008) further literature references to methods for structure line extraction from terrestrial laser scanning data, which go beyond the scope of this chapter, are provided.

4.3   Digital terrain model generation

As mentioned in the introduction to this chapter, a DTM is a model of the bare earth surface in digital form. This model is generated from a given set of terrain observations, i.e. points (the classified terrain echoes) and lines. Based on these observations, a certain function f can be determined that represents the terrain within a certain local area. Typically a DTM is limited to a 2.5D representation using the function $z = f(x,y)$, which maps a certain height value z to each 2D location $\mathbf{p}(x,y)$. 3D formulations (e.g. for overhangs and bridges) are not frequently used in practice and are typically used only for very limited areas. A discussion on the generation of 3D terrain models can be found in Pfeifer (2005). Within this section only methods for the generation and description of 2.5D DTMs are presented.

In order to make a DTM durable the function that describes the surface has to be stored in a certain data structure. One solution for this task is to store the function

$f(x,y)$, regardless of its description, and the area for which this function is valid. However, storing the function can cause problems in the exchangeability of the model, because every software package has to know how the digitally stored function has to be evaluated. Therefore, instead of storing the continuous function, the function is often evaluated at certain regular intervals and only the discretised function values are stored.

A very well known data format for the storage of DTMs is the raster representation. Within this data structure each raster value represents the height of the whole area covered by the raster element. As well as the raster format, grid formats are common. A grid is dual to the raster and represents the height information on discrete regularly arranged points. In contrast to the raster data structure, where the raster height is a certain representative value for the cell area, in a grid model the actual height value at the provided location is stored. In practice, the terms for the raster and grid model are often mixed and not always clearly distinguished. This is because for both data structures the same data formats can be used. The grid heights are typically determined by surface interpolation and approximation methods like inverse distance weighting, moving least squares, linear prediction, or kriging. An extension to the grid-based data format is the hybrid grid model, where linear feature lines are intermeshed with the grid points [Kraus, 2000; Ackermann and Kraus, 2004] (see Figure 4.12a). A different method for the generation of a bare earth representation is triangulation. In this process the neighbourhood topology of the given terrain dataset is derived according to a certain geometric rule (e.g. Delaunay triangulation). The resulting TIN can be stored in a certain data format. Furthermore, so-called constraint triangulations can be applied, which allow integration of structure lines into the TIN generation process.

For the representation of the DSM the same data structures and interpolation methods as for the DTM are typically in use. However, in contrast to the DTM the smoothing is usually reduced as much as possible in order to get a sharp representation of the object boundary. The biggest difference compared to DTM generation is the used input data. While for the DTM classified terrain points are necessary, the DSM is typically calculated based on all first echo points. However, if a high quality DSM from airborne laser scanning data is aimed for, the first echo input dataset should be refined. The first echo dataset might contain reflections from too high surfaces (gross errors, for example caused by birds or dust) and may include points on temporal objects that were present during the data acquisition (for example cars, tower cranes, people). Very few procedures are currently available for the automatic refinement of the first echo dataset. Nowadays, the DSM is typically determined fully automatically based on all first echoes. This DSM is usually just used for visualisation tasks or typically acts as input for the generation of an nDSM (normalised digital surface model), which is often used for building extraction and vegetation mapping (see Chapters 5 and 6, respectively).

4.3.1 Digital terrain model determination from terrestrial laser scanning data

The generation of local area 2.5D DTMs based on terrestrial laser scanning data works analogously to DTM generation from airborne laser scanning data [Briese *et al.*, 2002]. Prior to filtering of the terrestrial laser scanning point cloud all stations have to be transformed onto a common coordinate frame. Subsequently, the filter and DTM modelling process can follow. As already mentioned in the introduction, the main difference to airborne laser scanning data is the usually quite inhomogeneous 2D distribution of the terrestrial laser scanning data. This inhomogeneity is caused by the fixed and typically low observation height. This leads to a very high point density in the vicinity of the stations (typically more than 1000 points/m^2), whereas the point density at larger distances is significantly lower. Furthermore, one has to consider the effects of occlusions. Therefore, for practical use thinning out of the data, for example by taking the lowest point within a 0.25 m × 0.25 m cell, is recommended prior to filtering of the data. This helps to speed up the process and still allows determination of a high resolution DTM. However, for the generation of 3D DTM, for example of an overhang, a different modelling strategy has to be chosen. For this task 3D triangulation methods with a subsequent mesh refinement step (smoothing) can typically be used [Pfeifer, 2005].

4.3.2 Digital terrain model quality

DTMs are frequently used to make important decisions. In order to judge these decisions the DTM quality must be known. In general, the DTM quality description can be split into two parts. The first part, the data quality, describes the quality of the input dataset, whereas the second part considers the model quality. In the following sections an overview of several data and model quality layers is provided. Further details can be found in Karel *et al.* (2006).

Data quality

The following data quality measures allow the quality of the input data for DTM generation to be described. In order to derive these measures the whole input dataset that was used for the DTM generation is essential. The following data layers can be considered when describing data quality.

- Point density layer: the visualisation of the local point density in a raster image allows areas with few points to be investigated and gives information about interpolation and extrapolation areas.
- Point distance map: similar to the point density map, the point distance map allows detection of areas with a low amount of input data. In contrast to the point density map this layer provides information about the distance to the nearest input data point.

- Data class map: in the case of multiple data sources for one region the data class map visualises the presence of a certain input data class (for example airborne laser scanning, terrestrial laser scanning, photogrammetrically measured points) in a certain region.

- Data accuracy map: the data accuracy map presents a visualisation of the local best or worst data accuracy.

Furthermore, visualisations that give insight into data consistency are useful. One example in the case of airborne laser scanning data is strip difference visualisation based on strip-wise DSMs (see Figure 4.13). These difference models allow inspection of the relative georeferencing of the airborne laser scanning dataset (see Section 2.1.1.4). A further consistency check is displayed in Figure 4.14. This figure illustrates inconsistency within one airborne laser scanning strip. For this visualisation the airborne laser scanning data of an oscillating mirror scanner was split into two groups. The points that were acquired from a mirror moving from right to left were separated from those where the mirror was moving in the other direction. The figure illustrates the difference between DSMs derived from the two groups. The interesting differences that can be seen in Figure 4.14 are the result of an asynchronous registration of the angle and distance measurements (time lag).

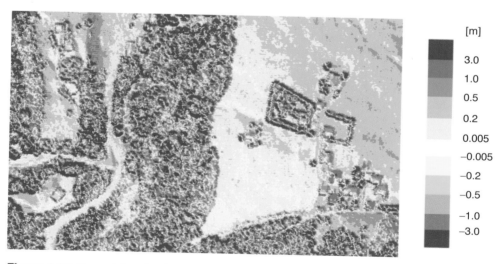

Figure 4.13 Inconsistencies between different data sets: discrepancies between two overlapping airborne laser scanner strips. The differences in smooth regions (for example on roofs and in open areas) are mainly a result of imprecise sensor orientation. Very large discrepancies are to be expected in non-smooth regions (for example vegetation and occlusion). These are not caused by georeferencing errors. (Karel *et al.*, 2006; courtesy the authors)

1.00
0.50
0.20
0.05
−0.05
−0.20
−0.50
−1.00

Figure 4.14 Inconsistencies within a single data set due to a sensor deficiency (units: metres). (Karel *et al.*, 2006; courtesy the authors)

Within the context of data quality, the quality of the filtering, i.e. the separation of terrain and off-terrain points, has to be considered, especially in the case of airborne laser scanning data. While this classification might be reliable in certain areas where the complexity of the terrain is low, significant classification errors can be present in built-up areas with a complex arrangement of buildings [Sithole and Vosselman, 2004]. Up to now none of the published filter algorithms has delivered a useful quality measure for the classification of airborne laser scanning points. The description of the filter quality is therefore limited and can best be described by practical experience. Another specific airborne laser scanning issue is the data quality in areas with low vegetation. As already mentioned airborne laser scanning sensors are not always able to estimate terrain echoes in these areas correctly. The resulting point cloud might therefore contain a systematic height shift in these areas. These problem areas are difficult to estimate without additional data.

Model quality

The model quality of a DTM can be separated into descriptions of interior and exterior quality. The internal quality describes the quality of the estimated DTM in contrast to the given terrain observation. For determining internal quality the law of error propagation can be applied or root mean square error (RMSE) maps can be calculated. These RMSE maps represent the difference between the input data and the derived DTM and help to detect areas where there is a significant difference between the input data and the models. These maps represent the precision of the DTM generation process.

The external quality of a DTM describes the quality of the estimated DTM in contrast to external control data. While the interior quality can be estimated based on the input data quality alone, for external quality documentation external reference data are essential. These external data must not have been used for DTM generation. In addition, these data need to have much better accuracy than the input data. The exterior quality of the model describes both the input data and the modelling process. In order to analyse the exterior quality of airborne laser scanning DTMs, several results of comparisons of airborne laser scanning DTMs with independently measured control data have been published. Briese (2000) and Pfeifer *et al.* (2001) report an average RMSE of an airborne laser scanning DTM of about ±0.11 m and highlight a strong dependency of the RMSE on the surface type. While the RMSE was below ±0.1 m in open terrain and even lower for well-defined areas on an asphalt street (RMSE of about ±0.03 m), the RMSE increased significantly in vegetated areas (±0.15 m). However, a RMSE of just ±0.03 m on an asphalt street requires high quality fine georeferencing of the airborne laser scanning data. Similar results for the analysis of the exterior DTM quality were reported by other authors [Hyyppä *et al.*, 2000; Wack and Stelzl, 2005].

In addition to these techniques for model quality description, empirical formulas are often used. These formulas provide a means of describing the model quality based on empirical analysis. For the estimation of the accuracy in height of DTMs derived from small footprint airborne laser scanning data Karel and Kraus (2006) propose the following formula

$$\sigma_z \, [\mathrm{cm}] = \pm \left(\frac{6}{\sqrt{n}} + 30 \tan(\alpha) \right) \tag{4.1}$$

where n (points/m^2) denotes the point density, and $\tan(\alpha)$ is the terrain slope. A more detailed *a posteriori* alternative for airborne laser scanning data is described by Kraus *et al.* (2006). It estimates DTM accuracy based on the original data and the DTM itself.

4.3.3 Digital terrain model data reduction

Airborne laser scanning sensor development has increased significantly in recent years. While ten years ago the measurement rate of airborne laser scanning systems was a few kilohertz, nowadays modern airborne laser scanning sensors utilise a measurement frequency of up to several hundred kilohertz (see Chapter 1). This development leads to very high point densities of one or more points per square metre, even for large areas. This increase in the volume of the original data leads to large area DTMs with high resolution that are often different to handle in subsequent applications. There is therefore a strong need for algorithms that reduce the large amount of data which is necessary for storing dense airborne laser scanning DTMs. Different algorithms are available for data reduction. Often they also depend on the given data structure. In general, one can distinguish between data reduction methods were the data structure type of the given high resolution DTM is not changed and others that result in a change of the DTM structure.

Resampling methods, which are standard methods in digital image processing, are often also applied to raster DTMs. In these approaches the reduced DTM is composed of bigger raster cells than the original DTM and the representative height value of this large cell is derived from all finer original raster cells within this large cell. The same reduction process can also be applied to grid models. By taking advantage of the simple topology these reduction processes are typically very fast. Furthermore, the result of this process has the same data structure as the input and can be stored in a similar way. However, the disadvantage of these procedures is that they are not adaptive (the cells used for the reduced representation are typically equal over the whole area) and therefore the approximation accuracy compared to the original high resolution DTM may vary significantly. While the approximation may be quite good in flat terrain, significant differences can occur in areas with large local height changes. These deficiencies may be solved by using local adaptive raster resp. grid cell sizes, but such data structures are only rarely used in practice.

For the decimation of TINs many different data reduction processes have been developed [Kobbelt *et al.*, 1997]. In general these approaches can be split into mesh decimation methods and subdivision approaches. While the mesh decimation methods thin out the TIN in a step-by-step procedure, the subdivision methods start with a rough TIN that is refined with a divide and conquer strategy until a certain approximation criterion is fulfilled. A detailed overview of several approaches of both types can be found in Heckbert and Garland (1997). However, due to the huge amount of data, these mesh decimation methods which build up a TIN of the whole dataset are not well suited to high density data. For the practical application of these methods, it would be necessary to split the huge DTM datasets into smaller parts. Furthermore, it has to be said that data reduction based on a TIN that is built up from the classified terrain points and available line information is often not the best choice for the derivation of reduced DTMs, because the thinning process will be directly influenced by random and systematic measurement errors that might be present within this dataset. In practice it can be observed that in areas with a strip overlap small errors in the relative georeferencing of the strips can lead to a very rough surface, which reduces the data reduction rate to a very high degree. It therefore has to be ensured that systematic measurement errors are reduced as much as possible prior to any modelling task. Furthermore, random measurement errors should be minimised during DTM generation as much as possible in order to guarantee higher compression rates during the reduction process. To achieve this, interpolation methods, such as linear prediction [Kraus and Mikhail, 1972; Kraus, 2000], with measurement noise reduction capabilities, are available. Subsequently, after the generation of the smoothed high resolution DTM using all input data, the data reduction process should be followed.

A procedure for data reduction of smoothed dense high quality hybrid grid DTMs that consists of regular grid points enhanced by structure lines (see Section 4.3) was introduced by Mandlburger (2006).

This reduction process is guided by a certain user-specified maximum height tolerance value $\Delta z_{\max}$ and by a maximum planimetric point distance value $\Delta xy_{\max}$ for the resulting reduced DTM representation. The process is initialised by the generation of a TIN that is composed of all structure lines and a coarse grid with the user-defined

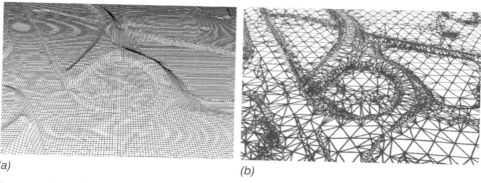

(a) (b)

Figure 4.15 (a) Hybrid grid airborne laser scanning DTM with a grid width of 2 m. (b) DTM after data reduction with irregular subdivision (Δz_{max} = 0.15 m, compression rate compared to the original 1 m DTM: 92%). The hybrid airborne laser scanning DTM was generated as part of the project "DGM-W Main-2" of the German Federal Institute of Hydrology (Bundesanstalt fuer Gewaesserkunde, BfG)

grid cell size of Δxy_{max} [Mandlburger, 2006; Mandlburger and Briese, 2007]. Subsequently, each coarse cell is subdivided as long as the criterion of the maximum height tolerance value Δz_{max} is not fulfilled. This subdivision can be performed in a hierarchical or irregular manner. In the hierarchical setup the cell is split into four parts and the height of the new finer cells is taken from the original high resolution grid. In the irregular case, however, just one grid point is added to the TIN at each step, i.e. the one with the largest difference compared to the previous representation. While a higher compression rate can be achieved with the irregular setup of this approach, the hierarchical subdivision produces a TIN with fewer inhomogeneous edge lengths and fewer varying triangle angles. In contrast to the other methods presented within this section, this approach changes the data structure of the model, because the resulting model is a TIN, whereas the original model was, for example, a hybrid grid. This TIN consists of an irregular subset of the grid points which are necessary to achieve the specified representation accuracy and it also contains all structure lines from the hybrid grid. This algorithm has two main advantages: (i) it adapts to the local surface characteristics; and (ii) the criteria for the subdivision can easily be adapted (for example instead of the height tolerance value a measure based on the surface curvature can be introduced). A practical result of this DTM data reduction process is displayed in Figure 4.15.

References

Ackermann, F. and Kraus, K., 2004. Grid based digital terrain models. *Geoinformatics*, 7(6), 28–31.

Axelsson, P., 1999. Processing of laser scanner data – algorithms and applications. *ISPRS Journal of Photogrammetry and Remote Sensing*, 54(2–3), 138–147.

Axelsson, P., 2000. DEM generation from laser scanner data using adaptive TIN models. *International Archives of Photogrammetry and Remote Sensing*, 33(Part B4), 110–117.

Bauerhansl, C., Rottensteiner, F. and Briese, C., 2004. Determination of terrain models by digital image-matching methods. *International Archives of Photogrammetry, Remote Sensing and Spatial Information Sciences*, 35(Part B4), 414–419.

Borkowski, A., 2004. Modellierung von Oberflächen mit Diskontinuitäten, Habilitation, Dresden University of Technology.

Bottelier, P., Briese, C., Hennis, N., Lindenbergh, R. and Pfeifer, N., 2005. Distinguishing features from outliers in automatic Kriging-based filtering of MBES data: a comparative study. *Geostatistics for Environmental Applications*, Renard, F., Demougeot-Renard, H. and Froidevaux, R. (Eds.), Springer-Verlag, Berlin and Heidelberg, Germany, 403–414.

Briese C. and Pfeifer, N., 2008. Line based reconstruction from terrestrial laser scanning data. *Journal of Applied Geodesy*, 2(2), 85–95.

Briese, C., 2000. Digitale Modelle aus Laser-Scanner-Daten in städtischen Gebieten (Digital models from laser-scanner-data in urban areas) Diploma Thesis, the Institute of Photogrammetry and Remote Sensing, Vienna University of Technology.

Briese, C., 2004a Three-dimensional modelling of breaklines from airborne laser scanner data. *International Archives of Photogrammetry, Remote Sensing and Spatial Information Sciences*, 35(Part B3), 1097–1102.

Briese, C., 2004b. Breakline Modelling from Airborne Laser Scanner Data. PhD Thesis, Institute of Photogrammetry and Remote Sensing, Vienna University of Technology.

Briese, C., Pfeifer, N. and Dorninger, P., 2002. Applications of the robust interpolation for DTM determination. *International Archives of Photogrammetry, Remote Sensing and Spatial Information Sciences*, 34(Part 3A), 55–61.

Brügelmann, R., 2000. Automatic breakline detection from airborne laser range data. *International Archives of Photogrammetry and Remote Sensing*, 33(Part B3), 109–115.

Brzank, A., Lohmann, P. and Heipke C., 2005. Automated extraction of pair wise structure lines using airborne laser scanner data in coastal areas. *International Archives of Photogrammetry, Remote Sensing and Spatial Information Sciences*, 36(Part 3/W19), 36–41.

Doneus, M. and Briese, C., 2006. Digital terrain modelling for archaeological interpretation within forested areas using full-waveform laserscanning. In *VAST 2006*, Ioannides, M., Arnold, D., Niccolucci, F. and Mania K. (Eds.), *The 7th International Symposium on Virtual Reality, Archaeology and Cultural Heritage VAST (2006)*, 1–4 November 2006, Nicosia, Cyprus, 155–162.

Doneus, M., Briese, C., Fera, M. and Janner, M., 2008. Archaeological prospection of forested areas using full-waveform airborne laser scanning. *Journal of Archaeological Science*, 35(4), 882–893.

Eckstein, W. and Munkelt, O., 1995. Extracting objects from digital terrain models. *Proceedings of the International Society for Optical Engineering: Remote Sensing and Reconstruction for Three-Dimensional Objects and Scenes*, 9–10 July 1995, San Diego, CA, USA, vol. 2572, 43–51.

Elmqvist, M., Jungert, E., Lantz, F., Persson, A. and Södermann, U., 2001. Terrain modelling and analysis using laser scanner data. *International Archives of Photogrammetry Remote Sensing and Spatial Information Sciences,* 34(Part 3/W4), 219–226.

Förstner, W., 1998. Image preprocessing for feature extraction in digital intensity, color and range images. In *Proceedings of the International Summer School on Data Analysis and Statistical Foundations of Geomatics*, 25–30 May 1998, Chania, Crete, Greece; *Geomatic Method for the Analysis of Data in the Earth Sciences, Springer Lecture Notes on Earth Sciences*, 95, 165–189.

Gomes-Pereira, L. and Wicherson, R., 1999. Suitability of laser data for deriving geographical information – a case study in the context of management of fluvial zones. *ISPRS Journal of Photogrammetry and Remote Sensing*, 54(2–3), 105–114.

Gomes-Pereira, L. and Janssen, L., 1999. Suitability of laser data for DTM generation: A case study in the context of road planning and design. *ISPRS Journal of Photogrammetry and Remote Sensing*, 54(4), 244–253.

Hansen, W. and Vögtle, T., 1999. Extraktion der Geländeoberfläche aus flugzeuggetragenen Laser-scanner–Aufnahmen. *Photogrammetrie Fernerkundung Geoinformation*, 1999(4), 229–236.

Haralick, R.M. and Shapiro, L.G., 1992. *Computer and Robot Vision*, Addison-Wesley, Longman Publishing Co. Inc., Boston, MA, USA.

Haralick, R.M., Sternberg, S.R. and Zhuang, X., 1987. Image analysis using mathematical morphology. *IEEE Transactions on Pattern Analysis and Machine Intelligence*, 9(4), 532–550.

Heckbert, P.S. and Garland, M., 1997. Survey of polygonal surface simplification algorithms. School of Computer Science, Carnigie Mellon University, Pittsburgh, Research report.

Höfle, B. and Pfeifer, N., 2007. Correction of laser scanning intensity data: data and model-driven approaches. *ISPRS Journal of Photogrammetry and Remote Sensing*, 62(6), 415–433.

Hyyppä, J., Pyysalo, U., Hyyppä, H. and Samberg, A., 2000. Elevation accuracy of laser scanning-derived digital terrain and target models in forest environment. *Proceedings of EARSeL-SIG-Workshop LIDAR*, 16–17 June 2000, Dresden, Germany; EARSeL eProceedings, No. 1, pp. 139–147.

Jacobsen, K. and Lohmann, P., 2003. Segmented filtering of laser scanner DSMs. *International Archives of Photogrammetry, Remote Sensing and Spatial Information Sciences*, 34(Part 3/W13), 87–92.

Karel, W., Pfeifer, N. and Briese, C., 2006. DTM quality assessment. *International Archives of Photogrammetry, Remote Sensing and Spatial Information Sciences*, 36(Part 2), 7–12.

Karel, W. and Kraus, K., 2006. Quality parameters of digital terrain models. In *Seminar on Automated Quality Control of Digital Terrain Models*, EuroSDR Seminar, 18–19 August 2005, Aalborg, Denmark; not included in proceedings but available from http://people.land.aau.dk/~jh/dtm_checking/SpecialContribution.pdf and http://www.ipf.tuwien.ac.at/publications/2006/Karel_Kraus_QualityPar4DTM.pdf.

Kerschner, M., 2003. Snakes für Aufgaben der digitalen Photogrammetrie und Topographie. Ph.D. thesis, Vienna University of Technology.

Kilian, J., Haala, N. and Englich, M., 1996. Capture and evaluation of airborne laser scanner data. *International Archives of Photogrammetry and Remote Sensing*, 31(Part B3), 383–388.

Kobbelt, L., Campagna, S. and Seidel, H.-P., 1998. A general framework for mesh decimation. *Proceedings of Graphics Interface*, 18–20 June 1998, Vancouver, BC, Canada, 43–50.

Kraus, K., 2000. *Photogrammetrie, Band 3, Topographische Informationssysteme*, Dümmler Verlag, Köln, Germany.

Kraus, K., Karel, W., Briese, C. and Mandlburger, G., 2006. Local accuracy measures for digital terrain models. *The Photogrammetric Record*, 21(116), 342–354.

Kraus, K. and Mikhail, E.M., 1972. Linear least squares interpolation. *Photogrammetric Engineering*, 38(10), 1016–1029.

Kraus, K. and Pfeifer, N., 1998. Derivation of digital terrain models in wooded areas. *ISPRS Journal of Photogrammetry and Remote Sensing*, 53(4), 193–203.

Kraus, K. 2000. *Photogrammetrie, Band 3, Topographische Informationssysteme*, Dümmler Verlag, Köln, Germany.

Kraus K. and Pfeifer, N., 2001. Advanced DTM generation from lidar data. *International Archives of Photogrammetry, Remote Sensing and Spatial Information Sciences*, 34(Part 3/W4), 23–30.

Kobler, A., Pfeifer, N., Ogrinc, P., Todorovski, L., Ostir, K. and Dzeroski, S., 2007. Repetitive interpolation: A robust algorithm for DTM generation from aerial laser scanner data in forested terrain. *Remote Sensing of Environment*, 108(1), 9–23.

Lindenberger, J., 1993. Laser-Profilmessungen zur topographischen Geländeaufnahme. Ph.D. thesis, http://elib.uni-stuttgart.de/opus/volltexte/2006/2712/ (accessed 30 March 2009).

Lohman, P., Koch, A. and Schaeffer, M., 2000. Approaches to the filtering of laser scanner data. *International Archives of Photogrammetry and Remote Sensing*, 33(Part B3), 540–547.

Mandlburger, G., 2006. Topographische Modelle für Anwendungen in Hydrauik und Hydrologie. Ph.D. thesis, Vienna University of Technology.

Mandlburger, G. and Briese, C., 2007. Using airborne laser scanning for improved hydraulic models. *Congress on Modeling and Simulation – MODSIM07*, 10–13 December 2007, Christchurch, New Zealand, 731–738.

Mandlburger, G., Briese, C. and Pfeifer, N. (2007) Progress in LiDAR sensor technology – chance and challenge for DTM generation and data administration. In *Proceedings of the 51st Photogrammetric Week*, 3–7 September 2007, Stuttgart, Germany, Fritsch D. (ed.), Wichmann, Karlsruhe, 159–169.

Mücke, W., 2008. Analysis of full-waveform airborne laser scanning data for the improvement of DTM generation. Diploma thesis, Institute of Photogrammetry and Remote Sensing, Vienna University of Technology.

Petzold, B., Reiss, P. and Stössel, W., 1999. Laser scanning – surveying and mapping agencies are using a new technique for the derivation of digital terrain models. *ISPRS Journal of Photogrammetry and Remote Sensing*, 54, S.95–104.

Pfeifer, N., 2003. Oberflächenmodelle aus Laserdaten. VGI – Österreichische Zeitschrift für Vermessung und Geoinformation, 91(4/03), 243–252.

Pfeifer, N., 2005. A subdivision algorithm for smooth 3D terrain models. *ISPRS Journal of Photogrammetry and Remote Sensing*, 59(3), 115–127.

Pfeifer, N., Gorte, B. and Elberink, S.O., 2004. Influences of vegetation on laser altimetry – analysis and correction approaches. *International Archives of Photogrammetry, Remote Sensing and Spatial Information Sciences*, 36(Part 8/W2), 283–287.

Pfeifer, N., Stadler, P. and Briese, C., 2001. Derivation of digital terrain models in the SCOP++ environment. *Proceedings of OEEPE Workshop on Airborne Laserscanning and Interferometric SAR for Detailed Digital Terrain Models*, 1–3 March 2001, Stockholm, Sweden; OEEPE Publication No. 40, Torlegard, K. and Nelson, J. (Eds.), on CD-ROM.

Rieger, W., Kerschner, M., Reiter, T. and Rottensteiner, F., 1999. Roads and buildings from laser scanner data within a forest enterprise. *International Archives of Photogrammetry and Remote Sensing*, 32(Part 3/W14), 185–191.

Roggero, M., 2001. Airborne laser scanning: Clustering in raw data. *International Archives of Photogrammetry, Remote Sensing and Spatial Information Sciences*, 34(Part 3/W4), 227–232.

SCOP++, 2007. The software package SCOP++, http://www.inpho.de/ (accessed 30 March 2009).

Sithole, G., 2001. Filtering of laser altimetry data using a slope adaptive filter. *International Archives of Photogrammetry, Remote Sensing and Spatial Information Sciences*, 34(Part 3/W4), 203–210.

Sithole, G., 2005. Segmentation and Classification of Airborne Laser Scanner Data, Dissertation, TU Delft, Publications on Geodesy of the Netherlands Commission of Geodesy, vol. 59.

Sithole, G. and Vosselman, G., 2004. Experimental comparison of filter algorithms for bare-Earth extraction from airborne laser scanning point clouds. *ISPRS Journal of Photogrammetry and Remote Sensing*, 59(3-4), 202–224.

Sithole, G. and Vosselman, G., 2005. Filtering of airborne laser scanner data based on segmented point clouds. *International Archives of Photogrammetry, Remote Sensing and Spatial Information Sciences*, 36(Part 3/W19), 66–71.

Sohn, G. and Downman, I., 2002. Terrain surface reconstruction by the use of tetrahedron model with the MDL criterion. *International Archives of Photogrammetry Remote Sensing and Spatial Information Sciences*, 34(Part 3A), 336–344.

Sui, L., 2002. Processing of laser scanner data and automatic extraction of structure lines. *International Archives of Photogrammetry, Remote Sensing and Spatial Information Sciences*, 34(Part 2), 429–435.

Vosselman, G., 2000. Slope based filtering of laser altimetry data. *International Archives of Photogrammetry and Remote Sensing*, 33(Part 3B), 935–942.

Wack, R. and Stelzl, H., 2005. Laser DTM generation for South-Tyrol and 3D-visualization. *International Archives of Photogrammetry, Remote Sensing and Spatial Information Sciences*, 36(Part 3/W19), 48–53.

Wagner, W., Hollaus, M., Briese, C. and Ducic, V., 2008. 3D vegetation mapping using small-footprint full-waveform airborne laser scanners. *International Journal of Remote Sensing*, 29(5), 1433–1452.

Wagner, W., Ullrich, A., Ducic, V., Melzer, T. and Studnicka, N., 2006, Gaussian decomposition and calibration of a novel small-footprint full-waveform digitising airborne laser scanner. *ISPRS Journal of Photogrammetry and Remote Sensing*, 60(2), 100–112.

Building Extraction | 5

Claus Brenner

Building extraction can be defined as the process to detect, structure, geometrically reconstruct and attribute building outlines and 3D models of buildings. When research started on the automation of building extraction more than 20 years ago, the use of aerial images was common, and research focused on the extraction of higher-level, 2D and 3D primitives from stereo images. However, primitives, such as line segments, proved to be hard to extract, since they are usually abundant in images, and many of them may not correspond to meaningful geometric features. In computer and machine vision, depth (or height) images were introduced as alternative or additional data sources, allowing direct access to geometric properties of the object without the need to first recover 3D features from 2D images. This research was mostly done using smaller objects or scenes, which could be acquired using active systems. During the 1990s, airborne laser scanning became widely available so that laser scans, or depth images, of huge scenes, cities, or even entire countries became feasible. As laid out in the first chapter, accuracy and point density have improved since then and are in fact still constantly improving. New acquisition systems, such as helicopter-based, terrestrial, or mobile mapping systems, have been further developed, adding not only greater geometrical accuracy and detail, but also terrestrial data, such as façade scans. Nowadays, dense point clouds are also obtained by image matching from highly overlapping image sequences, which can be used similarly to laser scan point clouds. Thus, from a building extraction perspective, the availability, density, and quality of raw data have never been better and in the future, extraction will benefit from general technical progress in the areas of acquisition systems and preprocessing.

The requirements for geometric and semantic modelling of building structures have long been underestimated. As traditional photogrammetric processing was based on the measurement of single points, the straightforward procedure for building complex objects was to add topology information describing line segments, faces and volumes. However, this does not sufficiently address questions of topological correctness, regularisation, and additional constraints between (sub)structures. Also, further processing of the derived building models, such as generalisation, often requires more information than just a topologic-geometric representation of the outer building hull in terms of a collection of faces. It is a huge step from points, which are not structured at all, to manmade objects such as buildings, which are highly structured and rich in semantics, and even for the extraction of 2D maps, issues of topology and constraint representation have not yet been tackled to a sufficient extent.

User acceptance for 3D building and city models has been evolving only slowly. From the beginning, there were numerous potential applications identified, such as the planning of antenna locations for cellular phone providers, tourism information systems, city marketing, intermodal navigation systems, microclimate and noise propagation simulations and computer games. However, it is clear that those applications often have very different requirements with regard to the data, so that, although it would be desirable, there will probably be no single 3D city model representation in the near future which suits all needs. Most national mapping agencies do not see sufficient interest in 3D city models to justify acquisition and updating of such data using public funding, but it is to be expected that private companies will do so for specialised use cases. Recently, projects such as Google Maps and Microsoft Virtual Earth, and also car navigation system map providers, have led to numerous activities. Compared to the situation a decade ago, due to general technical progress in the areas of processing, visualisation, interactive presentation (e.g. using "Web 2.0" interfaces), mobile devices and transmission bandwidths, applications for use by private consumers are now feasible, so there is an enormous potential for a mass market in the near future.

Although the production of city models has been underway for many years, building extraction remains a hot topic of research; not only are the raw data constantly improving, but so are our expectations, from simple building blocks to structured roofs, and nowadays to geometric and radiometric representations of façade details. This chapter is structured as follows. Section 5.1 covers aspects of building detection and Section 5.2 deals with the delineation of outlines. 3D building reconstruction is the topic of Section 5.3 and Section 5.4, including geometric modelling, examples of reconstruction systems, and current issues such as regularisation, constraint equations, interactivity and generalisation. Finally, Section 5.5 discusses representation and file formats.

5.1 Building detection

Building detection is a classification task which separates buildings from other objects such as natural and artificial ground (lawn, roads) and vegetation (bushes, trees). Since a common property of most buildings is that they are above ground, standard filter techniques can be used in the first place to identify above ground laser points. Popular methods are hierarchical robust interpolation [Kraus and Pfeifer, 1998; Pfeifer *et al.*, 2001], orthogonal polynomials [Akel *et al.*, 2004] and morphological filtering using sloped kernels [Vosselman, 2000]. Filtering is discussed in detail in the context of digital terrain model (DTM) extraction in the preceding chapter. The detection of buildings is not only a prerequisite for further interpretation steps, such as finding outlines or reconstructing the 3D geometry of buildings. It also has an important application in map updating, where laser scanning can be the basis for highly automated and reliable methods of spotting changes thereby reducing the amount of manual inspection needed [Matikainen *et al.*, 2003; Vosselman *et al.*, 2004].

Morphological operators have been particularly popular in the context of building extraction. Being well known from binary and grey level image processing [Haralick *et al.*, 1987], they were applied to laser scanner profile measurements by Lindenberger (1993). Weidner and Förstner (1995) used erosion followed by dilation (opening) to obtain an approximation of the DTM and subtracted this from the original surface model to obtain a normalised digital surface model (nDSM), which was subsequently height thresholded. The resulting binary image was then analysed for connected components of a certain minimum area, which removed smaller segments, which were presumably caused by trees. Many authors have used this or a similar approach in the past, since – especially if the laser data are interpolated to a regular raster – morphological operators are readily available in image processing libraries and the required parameters are easy to understand (these are the mask shape and size, the height threshold, and the minimum area of the connected components).

Obviously, enforcing a minimum area is not sufficient to reliably discriminate between buildings and other objects above ground. Hug (1997) notes that artificial objects usually consist of continuous surface segments of homogeneous material bound by discontinuous edges. At higher levels, artificial surfaces are often characterised by constant or slowly varying surface normals and certain specific characteristics such as geometric regularities or by the fact that they are assembled from simple geometric shapes, such as rectangles, triangles and circles. Consequently, he proposes analysing point clouds on four different aggregation/abstraction levels: surface points, grid cells, multipixel groups, and segments. Four features were defined: reflectance (intensity value from the laser scanner), elevation texture (which is the local variance of elevation values), local variance of the surface gradient, and directional distribution of the surface gradient (obtained by Fourier analysis of a histogram of normal vector directions inside a segment). It was found that the lowest classification errors were obtained by a combination of the two features "reflectance" and "directional distribution of the surface gradient". However, using only reflectance yields almost as good results. Presumably, the ineffectiveness of elevation and gradient variance features in that study was due to the relatively large interpolation grid size (1 m × 1 m), whereas the dominance of the reflectance feature was probably caused by the relatively high image quality obtained using a CW laser scanner.

Maas (1999) mentions that time-of-flight scanners usually yield a much lower intensity image quality. Using the three features "height data", "Laplace filtered height

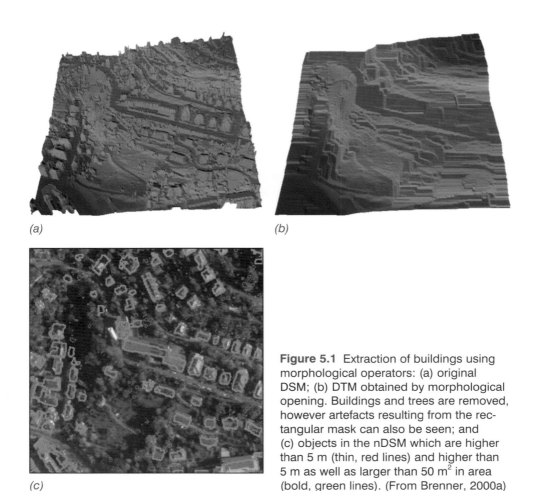

(a)

(b)

(c)

Figure 5.1 Extraction of buildings using morphological operators: (a) original DSM; (b) DTM obtained by morphological opening. Buildings and trees are removed, however artefacts resulting from the rectangular mask can also be seen; and (c) objects in the nDSM which are higher than 5 m (thin, red lines) and higher than 5 m as well as larger than 50 m² in area (bold, green lines). (From Brenner, 2000a)

data" and "maximum slope" as bands in a maximum likelihood classification, he is able to correctly classify 97% of tilted roofs and 94% of flat roofs with commission errors of 0.2% and 0.4%, respectively. While more than 99% of the trees were correctly classified, almost 9% of the pixels classified as trees actually belonged to flat terrain or roofs, the latter being caused by misclassification at roof edges. The test was carried out on an original dataset of about five scanned points per square metre, interpolated to a 0.5 m × 0.5 m grid. Using the intensity values from the scanner did not provide better results.

In order to improve classification, multispectral data can be used in addition. For example, Haala and Brenner (1999) use the height values of a normalised DSM in addition to the three bands of a CIR (false colour infrared) image to perform a pixel-wise classification into the classes street, grass, tree, building and shadow. Rottensteiner *et al.* (2003) do both a pixel-wise and a region-based classification within the framework of morphological filtering. To overcome the problem of selecting an appropriate mask size for the morphological filter (which should cut small terrain structures but preserve large buildings), they propose a hierarchical approach. This starts with a large mask size, which ensures large buildings are preserved but may also result in regions which are due to terrain or vegetation. These initial regions are improved by a *pixel-wise* examination of the normalised difference vegetation index (NDVI) and the height difference between first and last pulse return, followed by the following *region wide* criteria. First, using a texture measure, the number of "homogeneous", "linear" and "point-like" pixels are counted inside each region, corresponding to locally planar pixels, pixels on dihedral edges, and pixels of large and anisotropic variation of the normal vector, respectively. Buildings are filtered by setting thresholds on the percentage of homogeneous (which should be large) and point-like (which should be small) pixels. Second, a threshold on the average NDVI in the region is also used. Regions classified as buildings are then removed from further consideration by substituting their heights by the DTM heights obtained. Then, the hierarchical procedure continues using a smaller mask size. In addition to the mask size, the other thresholds are also varied, using tight bounds in the first and loose bounds in the last iteration. The procedure was later extended using a more elaborate classification scheme based on Dempster–Shafer theory [Rottensteiner *et al.*, 2005b]. As an infrared band was not available, in both cases a "pseudo-NDVI" image was computed using laser scanner intensities and the red channel of a true colour orthophoto. All operations were carried out on grids with a resolution of $1 \text{ m} \times 1 \text{ m}$.

Secord and Zakhor (2007) start from a point cloud which is preclassified into terrain and off-terrain. A segmentation is performed on a $0.5 \text{ m} \times 0.5 \text{ m}$ grid, using a weighted distance measure based on hue, saturation and value of a colour image and the height, height variation (local difference between maximum and minimum height) and x and y components of the surface normal vector. The weights are optimised using a learning method, which results in height variation and height having the largest weights, whereas hue and value have almost zero weight (i.e. no influence). After segmentation, a feature vector is computed (now for each segment instead of each pixel), consisting of the mean hue, saturation, value and height variation as well as the variance of the height. Finally, the classification into tree/non-tree is made using weighted support vector machines. Classification performance is characterised using receiver operating characteristic curves which allow assessment of the tradeoff between true positives and true negatives. Two major outcomes are that larger segments result in better classification results and segmentation followed by classification outperforms classification without prior segmentation, i.e. based on the point-wise features.

A disadvantage of using multispectral data is their limited availability. If they are obtained by a separate flight, errors may occur due to multitemporal acquisition. On the other hand, if the laser scanner and multispectral camera are used simultaneously, the overall system will be more dependent on weather conditions than a dedicated laser scanner system alone would be. Instead of using additional sensors, one can try to obtain more information from the laser scanner data themselves, in particular by analysing multiple pulses or the returned waveform. Tóvári and Vögtle (2004) use an interpolated grid (in this case 1 m × 1 m) and compute the features "first/last pulse differences", as well as "gradients on segment borders", "height texture" (Laplace operator), "shape and size of segment" and "laser pulse intensity" on segments which were obtained by region growing in the normalised surface model. As for the region shape, they propose a measure which evaluates the relative geometry of the four longest lines on the extracted boundary, being largest when the lines are parallel or orthogonal. The "gradients on segment borders" feature was intended to differentiate between buildings/vegetation and terrain objects, since the latter were expected to have smooth transitions to the surrounding terrain somewhere along their boundary. Their major findings were that first/last pulse differences had the greatest influence, substantially larger than "height texture", although these features are similar in visual appearance. Intensity was also found to be significant. Border gradients had no great influence and the shape parameter was found to work well only on large buildings. Both a fuzzy and a maximum likelihood classification were tested and no substantial differences were found. Classification rates for buildings were in the order of 93% to 96%.

Darmawati (2008) uses up to four pulse returns and a high density point cloud with an average of 10 points per square metre. Points are first grouped into planar segments using a Hough transform (see Chapter 2) for seed region selection, followed by region growing (both performed on the original points rather than a grid). This results in large building segments which contain mostly single echo points, and small, highly overlapping segments made of a high percentage of multiple-echo points which represent vegetation (Figure 5.2). The analysis shows that building segments usually contain more than 90% of single pulse return points (the exceptions being at the jump edges of segment borders), whereas vegetation segments contain less than 50%. Thus, in the proposed method, segments are classified according to thresholds on segment size and percentage of single pulse return points [similar to Wang, 2006]. This is subsequently further refined using analysis of neighbouring segments and a special detector for glass roofs. The results show that the high point density and the exploitation of multiple return information allow very good results to be obtained using a relatively straightforward approach (Figure 5.3).

As the first/last or multiple pulse criterion turns out to be a quite strong classification feature, it is worth investigating in more detail. Nowadays, several systems record the full-waveform of the pulse return, which allows it to be analysed using more elaborate (and time-consuming) methods. As Wagner *et al.* (2006) show, it is reasonable to view the returned waveform as a linear combination of Gaussian-shaped functions, each one resulting from the convolution of the nearly Gaussian-shaped system waveform of the scanner and the response of the scatterer (assumed to be Gaussian-shaped). Thus, given a pulse response, for each of the Gaussian-shaped peaks, the time

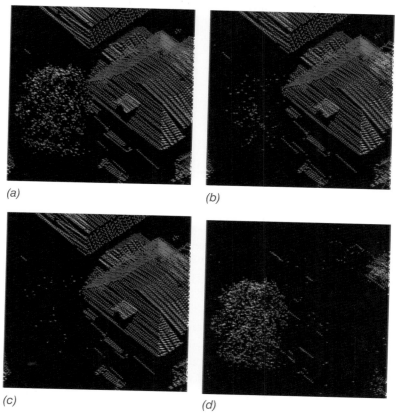

Figure 5.2 Comparison of pulse returns for buildings and trees: (a) first returns; (b) last returns; (c) points which belong to pulses with single returns; and (d) multiple returns. Points are coloured based on their assignment to planes, obtained by segmentation. Note the large planar segments on buildings and the small overlapping segments in trees. (From Darmawati, 2008; courtesy Arik Darmawati)

delay (corresponding to the range), amplitude and width can be determined. Using an analysis of 26 million waveforms, the authors successfully obtained a decomposition in 98% of cases. From the range, amplitude and width, the backscatter cross-section of the target can be computed. This target-specific value depends on the solid angle of the backscatter cone (which in turn depends on the surface roughness), the target reflectivity, and the target receiving area. Furthermore, it turns out that the pulse width in particular is a good indicator for separating buildings from vegetation, being narrow for artificial (continuous) surfaces and wide for trees [Rutzinger *et al.*, 2008].

Gross *et al.* (2007) use pulse width, amplitude and total number of echoes, and normalise each of them separately by a sigmoid function using empirically determined

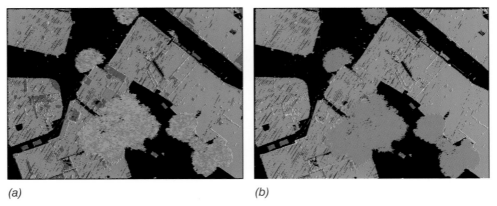

(a)   *(b)*

Figure 5.3 Classification results from Darmavati (2008). Colours are building (red), vegetation (green), others (dark blue), unclassified (light blue). (a) Initial classification based on percentage of single returns and segment size; and (b) final classification after analysis of neighbourhoods and glass roofs. (Courtesy Arik Darmawati)

parameters. The product of those values is then thresholded to obtain an indicator for tree regions. They find a high detection rate, but also a high false alarm rate. A second measure is based on the points in a local neighbourhood and multiplies averaged intensity, planarity and omnivariance (the geometric mean of the three eigenvalues of the scatter matrix). This is found to yield lower false alarm rates, but does not exceed a detection rate of around 65%. Again, the pulse width is found to be a good indicator for vegetation. From a visual inspection, the amplitude also seems to separate well between buildings and vegetation; however it can be seen that it is quite dependent on the incidence angle.

In a related approach using full-waveform data from the same sensor, Mallet *et al.* (2008) define the following eight features classifying points into the four classes building, vegetation, natural ground and artificial ground: (1) difference between range and largest range (lowest altitude) in a large environment (e.g. 20 m radius sphere); (2) residuals computed from a local plane estimation (0.5 m radius sphere); (3) deviation of local normal vector from the up direction; (4) first/last pulse difference; (5) number of echoes; (6) pulse amplitude; (7) width; and (8) shape. The shape parameter allows modification of the generalised Gaussian functions fitted to the returned waveform so that they are flat or peaked. It is observed that the amplitude is high for roofs, gravel, sand and cars, and low for vegetation and streets. Although contrast is low, different surfaces can be visually discriminated. Pulse width is found to be high for vegetation and low for building surfaces, even on sloped roofs. Support vector machines are used for the classification. Buildings are classified correctly in 87% of cases, with errors mainly due to a misclassification at building edges and superstructures. If the two terrain classes are merged, the overall classification rate rises to 92%.

Overall, the following observations and conclusions can be drawn.

- A major objective of building detection is differentiation between buildings, ground and vegetation. In particular, reliable separation between large trees or groups of trees and buildings, as well as trees close to or overlapping with buildings, has proven to be difficult.

- Useful criteria for the detection of buildings are the height relative to the terrain, the difference between minimum and maximum height in a local window, and texture, for example computed by Sobel or Laplace operators.

- Recently, it has been shown that multiple pulse returns, now available from modern scanners, improve the situation. For trees, there are usually large differences between the first and last pulse. Even the existence of multiple returns is a strong indicator for trees, although they may be present also at roof borders or at roof superstructures. When planar regions are considered instead of single points, regions on buildings usually consist of more than 90% single echo points, instead of 50% or lower for vegetation.

- Shape constraints on segmented regions, such as compactness, were found to be insufficient. However, with higher point densities, this may change. For example, a planar segmentation in a high density point cloud yields many small regions in trees, so that a simple threshold on the region size will give good results.

- The intensity of the returned pulse has been found to be of low quality and to be highly dependent on the incidence angle, and thus, in the past, it was concluded to be of little use. However, it seems that this has improved recently. In particular, the amplitude and width of the returned pulse, as obtained in full-waveform analysis, are strong candidates for a separation of buildings and trees.

- Multispectral data have been used in the past to aid separation between buildings and trees. However, the availability of such data may be a problem. It seems that to a certain extent higher point densities and multiple return/full-waveform analysis allow good results to be obtained without the need to acquire additional multispectral data.

- Several authors concluded that analysis has to be performed on several levels, for example the point/pixel level, the local neighbourhood level, and based on segmented regions. This is consistent with common practice in remote sensing.

- In general, it seems beneficial to explore well-known remote sensing approaches where several features, or "channels", are extracted from the data, followed by a standard classification, instead of building specialised extraction procedures using hand-tuned thresholds. If the features are directly computed from the point cloud, care must be taken to avoid dependency on the point density.

5.2   Outlining of footprints

After the detection of buildings, 2D outlines can be derived. These may be of interest on their own, for example to extract 2D cadastral maps automatically. They can also be used as an intermediate step for 3D building extraction, although they are not a necessary prerequisite. Three essential steps can be identified for finding the outline: (1) detection of pixels or points which make up the outline and formation of one polygon (or several polygonal rings); (2) reduction of points in the polygon so that only the "relevant" edges are preserved; and (3) enforcement of regularity constraints such as parallelism and rectangularity, which are typically present in manmade structures.

When a regularised DSM is used, the result of building detection, which is a set of connected components, can be easily converted to a polygonal outline, which, however, will follow the underlying grid structure [Weidner and Förstner, 1995]. Although a regular grid simplifies processing, this is especially critical for outline extraction, since interpolation may disturb the outline due to artefacts at height jumps. To extract the boundary from the raw point cloud directly, a 2D Delaunay triangulation (using points projected to the horizontal plane) can be computed in order to obtain neighbourhood relations among the points. Using the known point-wise classification of the vertices into building and ground, boundary triangles are characterised by having vertices of both classes. The building outline can then be defined using, for example, the "building" vertices of the triangles, the centre points of the triangles or the bisector of edges linking a building and ground vertex. Alternatively, the outline can be obtained by computing the α-shape [Edelsbrunner *et al.*, 1983] from the raw point cloud directly [Shen, 2008].

Boundaries extracted from a regular raster or point cloud usually contain a large number of edges and are substantially jagged. The next step is therefore to reduce the number of edges, keeping only "important" ones. A simple approach is to locally remove points which are lying on a line within a given maximum distance. For example, one can consider groups of three successive points and test whether the middle point can be removed.

A well-known algorithm for the simplification of polygons was given by Douglas and Peucker (1973). It reduces the number of points in a polyline by starting from a straight line connecting the first and last point (as an initial approximation) and recursively adding vertices which have a distance larger than a given threshold. The threshold for the algorithm can be set according to the point density of the original laser points. A fundamental problem of using this algorithm is that outliers will often be preserved. Also, the simplified polyline can only be composed of a subset of the original points. If important (corner) points are missing in the original data, they cannot be recovered by the algorithm. Finally, since the outline is a closed polygon instead of a polyline, suitable start points have to be found first. For example, Clode *et al.* (2004) first determine three "maximum" points of the initial outline, which span a large triangle partially covering the initial polygon. Then the three edges of this triangle are iteratively subdivided using the Douglas and Peucker algorithm. An additional filtering is proposed to reduce the effect of outliers. In order to recover correct corner points, each segment is adjusted by a weighted best fit, using the assignment of initial outline points to their respective line segments. The vertices of the final outline are then computed by intersection of the adjusted segments.

Sampath and Shan (2007) use a modified version of the well-known Jarvis march algorithm [Jarvis, 1973], the original being used to compute the convex hull. During each step, the successor point forming the smallest angle is only searched for in a local neighbourhood, thus leading to a simplified outline (instead of the convex hull). Another approach, which is global instead of local, is to use the Hough transform to detect major straight lines in the set of boundary points, which are then connected by intersection [Morgan and Habib, 2002]. However, some additional steps are required to ensure that a proper polygon is obtained. For example, if a short building edge does not lead to a detectable peak in Hough space, the polygon has to be closed by other means. As an alternative to the Hough transform, Sester and Neidhart (2008) use random sample consensus (RANSAC) [Fischler and Bolles, 1981; see also Chapter 2] to identify dominant lines in the outline polygon.

All these methods reduce the number of points according to some general criteria, mostly distances of points to lines. On the one hand, as Clode *et al.* (2004) point out, it is important not to impose further constraints for the sake of generality. For example, forcing line segments to form right angles in all cases will lead to bad results on curved buildings. On the other hand, most manmade objects follow certain construction principles and thus exhibit certain regularities. If there are no provisions made to include these in the simplification process, *most* of the reconstructed outlines will be wrong. This situation may improve with higher point densities, however experience shows that even small deviations from regularity conditions (such as parallel lines, repeating structures) easily attract attention and are found to be disturbing.

Many outlines consist of segments which are along or perpendicular to a single direction. Vosselman (1999) extracts a main building orientation using the intersection lines of adjacent planes (extracted in a previous step) or, alternatively, by analysing its contours. The outline is then constructed using an iterative algorithm similar to region growing. Two points are used as a seed and expanded until the distance of a candidate point to the current line, which is forced to be in the main direction or perpendicular to it, is larger than some bound. The next line then starts from this point in the perpendicular direction. In a second step, this is refined by changing point assignments at segment borders. Similarly, Alharthy and Bethel (2002) first determine the main building orientation using a histogram of angles computed by analysing binary image footprints. Using the outline, linear segments are extracted and connected in order to form a polygon, however they are only allowed to be in the main direction or orthogonal to it. Again using a histogram, major lines are detected, separately for the main and orthogonal direction. Segments close to these lines are moved so as to coincide. The approach thus implements the assumption that the outline is made only of segments in the main and orthogonal directions and that several disconnected segments are part of exactly the same line if they are close to it. Lahamy (2008) analyses reference outlines with regard to the angles between successive segments. Using segments of at least 1 m in length and a tolerance of ±5°, he finds 90% of right angles and 92% of multiples of 45°. He proposes applying three strategies in succession, first requiring multiples of 90°, then multiples of 45°, and finally without further constraints on the orientation.

Some outline simplification algorithms are based on rectangular structures. Gerke *et al.* (2001) start with an initial rectangle obtained using higher-order geometric moments of the original region. Differences between the region and the rectangle are then determined. If these are larger than a given minimum size, they are also described by rectangles and are added to or subtracted from the initial rectangle. This is iterated until there are no more differences larger than the minimum size. A similar approach has been used by Dutter (2007). She starts with a minimum enclosing rectangle (MER) and determines relevant deviations from the rectangle lines. This is done recursively, thus enabling different shapes of buildings, such as L-, T- or U-shapes, to be handled. Rectangles are also used in the detection part of the building extraction algorithm of Lafarge *et al.* (2008), discussed in Section 5.3.4.

An elegant way to introduce regularisation constraints is by integrating them into an overall adjustment process. For example, Sampath and Shan (2007) use a least-squares adjustment where all lines are forced to be parallel or perpendicular to one single direction. Similarly, Sester and Neidhart (2008), after having extracted initial outlines using RANSAC, first improve each segment separately by a least-squares adjustment, considering all connected points from the consensus set. Segments are then combined to form a closed outline. During this step, right angles and parallelism are detected and introduced as constraints (see Section 5.4.2) in an overall least-squares estimation. Since these constraints are only introduced if the angles are sufficiently close, non-rectangular building outlines can also be obtained (see Figure 5.4).

Although an overall adjustment – which considers line fitting and regularisation constraints at the same time – is a preferred solution, the problem of determining when to insert additional constraint equations remains. This is often decided by a simple threshold. For example, if two successive segments are found to form a right angle within ±10°, a right angle constraint equation is inserted in turn. In general, more elaborate criteria should be applied. For example, Weidner and Förstner (1995) propose using the minimum description length (MDL) principle to decide when to insert constraints. Groups of four points can be analysed if rectangle constraints can be imposed on one or both interior points or if they can be replaced by three points, also with a possible rectangle constraint. When using only local decisions, care has to be taken to prevent the insertion of constraints which are globally inconsistent.

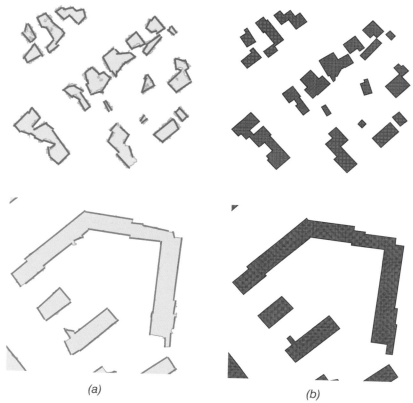

Figure 5.4 Outlining of footprints from laser scan data using the approach of Sester and Neidhart (2008). (a) Initial regions and line segments found by random sample consensus; and (b) final adjusted outlines. (Courtesy Monika Sester)

The problem of outline simplification has also been approached in cartography, where the objective is to generalise existing maps [Sester, 2005] (see also Section 5.4.5). Haunert and Wolff (2008) give an optimal solution which minimises the number of segments subject to a given distance tolerance. It is shown that the problem is (computationally) NP-hard when it is required that the result has no self-intersections. In general, methods like this cannot be applied directly because they simplify the polygon by selecting a subset of the original segments, whereas in outline extraction, due to the jagged appearance, usually no such subset can be found. Also, a number of generalisation methods work by first identifying typical patterns, such as right-angled protrusions and intrusions, which they subsequently remove. Again, in extracted (jagged) outlines, these patterns may be harder to find than in existing vector geodata.

Summing up, it is relatively straightforward to obtain an outline polygon. However, to reduce it to the minimum number of edges in such a way that typical regularities of manmade objects are preserved is a difficult problem. This is not surprising, since this step requires interpretation, which in turn has to be based on a suitable model. Model assumptions which have been made include the presence of one single main orientation, parallelism and right angles, and decomposability into rectangles. However, these are somewhat crude model assumptions and there seems to be plenty of room to explore search and machine-learning techniques. The latter could benefit from large reference (training) datasets, which could also be used for a more systematic, quantitative assessment of the proposed algorithms, which is currently lacking. However, comparing two outlines is not straightforward and, for example, simply computing the area difference is not sufficient [Ragia and Winter, 2000; Lahamy, 2008]. Beyond the analysis of single-building outlines, relations between adjacent buildings have also to be taken into account. For example, they may have the same main orientation, equal distances, alignment of segments facing the street, or form linear or repeating patterns. The simplification of outlines should lead neither to self-intersections nor to intersections with neighbouring objects. It is assumed that an increase in laser scan point density will improve the extraction of outlines. Even then, it has to be kept in mind that aerial scanning will usually obtain the roof outlines and not the walls, which are commonly present in cadastral maps.

5.3   Building reconstruction

5.3.1 Modelling and alternatives

When extracting building models from raw data, it is important to decide what is the most appropriate representation for them. For example, given a dense point cloud from laser scanning, one might argue that the best object representation would be a triangulated surface consisting of these points, or a subset of the points – because this best reflects what has actually been measured. This approach is typically used for free-form objects, where the original measured point cloud is triangulated, smoothed and simplified. Usually, this is not the description one strives for in the context of modelling manmade objects such as buildings, for various reasons:

- Even when used only for visualisation, densely triangulated meshes are not very suitable in the context of buildings, since the human eye is quite sensitive to deviations between the (often piece-wise flat) object and its approximation by the densely triangulated surface. Also, standard triangle reduction algorithms are usually not well suited to polyhedrons with sharp borders. A better visual impression can be obtained by adding texture [Früh and Zakhor, 2003]; however, the model might still look "jagged", especially at building borders.

- For graphics performance, storage and transmission, a surface description consisting of as few polygons as possible is often desirable.

- When building models are to become part of 3D geographic databases, an object-wise representation is more appropriate than a huge triangulated surface. As with 2D databases, different scales should be available, which correspond to different levels of detail (LOD). It is desirable to model buildings in such a way that different LODs can be derived automatically.

As a result, the challenge is to obtain an object instantiation from dense measurements which represents the major structure of an object, given a certain desired generalisation level, as expressed by an object model. Note that due to the random distribution of measured points, the "important" elements of the model instantiation, such as vertices and edges, are usually not even measured directly. Thus, in contrast to the reduction of point clouds or triangulated surfaces, an interpretation step is required. Defining an object model not only opens the way for further downstream usage, but is also important for the guidance and support of the extraction process itself.

It has to be noted that, especially for visualisation, there are alternative approaches which use the original point clouds directly. As Böhm and Pateraki (2006) note, in the domain of cultural heritage, objects to be modelled typically feature delicate and complex surfaces such as ornaments and statues, which are hard to represent using, for example, standard parts from a library. Usually, one would resort to a dense triangulated surface in such cases, but point splats have been shown to be an alternative [Kobbelt and Botsch, 2004]. They are discussed in more detail in Chapter 2.

It is possible to provide certain "object-based" operations, such as picking and highlighting of objects, using iconic surface representations. For example, Wahl and Klein (2007) use proxy objects to decouple rendering representations from semantic representations. While the interaction (such as clicking on an object) is based on coarsely modelled proxy geometries which are not directly visible to the user, visualisation uses high resolution (but uninterpreted) data. Thus, the system avoids the need for building reconstruction while at the same time also being useful for certain application scenarios.

5.3.2 Geometric modelling

Two main approaches for geometric modelling have evolved [Hoffmann, 1989]. The first constructs the boundary from measured points by adding topology information, thus building the boundary representation (B-rep) directly. The second obtains the object by combining volumetric primitives using Boolean operations (constructive solid geometry, CSG). From the CSG representation, B-rep can be obtained automatically and unambiguously. CSG modelling is used widely in computer-aided design (CAD), since it has a number of advantages: (1) modelling using primitives and Boolean operations is much more intuitive than specifying B-rep surfaces directly, because it frees the user from specifying geometric and topological constraints, which are implicit; (2) the primitives can be parameterised, thus enabling reuse and collection in libraries;

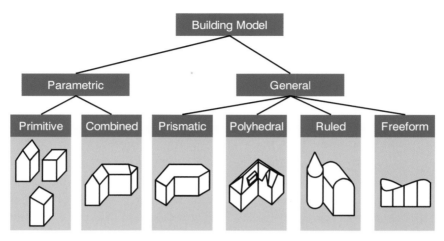

Figure 5.5 Standard geometric building models used in extraction systems.

(3) they can be associated with other, additional information, such as a link to a standard part table; and (4) the CSG modelling tree contains implicit information on the object hierarchy which can also be used for other purposes. However, it has to be kept in mind that since a conversion from a B-rep to a CSG representation is ambiguous, deriving a CSG representation from given (measured) data is also ambiguous. This means that a CSG tree obtained for the sole purpose of a representation of the outer hull of a building will not necessarily be useful for any other purpose which involves certain object semantics. However, in contrast to B-rep, CSG makes structured modelling and attachment of semantics easier in general.

Of course, even the most complex building shapes can be modelled using available CAD systems. Practical reconstruction systems, however, have often used only subclasses of building shapes to keep the systems simple (Figure 5.5). The first approaches for fully automatic building extraction were targeted at buildings consisting of a single primitive, such as buildings with a gabled roof. Also, prismatic building bodies are relatively easy to obtain, especially when ground plans are given. In such cases, only the determination of two scalars, ground and roof height, is necessary. Photogrammetric systems often prefer polyhedral models, as they can be easily integrated into a standard photogrammetric workflow by supplementing point measurement with an additional topology definition step. Finally, objects combined from simple parametric primitives are used, which is basically the standard CSG approach, providing application-specific primitives such as flat, gabled, hipped and desk roofs.

5.3.3 Formal grammars and procedural models

As mentioned, a CSG modelling tree has the advantage that besides defining a geometric object representation, it also encodes valuable information about the object's (hierachical) structure. However, the CSG tree by itself does not contain or enforce semantics. Rather, the basic building blocks (volumetric primitives) and operations (alignment, Boolean operations) are strictly geometric in nature. While this is very general, it does not constrain the possible building shapes.

Formal grammars can be used to introduce semantics. A formal grammar is a four-tuple (N, Σ, P, S), consisting of a set of variable symbols, terminal symbols, production rules and a start symbol [Hopcroft and Ullman, 1979]. Beginning with the start symbol, words can be derived by successively applying production rules until all of the symbols in the word are terminals. The resulting word is an element of the formal language described by the grammar. Depending on the characteristics of the production rules, formal languages can be ordered according to their "complexity", known as the Chomsky hierarchy. Often, context-free languages are used where all rules are of the form $A \rightarrow \alpha$, with A being a non-terminal symbol and α being zero or more terminal or non-terminal symbols.

While formal grammars have been used extensively to specify and analyse computer programming languages, they have also been applied in other areas. For example, Prusinkiewicz and Lindenmayer (1990) use Lindenmayer systems (L-systems) to describe plants. Describing a growth-like process, L-systems have also been used by Parish and Müller (2001) to generate entire cities automatically. They use a hierarchical system which first generates a pattern of roads, subdivides the space in between into lots, and finally generates buildings inside these lots. Both the road pattern and building generation are controlled by L-systems. The approach is targeted at the creation of purely virtual city models, for example for the movie and games industries. This approach needs very little input data such as coarse elevation and population density maps.

Wonka *et al.* (2003) note that plants usually grow in open spaces, while this is not the case for buildings, which rather rely on a partitioning process. They combine a split grammar, which is used for the spatial partitioning of façades, and a control grammar, which is responsible for attribute propagation. The rules are selected based on a deterministic attribute matching and a stochastic selection function. The system is shown to be especially useful for the production of virtual façades. Later, Müller *et al.* (2006) found that while split grammars are useful for façades, they are not optimal for initially generating the crude volumetric building model. They propose a set grammar called 'CGA Shape' with production rules of the form *predecessor : condition → successor : probability*, where predecessor is replaced by successor, condition has to evaluate to true in order for the rule to be applicable, and probability is the probability with which the rule is selected. An important aspect of this approach is the transition between dimensions using production rules. By applying the so-called "component split" rule, the dimension can be reduced, for example splitting a volumetric primitive into its faces, edges, or vertices. Conversely, using size production enables a face to be extruded to obtain a 3D volume again.

In general, formal grammars are a very powerful mechanism for encoding knowledge about building structures. Ideally, the grammar reflects typical architectural patterns; however, in practice, informal rules such as those formulated by Alexander *et al.* (1977) will still be hard to formalise. Nevertheless, there is a general tendency to generate graphics procedurally, not only by the application of grammars, but also by embedding procedures, such as in generative modelling [Berndt *et al.*, 2004].

Although these approaches are aimed at the generation of purely artificial (but plausible) building and city models, they are nevertheless very promising for building reconstruction. Not only do they introduce semantics, but they can also be directly used as part of a hypothesise-and-test reconstruction approach.

5.3.4 Building reconstruction systems

The design and implementation of building reconstruction systems has been a research topic for many years [Grün *et al.*, 1995, 1997; Baltsavias *et al.*, 2001; Brenner, 2005a]. As noted above, early research focused on monoscopic and stereoscopic aerial image setups. However, with the advent of laser scanning, the usefulness of dense point clouds became obvious [Mayer, 1999].

The extraction of manmade objects is a special case of *object recognition*. As such, it is part of a large research field [Grimson, 1990]. However, as techniques are often driven by the data available, the special characteristics of the objects to be extracted and the expected end products, the extraction of manmade objects is a large research branch in its own right. Building reconstruction approaches can be classified according to the following criteria.

Control Typically, approaches are seen as examples of *bottom-up*, *top-down*, or *hybrid* control. Bottom-up, or data-driven, means that only a few assumptions are made about the object. For example, prominent points are extracted, which are then grouped into lines, which in turn are grouped into rectangles, then to clusters of rectangles, etc. As one can see, there are still model assumptions, namely, that the object has prominent points, that it is (mostly) described by lines, which actually form rectangles, which usually occur in clusters, etc. However, this knowledge of the model is introduced only "little-by-little" into the reconstruction process. On the other hand, top-down, or model-driven, approaches usually try to find a "complex" model in the data. For example, searching for L-shaped buildings directly by placing a geometric model in the scene, and then verifying it, would be considered a top-down approach. In practice, systems are often hybrid, i.e. a mixture of bottom-up and top-down.

Data sources These have been traditionally single-, stereo- or multi-image sources of varying image resolution, panchromatic, coloured or false colour infrared. In laser scanning, they are point clouds obtained by fixed-wing aircraft, helicopters, and terrestrial and mobile mapping systems with corresponding characteristics regarding view angle, point density and multiple-pulse/full-waveform capabilities. Also, many systems allow images to be taken simultaneously, so that colour (and possibly infrared) may be used (in addition to the intensity of the returned laser beam). Apart from the directly captured data, data captured previously and/or with different platforms may be used, such as aerial or satellite images. Often, the use of existing data from Geographic Information Systems (GIS) is also considered. For example, the 3D reconstruction of buildings is sometimes required to fit the 2D ground plans exactly as given in the cadastral map (whether or not they are correct). Apart from such "hard" constraints, using existing GIS data can be useful since it directs the solution towards a result which is compatible with existing data. Seen from a more general perspective, GIS data "injects" data which are already interpreted (usually in two dimensions) into the reconstruction process.

Reconstructed models These can be differentiated into geometric (CSG, B-rep), topological, semantical description and additional attributes (e.g. spectral response, material).

Automatic and semi-automatic methods Many of the early approaches were devised for fully automatic processing. However, as it turned out that this does not provide the required reliability, post- or pre-editing steps were added. Baltsavias (2004) points out that the term "semi-automated system" is often used in a positivistic manner for automatic systems which sooner or later fail during processing and thus require manual intervention. It is certainly true that interactive reconstruction systems nowadays use quite traditional interaction techniques from measurement or CAD systems.

In the following, some proposed reconstruction systems which are based on laser scanning are reviewed. It is not the intention to provide an exhaustive list of all suggested approaches, but rather to provide some insight into how practical systems achieve the above criteria.

Vosselman (1999, 2001) extracts faces from non-regularised laser scan data using a Hough transform, followed by connected component analysis [Vosselman, 1999] (see Figure 5.6). From this, edges are found by intersection of faces and analysis of height discontinuities. No ground plans are used as additional information, however since jump edges cannot be determined very well from laser scanning, a "main building orientation" is derived and used as a constraint for the edge orientation. The roof topology is built by bridging gaps in the detected edges. The use of geometric constraints is proposed in order to enforce building regularities.

In a similar approach ground plans are introduced as additional information to prevent spurious roof faces [Vosselman and Dijkman, 2001]. Concave ground plan corners are extended to cut the building area into smaller regions. The Hough-based plane extraction is constrained to these regions. Split-and-merge is used to obtain the final faces. Thus, while the reconstruction is guided by the ground plan, the results are to a certain extent independent of it.

Maas and Vosselman (1999) describe an interesting approach where all seven parameters, which determine a simple gabled roof primitive, are computed in closed form

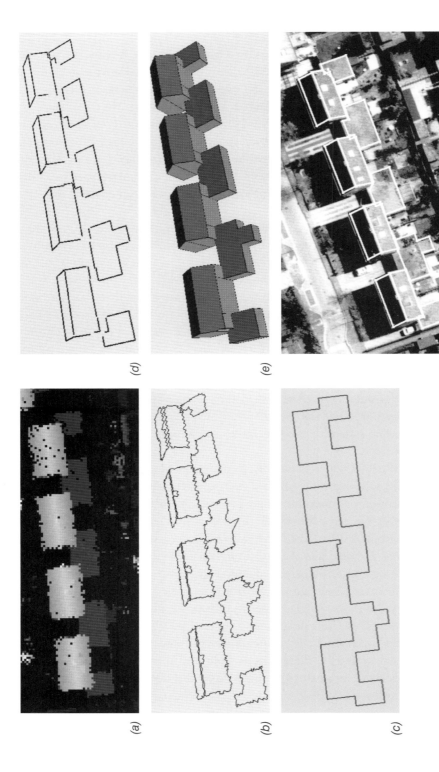

Figure 5.6 (a) DSM of the scene (the regular raster is used for visualisation only); (b) roof faces as detected by the Hough transform and connected components analysis; (c) building outline, obtained using a main building orientation; (d) detected roof face edges; (e) final 3D reconstruction; and (f) results projected to aerial image. (Results from Vosselman, 1999; courtesy George Vosselman)

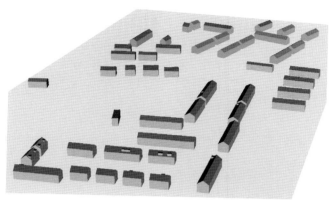

Figure 5.7 Reconstruction results from Maas and Vosselman (1999). (Courtesy ISPRS)

using first-order and second-order moments of the original laser points (Figure 5.7). After the model has been determined, it is projected to the point cloud and residuals are computed. Some of the outliers are used to model dormers on top of the gabled roof, their parameters also being computed using moments. However, since there are only a few (mostly less than 20) laser points on a single dormer, a simplified model with a flat roof and main orientation according to the orientation of the gabled roof is applied. Obviously, the approach benefits from a high point density of five points per square metre, which, however, is not unusual today.

Brenner (1999–2000) investigated several approaches to including existing 2D ground plans into the reconstruction.

A first approach uses a hypothesise-and-test scheme [Haala and Brenner, 1997] (see Figure 5.8). First, the straight skeleton [Aichholzer *et al.*, 1995] is computed using only the 2D ground plan, assuming identical roof slopes. The resulting roof faces, in 2D, are used to define areas of interest. Inside each of these areas, normal vectors computed for each point in the DSM are compared to the normal vector from the constructed roof face. If there is a sufficient number of compatible vectors, the face is considered confirmed and its slope is set according to the normal vectors from the DSM. If not, the face is set upright. Then, the straight skeleton computation is invoked for a second time, using the new slopes. Finally, a least-squares adjustment is performed and vertical walls are added.

In a second approach, ground plans were divided into rectangular primitives using a heuristic algorithm [Brenner, 1999]. For each of the 2D primitives, a number of different 3D parametric primitives from a fixed set of standard types are instantiated and their optimal parameters are estimated using a DSM obtained by laser scanning. The best instantiation is selected based on area and slope thresholds and the final-fit error. The 3D primitive selection and parameters can be changed later using a semi-automatic

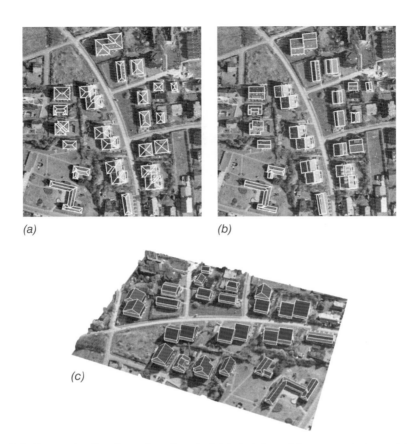

Figure 5.8 Reconstruction based on the straight skeleton of the ground plans: (a) initial reconstruction using ground plans only; (b) final reconstruction, after second invocation of the straight skeleton algorithm; and (c) result in a 3D view. (From Brenner, 2000a)

modelling tool, which allows interactive modifications in 2D, while the height and slope parameters (and thus the 3D shape) are estimated in real time during user interaction (Figure 5.9). The final object representation is obtained by merging all 3D primitives. Orthoimages can be used to facilitate interpretation by a human operator during inter-active editing, however they are not used by the automatic measurement. One advantage of this method is that, due to the fixed primitives, it is quite robust. The major disadvan-tage is that the subdivision into primitives is based only on the ground plan, so that roof structures can only be detected by the automatic method if they can be inferred from the ground plan shape. Consequently, while the method works well for simple buildings, it fails more often on complex roofs, such as roofs in densely built-up areas.

A third approach tried to lessen the tight coupling between ground plan and roof reconstruction [Brenner, 2000b]. It uses three steps: (1) the extraction of planar faces;

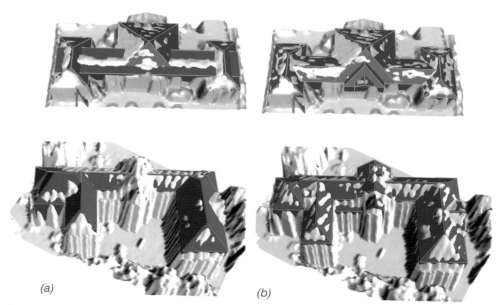

(a)   (b)

Figure 5.9 Results from rectangle-based reconstruction for two complex roofs: (a) result of automatic rectangle subdivision and primitive type selection algorithm; and (b) result of semi-automatic post-editing. (From Brenner, 2000a)

(2) the removal of spurious faces which are either due to segmentation errors or cannot be explained by the model; and (3) the formation of a correct roof topology using the remaining planar faces. In the first step, planar faces are extracted from a regularised DSM using random sample consensus (RANSAC). The second step links the faces to an existing ground plan by accepting or rejecting them based on a set of rules, which express possible relationships between faces and ground plan edges. Finally, the roof topology is obtained from all accepted regions by a global search procedure, which uses discrete relaxation and constrained tree search to cut down search space (Figure 5.10). The additional introduction of constraints and a least-squares adjustment to enforce regularity is described in Brenner (2000a).

Rottensteiner *et al.* (2002–2006) generate DTM and DSM by robust interpolation, and an initial building mask by thresholding the difference [Rottensteiner and Briese, 2002]. Further filtering removes remaining tree regions (Figure 5.11). Subsequently, homogeneous pixel regions are searched for and those with the best fit to a local plane are used as seed regions for region growing. A Voronoi diagram is then computed to assign the remaining pixels to their closest region. Finally, a 3D model is created from the 2D Voronoi region outlines, vertically bound by the roof plane and the DTM. Although this gives a first visual impression, the outlines look jagged and neighbour-ing planes do not intersect correctly. In further work, adjoining regions are classified

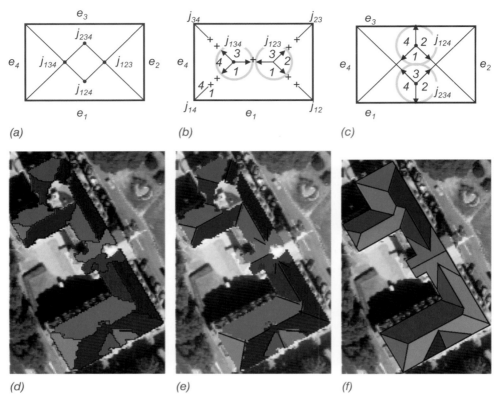

Figure 5.10 Results from Brenner (2000). Assume a simple, four-sided ground plan, each edge defining a roof plane. (a) Four intersection points are possible by intersecting these four planes. The correct roof topology is to be selected from the corresponding graph. This is a search problem. However, by labelling edges (two possibilities: convex (+) and concave (−)) and nodes (eight possibilities), it can be found by discrete relaxation that (b) is a solution, while (c) is not. In this case, this is the only solution that remains, whereas in general (for more complex buildings) a search is required in addition to discrete relaxation. (d) Regions obtained by segmentation, after selection by the rule-based approach. (e) Region adjacencies and intersection lines. (f) Final roof structure, obtained after relaxation and search.

into whether there is an intersection or jump edge between them and step edges are established based on statistical tests and robust estimation [Rottensteiner *et al.*, 2005a]. In addition, a "constraint generator" is proposed to identify and add further regularisation equations to be used in a final overall adjustment. The adjustment model was later improved in Rottensteiner (2006).

Verma *et al.* (2006) use a combined bottom-up and top-down approach where a region adjacency graph is set up based on a planar segmentation, in which subgraphs representing simple primitives are subsequently searched.

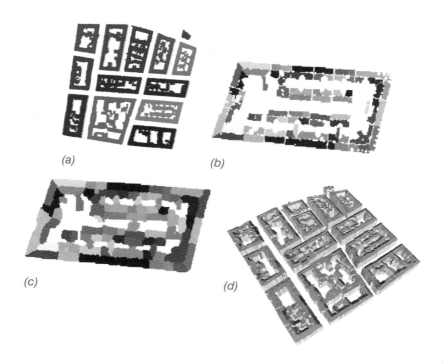

Figure 5.11 (a) Final detected building regions; (b) label image of segmented planes for one of the building regions; (c) Voronoi diagram for the same region; and (d) 3D visualisation generated from the Voronoi regions. (Results from Rottensteiner and Briese, 2002; courtesy Franz Rottensteiner)

First, a principal component analysis in local point neighbourhoods of fixed radius is used to determine whether points are flat. Non-flat points are removed to ensure sufficient gaps between roof and ground points. Connected components are computed using a voxel data structure, which ideally results in a connected component for each roof and one large component representing the ground. Connected components inside trees are removed by a threshold on the minimum number of points.

Each complete roof is subdivided into patches using a planar segmentation. A roof topology graph is obtained where each graph vertex corresponds to a planar patch. Whenever two planar patches are adjacent, a corresponding edge is inserted into this graph (Figure 5.12a). Similar to Brenner (2000b), edges are labelled as convex and concave based on the projection of their normal vector into the plane, using $O+$ for convex and orthogonal, $O-$ for 90° concave and orthogonal, $S+$ for convex and pointing away from each other (180°), and N for all other graph edges.

The topology graph is then searched for simple subgraphs representing U-shaped, L-shaped and hipped primitives, starting with the most complex (U-shaped) primitive (Figure 5.12b). For each substructure found, the corresponding points are labelled

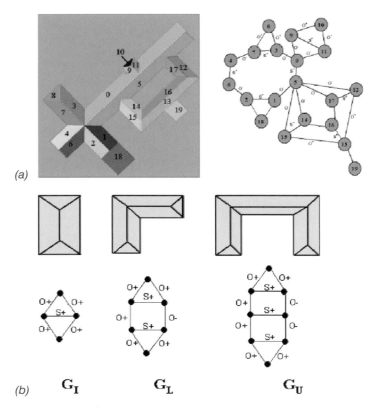

Figure 5.12 (a) Segmented regions of a complex roof shape and derived roof topology graph; and (b) simple shape models and corresponding topology graphs. These graphs are searched for by subgraph matching. (Results from Verma *et al.*, 2006)

accordingly. Points which remain unlabelled are considered "unexplained" and are fitted rectilinear primitives which use the main orientation of the building. Finally, the geometry of the building is computed by a least-squares estimation which uses the distance of the points to their assigned planes. Additional constraints are inferred as follows. Pairs of faces which share a ridge segment are forced to have opposing gradient orientations. Moreover, if their slopes are similar, they are forced to be the same. Also, a main building orientation is estimated and (projected) normal vectors which are close to this orientation or its orthogonal are forced to this common orientation.

Lafarge *et al.* (2008) use a method based on reversible jump Markov chain Monte Carlo (RJMCMC) to extract buildings. Original data are a DSM (in this case computed by image matching, however this could also be obtained using laser scanning). The method uses the following steps (Figure 5.13). First, a RJMCMC process is run to find a set of rectangles which tightly cover the building structures, as present in the DSM. At this stage, there will still be overlaps of, or gaps between, rectangles. In a second step,

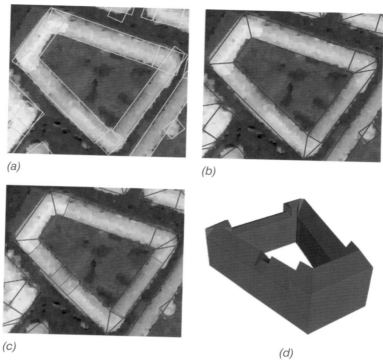

(a)

(b)

(c)

(d)

Figure 5.13 (a) Rectangles found by the RJMCMC process; (b) after merging of rectangles; (c) after detection of roof height discontinuities; and (d) final 3D result. (Results from Lafarge *et al.*, 2008; courtesy ISPRS)

this is corrected by merging neighbouring rectangles, yielding a connected area. A third step detects roof height discontinuities and inserts additional (jump) edges into this connected area. Finally, rooftop heights are estimated and 3D models are obtained.

The first step of the algorithm is detailed in the following. If a single rectangle is placed (arbitrarily) in the scene, with a certain position, width, depth and orientation, an energy function may be computed which measures how well the proposed rectangle matches the structures present in the DSM. Lafarge *et al.* (2008) call this "external" energy, which is made up of a volume, moment and localisation rate. Additionally, if another rectangle is in the vicinity, terms for "internal" and "exclusion" energy are used to favour "good" configurations, such as aligned and non-overlapping rectangles. Combining all energy functions, an overall objective, or density, function is obtained. The best selection of rectangles then corresponds to the peak, or mode, of the density function, and the goal is to find this mode. Due to the high dimensionality, an exhaustive search of this space is not possible. Instead, it is explored by random sampling, using a Markov chain which converges to the desired density function. The rationale behind this is that since the samples are drawn according to the target density, there is a greater chance of ending in (and thus finding) peaks. Five parameters (position, width, depth and orientation) are required to describe a single rectangle, but since the number of rectangles in the scene is not known in advance, the dimension of the parameter vector is variable. The process thus has to change dimensions, which is done using the "reversible jump" approach of Green (1995).

MCMC methods are computation intensive, however they are very interesting for several reasons. First, changes are driven by a kernel, and partial knowledge of the scene or the model can be used to favour certain changes. For example, if a "main building direction" has been extracted by scene analysis, or if a building size range is known, this can be incorporated into the kernel function, which then yields a higher rate of "good" samples. Second, MCMC allows a "hypothesise-and-test" approach, where there is no need to actively construct a solution, but only to verify a given one (by computing an energy, or score, function). Third, MCMC can be used to "forward run the model", if only the "internal energy" terms are used. Since the knowledge of the model is contained in the internal energy terms, running MCMC without external energy terms (i.e. disregarding actual measurements) will give simulation results which only represent the model knowledge. This can be used to detect errors in the model, but also to identify possible additional constraints which could be introduced.

MCMC has also been used by Dick *et al.* (2004) for the reconstuction of buildings from multiple images. Ripperda and Brenner (2006) suggested an approach to reconstructing building façades which combines a formal grammar to describe the structural properties and RJMCMC to explore the solution space.

5.4   Issues in building reconstruction

5.4.1 Topological correctness, regularisation and constraints

There is a huge difference between object *construction* and *reconstruction*, which unfortunately has not yet been addressed to a sufficient extent in reconstruction systems. In CAD construction, objects are usually built based on idealised assumptions. For example, exact dimensions are set, objects are copied, leading to identical clones, and snap routines ensure that objects meet at a single point along a line or plane, or are in line or orthogonal. In reconstruction, however, the task is not to construct an envisioned object from scratch, but rather one that has already been described by measured data, possibly millions of 3D points. This has consequences for ensuring correct *topology* and *regularisation*.

When objects are constructed using a CAD system, correct *topology* is usually ensured by the available operations. For example, all CSG primitives will have correctly adjoined, "watertight" faces with correctly oriented face normal vectors. Combining primitives using Boolean operations will preserve this property. Snap operations allow separate CSG primitives to be perfectly joined. In reconstruction, however, one may start with single faces fitted to the point cloud, which then have to be modified to obtain a watertight structure. What seems to be straightforward is actually a difficult problem since snapping is a good alternative in construction, but not in reconstruction. Assume, for example, that four planes have been reconstructed, each one by a fit to part of the laser scan point cloud. If it is now additionally required that those four planes meet at a single 3D point, there are two alternatives. First, performing a snapping operation, which corrects the topology, but destroys the best-fit property of (at least) one plane. Or second, keeping the best-fit property, which will result in a wrong topology containing an extra edge. What is actually required is to estimate the geometry of all four planar patches, using all laser scan points *under the constraint* that they meet at a single point.

A similar situation holds for the problem of *regularisation*. For example, right angles, parallelism, identical roof slopes, etc. are properties which are often encountered. Again, in CAD, they are usually fulfilled by construction, whereas in reconstruction they have to be enforced by additional measures.

In both cases, it is obvious that the best solution is to impose constraints during the reconstruction process. As for the topology, one would usually like "hard" constraints. In the case of four planes meeting at one point, it may not be tolerable if they meet at two separate points which are "close". In contrast, "soft" constraints will usually be sufficient for regularisation conditions. For example, if two roof faces are required to have the same slope, a small deviation from this condition will be tolerable. In fact, weights could be set according to the confidence in the assumption of identical slopes.

5.4.2 Constraint equations and their solution

A geometric constraint problem consists of a set of geometric primitives and a set of constraints which they should fulfil. Geometric primitives are points, lines, circles in two dimensions, or points, planes, cylinders, spheres, cones in three dimensions, etc. Constraints may be logical, such as incidence, perpendicularity, parallelism, tangency, etc., or metric, such as distance, angle, radius, etc. Table 5.1 lists some objects and constraints. From this, one can see that the constraint equations are mostly quadratic (such as fixed point to point distance) or bilinear (such as fixed point to plane distance); note that a, b, c and x, y, z are unknowns. See Heuel (2004), for example, for a more complete list of relationships.

In general, constraint equations are algebraic equations of the form $f_i(\mathbf{x}) = 0$, where $\mathbf{x} \in \mathbb{R}^n$ is the parameter vector describing the scene. For example, $\mathbf{x}$ can directly contain the coefficients describing primitive objects such as points, lines, planes or other surfaces, but can also describe "higher level" parameters such as the transformation parameters of a substructure. Putting all the functions f_i, $1 \le i \le m$ together, the problem is to find a parameter vector $\mathbf{x}$ which satisfies the implicit (and usually non-linear) constraint equation

$$\mathbf{f}(\mathbf{x}) = \mathbf{0} \qquad (5.1)$$

Table 5.1 Examples of geometric objects, logical and metric constraints (U: unknowns, C: constraints, HNF: Hesse normal form)

Object/constraint	Unknowns/equations	U	C
Point (in 3D)	x, y, z	3	–
Line (in 2D)	HNF (a, b, c), $a^2 + b^2 - 1 = 0$	3	1
Plane (in 3D)	HNF (a, b, c, d), $a^2 + b^2 + c^2 - 1 = 0$	4	1
Point on plane	$ax + by + cz + d = 0$	–	1
Two lines perpendicular	$a_1 a_2 + b_1 b_2 + c_1 c_2 = 0$	–	1
Four planes meeting at a point	$a_i x + b_i y + c_i z + d_i = 0$, $1 \leq i \leq 4$	3	4
Fixed distance ϵ between points	$(x_1 - x_2)^2 + (y_1 - y_2)^2 + (z_1 - z_2)^2 - \epsilon^2 = 0$	–	1
Fixed (signed) distance ϵ of point to plane	$ax + by + cz + d - \epsilon = 0$	–	1

As noted in the previous section, reconstruction involves not only hard constraints (for example to ensure a correct topology), but also soft constraints (for example for regularisation) and equations which fit the model to the (noisy) data. Thus, the problem can be formulated as

$$||\mathbf{g}(\mathbf{b}, \mathbf{x})|| \overset{!}{=} \min, \quad \text{subject to } \mathbf{f}(\mathbf{x}) = 0 \tag{5.2}$$

where $\mathbf{g}$ subsumes the (possibly contradictory) constraints imposed by some measurement data $\mathbf{b}$ and the soft constraints, whereas $\mathbf{f}$ represents the hard constraints. The equation system $\mathbf{f}$ will normally be underconstrained (as the measurements would otherwise have no effect on the solution), whereas $\mathbf{g}$ will be typically overconstrained (since redundant measurement data are used), which leads to a system which is both locally overconstrained and globally well-constrained or underconstrained. The globally underconstrained case is interesting since it leaves room for interactive manipulation of object parts which are not fixed by either measurements or model constraints (see the following section).

In reconstruction, there is usually an initial solution available, which needs to be refined. Equation (5.2) can thus be solved by linearisation and iterative solution. This results in the equation system

$$||\mathbf{Ax} - \mathbf{b}|| \overset{!}{=} \min, \quad \text{subject to} \tag{5.3}$$

$$\mathbf{Cx} = \mathbf{d} \tag{5.4}$$

where $\mathbf{A}$ and $\mathbf{C}$ are the design and constraint matrices, respectively (a weight matrix is omitted for clarity). If $||\cdot||$ in Equation (5.3) is the L_2 norm, this can be solved in numerous ways [Förstner, 1995; Brenner, 2005b]: (1) by adding the constraint equations to the observation equations, using large weights; (2) by solving for an extremum with auxiliary conditions, which introduces Lagrange multipliers; or (3) by an explicit parameterisation of the null space of $\mathbf{C}$, using the Moore–Penrose pseudoinverse $\mathbf{C}^+$ to compute the solution directly [Lawson and Hanson, 1995]. If alternative (1) is chosen, care has to be taken, since the constraint equations are only enforced to a certain degree, depending on the weight. However, they may compete with observation equations – actually, with mass data from laser scanning, there can be many thousands "against" a single constraint equation. Consequently, the weight has to be very large, running the risk that the equation system becomes poorly conditioned.

The rank of the matrices is also an issue. The equation system (5.3) and (5.4) only has a unique solution if $[\mathbf{A}^{\mathrm{T}}, \mathbf{C}^{\mathrm{T}}]$ has full rank. If not, solutions based on normal equations and inversion will fail, in contrast to pseudoinverse or singular value decomposition (SVD) solutions. In the context of a constraint modeller, a newly placed object will usually be underconstrained, and in fact it will usually stay underconstrained as long as it is intended to be modified interactively.

5.4.3 Constraints and interactivity

As fully automatic reconstruction systems may fail, production systems have to include the possibility of modifying reconstructed objects interactively. Imagine the situation where a (complex) building object O is instantiated and partially fitted to a point cloud, and the user subsequently decides to modify it interactively, for example intending to drag a corner of the model to another (presumably correct) position. Now it can happen that (1) O is completely free with respect to the selected modify operation and no other objects are involved when it is modified; (2) O is free, however it is linked by constraints to other objects, which also have to be modified as a result of modifying O; or (3) O is partially or fully constrained, i.e. the user's wish to modify O can be only partially fulfilled or cannot be fulfilled at all.

The interesting case is when the modification is feasible, however, in response to the user's request, other objects (variables) also have to be changed, and there is a choice as to which variables can be changed and how the variables can be changed. It turns out that this is less a mathematical question than a question of what we expect from the behaviour of objects. For example, consider a 2D drawing program. If a "polygon primitive" is placed in the scene and a single polygon vertex is selected and dragged, one would usually expect that only the vertex itself and the incident edges (i.e. a total of three objects) are actually changed (see Figure 5.14a). On the other hand, if the vertex of a "rectangle primitive" is moved, it is not so clear what a user would expect. For example, it would be sensible that the diagonally opposite vertex and the orientation stay fixed, while the width and height are adjusted (Figure 5.14b); or that the midpoint of the rectangle and the orientation stay fixed, while width and height are adjusted (Figure 5.14c); or that orientation and scale are changed (Figure 5.14d). A drawing

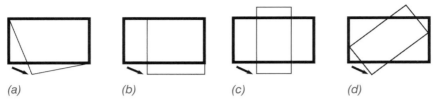

Figure 5.14 Different user interface behaviour when (a) a point of a polygon, or (b)–(d) a rectangle, is changed.

program usually has different interactive modification operators which implement a certain, object-specific behaviour. In the context of building reconstruction, Lang and Förstner (1996) have coded the behaviour of wireframe primitives in a so-called "association table". Depending on which two points of the wireframe model are clicked in succession, either the model is moved as a rigid body or parameters such as width and height are adapted.

It is not obvious how such an approach can be extended to a large number of objects and constraints. Intuitively, we could require that there is "not much change" among the objects which are not directly modified. To see what happens, consider the example in Figure 5.15, again in 2D for simplicity. There are five objects (three points, two lines), for a total of 12 unknowns, as well as two "normal vector length" and four "point on line" constraints, so that six degrees of freedom remain (i.e. the freedom to move any of the points). If $\mathbf{p}_1$ is moved to the right, one would intuitively expect Figure 5.15(b) to result. However, if "not much change" is formulated in the least-squares sense, the (unexpected) result of Figure 5.15(c) is obtained. The reason is that not only the changes to the points, but also the changes in line parameters are minimised, which suppresses rotation of $\mathbf{l}_2$ and forces $\mathbf{p}_3$ to move. Similarly, if a right angle

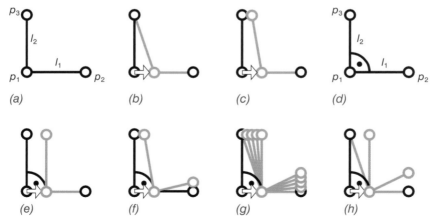

Figure 5.15 Behaviour of a simple constraint system under different optimisation criteria (see text).

is added (Figure 5.15d), instead of Figure 5.15(e), the result shown in Figure 5.15(f) is obtained, where the movement leads to a rotation of the object. In general, a least-squares solution will change many variables by a small amount rather than a few variables by a large amount. Vosselman and Veldhuis (1999) have used such an approach for the interactive modification of building wireframe models.

Using the L_1 norm instead, any solution shown in Figure 5.15(g) is valid, i.e. there is no unique result. Another approach would be to demand that as few as possible variables are affected. In this example, this leads to two solutions as shown in Figure 5.15(h). Note again that in all these cases, the movement of $\mathbf{p}_1$ is not constrained and the results differ only in how connected elements are affected by the interaction. A more detailed treatment can be found in Brenner (2005b).

5.4.4 Structured introduction of constraints and weak primitives

The need to introduce constraints into the reconstruction process of manmade objects was recognised at an early stage. For example, Weidner (1997) proposed automatically deriving mutual relationships between the extracted faces, such as "same slope", "symmetry", and "antisymmetry", in order to insert them as constraints into a global robust adjustment. Although this has been proposed by several authors [Vosselman, 1999; Rottensteiner *et al.*, 2005a; Verma *et al.*, 2006], it is still unclear if this approach is feasible in practice. If the system is allowed to establish constraint equations automatically, a large number of constraints may result, probably with circular dependencies, from which a "spaghetti constraint" situation may arise. However, it is clear that the introspection of large, non-linear equation systems is non-trivial. Thus, in practical situations, it may be the case that the behaviour of a set of objects and constraints is not as expected, and it is not easy to find out which constraints are responsible for this.

A practical way to approach this problem is to set up constraint equations in a structured fashion so as to maintain a situation where their role is clear and circular dependencies tend to be avoided. In a sense, CSG modelling is such an approach, as CSG primitives (imagine a cube) can be seen as packages containing sets of objects and sets of implicit constraints. The drawback is that these strong constraints cannot be "relaxed" when necessary, since they are an inherent property of the primitives. However, real-world manmade objects often have deviations from the idealised shape, which are cumbersome to model using ideal primitives.

To solve this, Brenner (2004) suggested *weak primitives*. They consist of a set of unknowns and a set of constraints, packaged as a single "primitive". Using a weak primitive is the same as using a "real" (CSG) primitive, although the constraints are enforced internally by equations rather than by an implicit model. The difference becomes obvious when constraints are to be relaxed. Weak primitives allow constraints to be

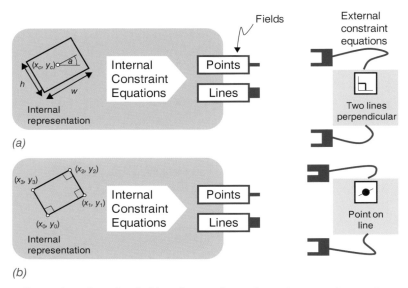

Figure 5.16 Illustration of weak primitives in two dimensions. A rectangle may be represented either (a) by its minimal, five-parameter representation, or (b) by four points and appropriate constraints. In each case, its fields allow access to all (four) points and lines, making it possible to connect external constraints. The field types allow for type checking.

"switched off" and thus are able to deviate from their ideal shape. In order to enforce regularities between different primitives, additional constraints have to be imposed. To achieve this, weak primitives expose an interface consisting of fields. Fields of different primitives can be connected by constraints, where the connections themselves allow for type checking (Figure 5.16). The use of weak primitives, primitive fields and connecting constraints has been inspired by scene graph concepts from computer graphics [Wernecke, 1994; OpenSceneGraph, 2008], where the corresponding items are termed node kits, node fields and engines, respectively.

5.4.5 Reconstruction and generalisation

Apart from supporting the extraction process, a strong model is also important for the downstream usability of the data, especially for the automatic derivation of coarser levels of detail from detailed models (a process called generalisation). Being able to deliver different LODs tailored to different customers' needs, to context-adapted visualisations, such as on mobile displays, or simply to cut down rendering time of large models is essential for 3D models to enter the market. The Sig3D group has defined five levels of detail for building models: LOD 0 is essentially a 2.5D DTM; LOD 1 represents buildings as simple blocks; LOD 2 adds roof structure and texture; LOD 3 consists of architectural models with detailed wall and roof structures; and LOD 4 adds interior structures such as rooms, doors, stairs and furniture [Kolbe *et al.*, 2005]. However, the definition of discrete LODs does not imply any path for deriving one level from the other in an automated way.

Automation of (manual) map generalisation procedures has been a topic in cartography for several decades. There are now operational systems available, which usually start from a scene description in the form of 2D primitives such as polygons or polylines. From this, implicit relationships are discovered, such as adjacency, parallel and rectangular structures, distances, protrusions, etc., which are to be modified or preserved during generalisation. The final outcome is again a description of the objects in terms of their geometry alone. Since the discovered structures are not being made explicit, they cannot be modified, which frequently leads to the need to check and correct the outcome of the automatic generalisation step manually.

In cartography, methods have been developed which aim at the recognition of important structures that are needed as a basis for generalisation, e.g. parallelism, linear arrangement, and clusters [Anders and Sester, 2000; Christophe and Ruas, 2002]. However, experience shows that generalisation is already a difficult problem in 2D and gets much worse in 3D. Kada (2005) uses space partitioning to obtain a generalised building volume from a CAD model. In contrast to this strictly geometric approach, Thiemann and Sester (2004) first try to interpret the scene by identifying prominent substructures such as chimneys (protruding) or windows (intruding). In a second step, these structures can be selectively removed in order to obtain a generalised building description.

However, it seems that the key to successful generalisation systems lies in the modelling itself. Instead of trying to generalise a purely geometric structure, one should rather start from a semantically rich description. For example, in 2D, Brenner and Sester (2005) apply the "weak primitive" approach to generalisation and extend it by containers and discrete behaviour. Primitives are defined as the combination of geometric description (for example polygons), sets of constraints (for example all line segments aligned horizontally or vertically), and discrete behaviour (for example boundary simplification rules). Containers provide the ability to spatially layout primitives, with dedicated interface slots which allow connection of primitives to containers. An even more powerful concept would be to exploit descriptions based on formal languages (discussed in Section 5.3.3) for generalisation.

5.4.6 Lessons learnt

Knowledge and learning are arguably the most critical aspects of today's reconstruction systems. Often, systems rely heavily on parameters and weights, which are either assumed, or determined experimentally while the system is developed. Thus, the required knowledge is not derived in a systematic manner. It is also usually not separated properly from the inference mechanism but rather both (mechanism and knowledge) are included in one monolithic system. One major problem is that there is no database consisting of thousands of buildings, each annotated with the correct reconstruction result, from which such knowledge could be derived. If there was one, it would probably only be for one special type of modelling, for example polyhedral geometric representations, and a certain generalisation level. This makes it much harder than, for example, the case of handwritten character recognition, where such databases have been used extensively.

Reliability and reliability indicators are perhaps the second most critical points. Automatic systems nowadays do not reach the reliability which would be required to integrate them into geodata production lines. The situation would be not so bad if it was possible to clearly see when the system succeeded and when it failed. A 90% success rate would then still yield substantial productivity increases, pointing an operator only to the cases where the automatic procedure failed. However, this has not yet been achieved, which is one reason why semi-automatic systems are not prevalent in geodata production.

The machinery used for interpretation can be drawn from any general technique used in computer vision and artificial intelligence. For example, evidence theory, fuzzy logic, Bayes nets, Markov chain Monte Carlo, simulated annealing, or artificial neural networks may be used [Russell and Norvig, 2002]. However, existing systems are often found to use many heuristics which are derived from special, known properties of the data or the model. Also, it should be kept in mind that, in general, recovering the building structure is a search problem (i.e. where a search space is to be explored).

Existing geodata are often considered to be useful as part of a data fusion reconstruction approach. However, although "more information" should lead to better results, there are some obstacles for practical systems. First, a system relying on multiple data sources can only be applied if they are all available, so it is more limited in applicability. Second, existing data always have some deficiencies, for example data are inaccurate or incomplete, or multitemporal data depict different object states. GIS data may be generalised or may not exactly reflect the required information. For example, cadastral maps usually contain ground plans which are in most cases different from the gutter lines which can be obtained from aerial observation. Third, there are also practical reasons which hinder the use of existing geodata, such as lack of producers' coordination, missing data, warehouse/geodata infrastructure, or prohibitive pricing [Baltsavias, 2004]. There is of course hope that this will change when geodata infrastructure becomes more developed.

Data fusion is a term used for any system which uses heterogeneous input data in some way. However, it is very complex from a technical viewpoint, since a weighting between data sources has to be conducted. Today, data fusion is often carried out on a much more discrete (and simple) level. For example, lines are extracted from a first dataset (for example a laser scan) and subsequently refined in position using a second dataset (for example an aerial image).

Bottom-up systems often run the risk of accumulating many thresholds on the way up. Considering the example above, one would grow extracted points to lines if they are within a certain distance and form a certain maximum angle. One would then group lines into rectangles according to certain buffer, angle and minimum line length constraints, etc. In many cases, controlling all the thresholds becomes intractable, especially if it is not very clear how intermediate products (like lines) influence the result; for example, should there be fewer or more (but less reliable), shorter or longer lines for the subsequent steps to succeed?

The focus of a system should be stated clearly, especially in research. Is it targeted at the recognition process itself, i.e. does it explore how a performance similar to humans can be reached, using the same input data? Or is its main purpose to engineer a technical system, which probably uses much more and/or different data than a human would need to derive the same results? Related to this is the question of whether the best available input data are used. Often, especially in research, outdated data are used due to budget constraints. However, technical progress has yielded higher laser pulse rates and multiple echo detection, so that much denser point clouds essentially come free, easing interpretation tasks substantially. It is also beneficial if it is clearly stated which stage of the reconstruction chain is addressed by a proposed algorithm, be it detection, topology building, estimation, regularisation, reconstruction or visualisation.

Updating has not been fully treated so far. 3D city models are often updated by a complete reacquisition, basically at initial acquisition costs. As Vosselman *et al.* (2004) have shown, change detection based on laser scanning is highly feasible. However, it also became clear that some erroneous results had been caused by the object selection rules on which the cadastral map (to be updated) was based. In general, the existing data to be updated should be in an appropriate representation. For example, it may be useful not only to represent where buildings have been detected, but also where trees were found which can be expected to change in height over time. As for the updating of individual building structures, the area which was assigned to a certain face in the reconstructed geometry could be memorised. Updating may therefore be more successful if a description of the "world state" is kept which is richer than just the reconstructed buildings. Also, consistent upgrading is a topic which is closely related to generalisation (Section 5.4.5).

Systematic evaluation of reconstruction approaches is important and has recently been reported by Kaartinen and Hyyppä (2006). The investigation included photogrammetric as well as laser scanning methods. Assessment was performed using reference points and ground plans. For the points, deviations of location, length and roof inclination were determined. The reference ground plans were rastered and area differences between ground plans and reconstructed building outlines were computed. The major findings were that (1) laser scanning yielded four times more errors than photogrammetric modelling; (2) laser scanning methods had a higher automation level so that the time spent was only 5% of the time spent for photogrammetric measurement; (3) laser scanning is good at height and roof inclination determination, whereas photogrammetry is good at measuring the outline; and (4) laser scanning point density has a significant impact on the result. One major problem is the assessment of the reconstruction quality beyond point measurements and area differences. Ragia and Winter (2000) and Lahamy (2008) present more complex criteria in the 2D case, for example there is a difference if the (scalar) deviation of two volumes is due to a small displacement affecting the entire object surface rather than to a large deviation which is confined to only a small surface area.

5.5   Data exchange and file formats for building models

Future geodata products will be combined from different vendors using a geodata infrastructure, web feature services and web processing services to a much greater extent than is currently the case. Demand for the interoperability of such data is rising.

Modelling 3D buildings touches several areas, each of which has established its own standards [Kolbe *et al.*, 2005]. GIS have their strengths in the management and representation of large area, georeferenced datasets. After being targeted mainly at 2D information for many years, they can nowadays be used to represent and visualise 3D information. A widely used GIS file format (and pseudo standard) is the "shapefile" format from ESRI, Inc., which allows 2D and 3D geometry and associated attributes to be represented. GIS do not have the same 3D modelling possibilities as CAD systems, which have their own file formats, for example the well-known drawing interchange format DXF. Computer graphics still use other tools and formats, such as VRML97, X3D and COLLADA.

In these formats there is mostly a lack of thematic and semantic information. In the area of architecture, engineering construction and facility management (AEC/FM), the international alliance for interoperability has developed the industry foundation classes (IFC; an ISO/PAS 16739 standard), which include semantics. The focus is on detailed building models, especially for the purposes of construction and maintenance, however lacking spatial objects such as streets, vegetation and water bodies. Kolbe *et al.* (2005) present CityGML, which implements an interoperable, multifunctional, multiscale and semantic description for 3D city models. It concentrates on a multi level-of-detail representation of buildings, but also contains other relevant objects, such as streets, traffic lights and vegetation. CityGML is a profile of GML3 and has been adopted as an official standard by the open geospatial consortium (OGC).

References

Aichholzer, O., Aurenhammer, F., Alberts, D. and Gärtner, B., 1995. A novel type of skeleton for polygons. *Journal of Universal Computer Science*, 1(12), 752–761.

Akel, N.A., Zilberstein, O. and Doytsher, Y., 2004. A robust method used with orthogonal polynomials and road network for automatic terrain surface extraction from LIDAR data in urban areas. *International Archives of Photogrammetry, Remote Sensing and Spatial Information Sciences*, 35(Part B3), 274–279.

Alexander, C., Ishikawa, S. and Silverstein, M., 1977. *A Pattern Language: Towns, Building, Construction*. Oxford University Press, New York, NY, USA.

Alharthy, A. and Bethel, J., 2002. Heuristic filtering and 3D feature extraction from LIDAR data. *International Archives of Photogrammetry, Remote Sensing and Spatial Information Sciences*, 34(Part 3A), 29–34.

Anders, K.-H. and Sester, M., 2000. Parameter-free cluster detection in spatial databases and its application to typification. *International Archives of Photogrammetry and Remote Sensing*, 33(Part B4/1), 75–82.

Baltsavias, E., Grün, A. and Gool, L.V. (Eds.), 2001. *Automatic Extraction of Man-Made Objects from Aerial and Satellite Images (III)*. Balkema Publishers, Leiden, The Netherlands.

Baltsavias, E.P., 2004. Object extraction and revision by image analysis using existing geodata and knowledge: current status and steps towards operational systems. *ISPRS Journal of Photogrammetry and Remote Sensing*, 58(3–4), 129–151.

Berndt, R., Fellner, D.W. and Havemann, S., 2004. Generative 3D models: A key to more information within less bandwidth at higher quality. Tech. Rep. TUBS-CG-2004-8, Institute of Computer Graphics, Technical University of Braunschweig.

Böhm, J. and Pateraki, M., 2006. From point samples to surfaces – on meshing and alternatives. *International Archives of Photogrammetry, Remote Sensing and Spatial Information Sciences*, 36(Part 5), 50–55.

Brenner, C., 1999. Interactive modelling tools for 3D building reconstruction. In Fritsch, D. and Spiller, R. (Eds.), *Photogrammetric Week 99*, 20–24 September 1999, Stuttgart, Germany, Wichmann Verlag, 23–34.

Brenner, C., 2000a. Dreidimensionale Gebäuderekonstruktion aus digitalen Oberflächenmodellen und Grundrissen. Ph.D. thesis, Universität Stuttgart, Deutsche Geodätische Kommission, DGK Reihe C, Nr. 530.

Brenner, C., 2000b. Towards fully automatic generation of city models. *International Archives of Photogrammetry and Remote Sensing*, 33(Part B3), 85–92.

Brenner, C., 2004. Modelling 3D objects using weak primitives. *International Archives of Photogrammetry, Remote Sensing and Spatial Information Sciences*, 35(Part B3), 1085–1090.

Brenner, C., 2005a. Building reconstruction from images and laser scanning. *International Journal of Applied Earth Observation and Geoinformation*, 2005(6), 187–198.

Brenner, C., 2005b. Constraints for modelling complex objects. *International Archives of Photogrammetry, Remote Sensing and Spatial Information Sciences*, 36(Part 3/W24), 49–54.

Brenner, C. and Sester, M., 2005. Cartographic generalization using primitives and constraints. In *Proceedings of the 22nd International Cartographic Conference*, 9–16 July 2005, La Coruna, Spain.

Christophe, S. and Ruas, A., 2002. Detecting building alignments for generalisation purposes. In Richardson, D. and van Oosterom, P. (Eds.), *International Symposium on Spatial Data Handling*, 9–12 July 2002, Ottawa, Canada. Springer Verlag, Heidelberg, Germany, 419–432.

Clode, S.P., Kootsookos, P.J. and Rottensteiner, F., 2004. Accurate building outlines from ALS data. *Proceedings of the 12th Australasian Remote Sensing and Photogrammetry Conference*, 18–22 October, 2004, Fremantle, Perth, Australia. On CD-ROM.

Darmawati, A.T., 2008. Utilization of multiple echo information for classification of airborne laser scanning data. Masters thesis, International Institute for Geo-Information Science and Earth Observation, Enschede, The Netherlands.

Dick, A., Torr, P., Cipolla, R. and Ribarsky, W., 2004. Modelling and interpretation of architecture from several images. *International Journal of Computer Vision*, 60(2), 111–134.

Douglas, D. and Peucker, T., 1973. Algorithms for the reduction of the number of points required to represent a digitized line or its caricature. *The Canadian Cartographer*, 10(2), 112–122.

Dutter, M., 2007. Generalisation of building footprints derived from high resolution remote sensing data. Diploma Thesis, Technical University of Vienna.

Edelsbrunner, H., Kirkpatrick, D. G. and Seidel, R., 1983. On the shape of a set of points in the plane. *IEEE Transactions on Information Theory*, 29(4), 551–559.

Fischler, M.A. and Bolles, R.C., 1981. Random sample consensus: A paradigm for model fitting with applications to image analysis and automated cartography. *Communications of the ACM*, 24(6), 381–395.

Förstner, W., 1995. Mid-level vision processes for automatic building extraction. In Grün, A., Kübler, O. and Agouris, P. (Eds.), *Automatic Extraction of Man-Made Objects from Aerial and Space Images, Ascona Workshop*, 24–28 April 1995, Ascona, Switzerland, Birkhäuser, Basel, 179–188.

Früh, C. and Zakhor, A., 2003. Constructing 3d city models by merging ground-based and airborne views. *IEEE Computer Graphics and Applications*, 23(6), 52–61.

Gerke, M., Heipke, C. and Straub, B.-M., 2001. Building extraction from aerial imagery using a generic scene model and invariant geometric moments. *Proceedings of the IEEE/ISPRS joint workshop on remote sensing and data fusion over urban areas*, 8–9 November 2001, Rome, Italy, 85–89.

Green, P.J., 1995. Reversible jump markov chain monte carlo computation and bayesian model determination. *Biometrika*, 82(4), 711–732.

Grimson, W.E.L., 1990. *Object Recognition by Computer*. Series in Artificial Intelligence. MIT Press, Cambridge, MA, USA.

Gross, H., Jutzi, B. and Thoennessen, U., 2007. Segmentation of tree regions using data of a full-waveform laser. *International Archives of Photogrammetry, Remote Sensing and Spatial Information Sciences*, 36(Part 3/W49A), 57–62.

Grün, A., Baltsavias, E. and Henricsson, O. (Eds.), 1997. *Automatic Extraction of Man-Made Objects from Aerial and Space Images (II)*. Birkhäuser, Basel.

Grün, A., Kübler, O. and Agouris, P. (Eds.), 1995. *Automatic Extraction of Man-Made Objects from Aerial and Space Images*. Birkhäuser, Basel.

Haala, N. and Brenner, C., 1997. Interpretation of urban surface models using 2D building information. In Grün, A., Baltsavias, E. and Henricsson, O. (Eds.), *Automatic Extraction of Man-Made Objects from Aerial and Space Images (II)*. Birkhäuser, Basel, 213–222.

Haala, N. and Brenner, C., 1999. Extraction of buildings and trees in urban environments. *ISPRS Journal of Photogrammetry and Remote Sensing*, 54(2–3), 130–137.

Haralick, R., Sternberg, S. and Zhuang, X., 1987. Image analysis using mathematical morphology. *IEEE Transactions on Pattern Analysis and Machine Intelligence*, 9(4), 532–550.

Haunert, J.-H. and Wolff, A., 2008. Optimal simplification of building ground plans. *International Archives of Photogrammetry, Remote Sensing and Spatial Information Sciences*, 37(Part B2), 373–378.

Heuel, S., 2004. *Uncertain Projective Geometry: Statistical Reasoning for Polyhedral Object Reconstruction*. Vol. 3008 of Lecture Notes on Computer Science. Springer, Heidelberg, Germany.

Hoffmann, C.M., 1989. *Geometric and Solid Modeling: An Introduction*. Morgan Kaufmann, San Francisco, CA, USA.

Hopcroft, J.E. and Ullman, J.D., 1979. *Introduction to Automata Theory, Languages, and Computation*. Addison-Wesley Publishing Co., Reading, MA, USA.

Hug, C., 1997. Extracting artificial surface objects from airborne laser scanner data. In Grün, A., Baltsavias, E. and Henricsson, O. (Eds.), *Automatic Extraction of Man-Made Objects from Aerial and Space Images (II)*. Birkhäuser, Basel, 203–212.

Jarvis, R.A., 1973. On the identification of the convex hull of a finite set of points in the plane. *Information Processing Letters*, 2(1), 18–21.

Kaartinen, H. and Hyyppä, J., 2006. Evaluation of building extraction. Tech. rep., *European Spatial Data Research*, Publication No. 50.

Kada, M., 2005. 3D building generalisation. *Proceedings of the 22th International Cartographic Conference*, 9–16 July 2005, La Coruna, Spain.

Kobbelt, L. and Botsch, M., 2004. A survey of point-based techniques in computer graphics. *Computers and Graphics*, 28(6), 801–814.

Kolbe, T.H., Gröger, G. and Plümer, L., 2005. CityGML – interoperable access to 3D city models. In van Oosterom, P., Zlatanova, S. and Fendel, E. M. (Eds.), *Proceedings of the 1st International Symposium on Geo-information for Disaster Management*, Delft, The Netherlands. Springer Verlag, 883–899.

Kraus, K. and Pfeifer, N., 1998. Determination of terrain models in wooded areas with airborne laser scanner data. *ISPRS Journal of Photogrammetry and Remote Sensing*, 53(4), 193–203.

Lafarge, F., Descombes, X., Zerubia, J. and Pierrot-Deseilligny, M., 2008. Automatic building extraction from DEMs using an object approach and application to the 3D-city modeling. *ISPRS Journal of Photogrammetry and Remote Sensing*, 63(3), 365–381.

Lahamy, H., 2008. Outlining buildings using airborne laser scanner data. Masters thesis, International Institute for Geo-Information Science and Earth Observation, Enschede, The Netherlands.

Lang, F. and Förstner, W., 1996. 3D-city modeling with a digital one-eye stereo system. *International Archives of Photogrammetry and Remote Sensing*, 31(Part B3), 415–420.

Lawson, C.L. and Hanson, R.J., 1995. *Solving Least Squares Problems*. SIAM Society for Industrial and Applied Mathematics, Philadelphia, PA, USA.

Lindenberger, J., 1993. Laser-Profilmessungen zur topographischen Geländeaufnahme. Ph.D. thesis, Universität Stuttgart, Deutsche Geodätische Kommission, DGK Reihe C, Nr. 400.

Maas, H.-G., 1999. The potential of height texture measures for the segmentation of airborne laserscanner data. *Fourth International Airborne Remote Sensing Conference and Exhibition/*

21st Canadian Symposium on Remote Sensing, 21–24 June 1999, Ottawa, Ontario, Canada, Vol. I., 154–161.

Maas, H.-G. and Vosselman, G., 1999. Two algorithms for extracting building models from raw laser altimetry data. *ISPRS Journal of Photogrammetry and Remote Sensing*, 54(2–3), 153–163.

Mallet, C., Soergel, U. and Bretar, F., 2008. Analysis of full-waveform LIDAR data for classification of urban areas. *International Archives of Photogrammetry, Remote Sensing and Spatial Information Sciences*, 37(Part B3a), 85–91.

Matikainen, L., Hyyppä, J. and Hyyppä, H., 2003. Automatic detection of buildings from laser scanner data for map updating. *International Archives of Photogrammetry, Remote Sensing and Spatial Information Sciences*, 34(Part 3/W13), 218–224.

Mayer, H., 1999. Automatic object extraction from aerial imagery – a survey focusing on buildings. *Computer Vision and Image Understanding*, 74(2), 138–149.

Morgan, M. and Habib, A., 2002. Interpolation of LIDAR data and automatic building extraction. *ACSM-ASPRS 2002 annual conference proceedings*, 12–14 November 2002, Denver, CO, USA.

Müller, P., Wonka, P., Haegler, S., Ulmer, A. and Gool, L.J.V., 2006. Procedural modeling of buildings. *ACM Transaction on Graphics*, 25(3), 614–623.

OpenSceneGraph, 2008. http://www.openscenegraph.org (accessed December 2, 2008).

Parish, Y. and Müller, P., 2001. Procedural modeling of cities. In Fiume, E. (Ed.), *Proceedings of ACM SIGGRAPH*, 12–17 August 2001, Los Angeles, CA, USA. ACM Press, 301–308.

Pfeifer, N., Stadler, P. and Briese, C., 2001. Derivation of digital terrain models in the SCOP++ environment. *Proceedings OEEPE Workshop on Airborne Laserscanning and Interferometric SAR for Digital Elevation Models*, 1–3 March, 2001, Stockholm, Sweden, OEEPE Publication No. 40 (on CD-ROM).

Prusinkiewicz, P. and Lindenmayer, A., 1990. *The Algorithmic Beauty Of Plants*. Springer, New York, NY, USA.

Ragia, L. and Winter, S., 2000. Contributions to a quality description of areal objects in spatial data sets. *ISPRS Journal of Photogrammetry and Remote Sensing*, 55(3), 201–213.

Ripperda, N. and Brenner, C., 2006. Reconstruction of facade structures using a formal grammar and rjMCMC. *Proceedings of the 28th Annual Symposium of the German Association for Pattern Recognition, Springer Lecture Notes on Computer Science*, 4174, 750–759.

Rottensteiner, F. and Briese, C., 2002. A new method for building extraction in urban areas from high-resolution LiDAR data. *International Archives of Photogrammetry, Remote Sensing and Spatial Information Sciences*, 34(Part B3a), 295–301.

Rottensteiner, F., Trinder, J., Clode, S. and Kubik, K., 2003. Building detection using LIDAR data and multi-spectral images. In Sun, C., Talbot, H., Ourselin, S. and Adriaansen, T. (Eds.), *Proceedings of the VIIth Digital Image Computing: Techniques and Applications*, 10–12 December, 2003, Sydney, Australia. 673–682.

Rottensteiner, F., Trinder, J., Clode, S. and Kubik, K., 2005a. Automated delineation of roof planes from LIDAR data. *International Archives of Photogrammetry, Remote Sensing and Spatial Information Sciences*, 36(Part 3/W19), 221–226.

Rottensteiner, F., Trinder, J., Clode, S. and Kubik, K., 2005b. Using the Dempster-Shafer method for the fusion of LIDAR data and multi-spectral images for building detection. *Information Fusion*, 6(4), 283–300.

Rottensteiner, F., 2006. Consistent estimation of building parameters considering geometric regularities by soft constraints. *International Archives of Photogrammetry, Remote Sensing and Spatial Information Sciences*, 36(Part 3), 13–18.

Russell, S. and Norvig, P., 2002. *Artificial Intelligence: A Modern Approach*. Prentice Hall, Englewood Cliffs, NJ, USA.

Rutzinger, M., Höfle, B., Hollaus, M. and Pfeifer, N., 2008. Object-based point cloud analysis of full-waveform airborne laser scanning data for urban vegetation classification. *Sensors*, 8(8), 4505–4528.

Sampath, A. and Shan, J., 2007. Building boundary tracing and regularization from airborne lidar point clouds. *Photogrammetric Engineering and Remote Sensing*, 73(7), 805–812.

Secord, J. and Zakhor, A., 2007. Tree detection in urban regions using aerial lidar and image data. *IEEE Geoscience and Remote Sensing Letters*, 4(2), 196–200.

Sester, M., 2005. Optimising approaches for generalisation and data abstraction. *International Journal of Geographic Information Science*, 19(8–9), 871–897.

Sester, M. and Neidhart, H., 2008. Reconstruction of building ground plans from laser scanner data. *Proceedings AGILE*, 5–8 May 2008, Girona, Spain, 11.

Shen, W., 2008. Building boundary extraction based on LIDAR point clouds data. *International Archives of Photogrammetry, Remote Sensing and Spatial Information Sciences*, 37(Part B3b), 157–161.

Thiemann, F. and Sester, M., 2004. Segmentation of buildings for 3D generalisation. *Working paper of the ICA workshop on generalisation and multiple representation*, 20–24 August 2004, Leicester, UK (on CD-ROM).

Tóvári, D. and Vögtle, T., 2004. Classification methods for 3D objects in laser-scanning data. *International Archives of Photogrammetry, Remote Sensing and Spatial Information Sciences*, 35(Part B3), 408–413.

Verma, V., Kumar, R. and Hsu, S., 2006. 3D building detection and modeling from aerial LIDAR data. *Proceedings of the IEEE Computer Society Conference on Computer Vision and Pattern Recognition* (CVPR'06), 2213–2220.

Vosselman, G., 1999. Building reconstruction using planar faces in very high density height data. *International Archives of Photogrammetry, Remote Sensing and Spatial Information Sciences*, 32(Part 3/2W5), 87–92.

Vosselman, G., 2000. Slope based filtering of laser altimetry data. *International Archives of Photogrammetry and Remote Sensing*, 33(Part B3), 935–942.

Vosselman, G. and Dijkman, S., 2001. 3D building model reconstruction from point clouds and ground plans. *International Archives of Photogrammetry and Remote Sensing*, 35(Part 3/W4), 37–44.

Vosselman, G., Gorte, B.G.H. and Sithole, G., 2004. Change detection for updating medium scale maps using laser altimetry. *International Archives of Photogrammetry, Remote Sensing and Spatial Information Sciences*, 35(Part B3), 207–212.

Vosselman, G. and Veldhuis, H., 1999. Mapping by dragging and fitting of wire-frame models. *Photogrammetric Engineering and Remote Sensing*, 65(7), 769–776.

Wagner, W., Ullrich, A., Ducic, V., Melzer, T. and Studnicka, N., 2006. Gaussian decomposition and calibration of a novel small-footprint full-waveform digitising airborne laser scanner. *ISPRS Journal of Photogrammetry and Remote Sensing*, 60(2), 100–112.

Wahl, R. and Klein, R., 2007. Towards semantic interaction in high-detail realtime terrain and city visualization. *International Archives of Photogrammetry, Remote Sensing and Spatial Information Sciences*, 36(Part 3/W49A), 179–184.

Wang, O., 2006. Using aerial LIDAR data to segment and model buildings. Masters thesis, University of California Santa Cruz, USA.

Weidner, U., 1997. Digital surface models for building extraction. In Grün, A., Baltsavias, E. and Henricsson, O. (Eds.), *Automatic Extraction of Man-Made Objects from Aerial and Space Images (II)*. Birkhäuser, Basel, 193–202.

Weidner, U. and Förstner, W., 1995. Towards automatic building extraction from high resolution digital elevation models. *ISPRS Journal of Photogrammetry and Remote Sensing*, 50(4), 38–49.

Wernecke, J., 1994. *The Inventor Mentor: Programming Object-Oriented 3-D Graphics with Open Inventor*. Addison-Wesley, Reading, MA, USA.

Wonka, P., Wimmer, M., Sillion, F. and Ribarsky, W., 2003. Instant architecture. *ACM Transactions on Graphics*, 22(3), 669–677.

Forestry Applications | 6

Hans-Gerd Maas

6.1 Introduction

Airborne laser scanning had already been adopted and accepted as a very valuable tool in forestry applications shortly after its advent as a commercially available measurement technique in the 1990s. The 3D nature of laser scanner data makes it especially suited to applications in forestry, which has a wide range of rather complex 3D modelling tasks in fields such as forest inventory, forest management, carbon sink analysis, biodiversity characterisation and habitat analysis. While photogrammetric image matching based 3D object reconstruction will often fail in forestry applications due to the complexity of the scene and the lack of a continuous object surface, the laser scanner pulse penetration characteristics and the multi-echo recording capabilities of many laser scanner systems allow for high quality 3D representations of forest geometries.

Figure 6.1 shows a profile through an airborne laser scanner dataset, which clearly shows the potential of the technique in capturing 3D forest structure data. The interference of laser scanner pulses and forest canopies is characterised by reflection from the canopy and penetration through the canopy. Typical airborne laser scanner instruments have a beam divergence between 0.25 mrad and 1.0 mrad, resulting in a laser spot diameter of 25 cm to 1 m at a flying height of 1000 m above ground. Modern laser scanner instruments are able to record the time-of-flight for at least two echoes per emitted laser pulse (first and last echo); many current systems record up to four

Figure 6.1 Airborne laser scanner data profile. (Courtesy TU Dresden, Germany)

to five echoes per pulse. In the case of vegetation, most first echoes will come from the surface of the forest canopy. Depending on the laser spot size and the gap structure of the forest canopy, a significant percentage of the laser scanner pulses will penetrate through gaps in the canopy and will be reflected by lower parts of the canopy, the stem or the forest ground (see Figure 6.2). These points, which hit the ground, form a very good basis for the generation of forest region digital terrain models (Section 6.2). The canopy height model, basically obtained as the height difference between the first pulse echo and the digital terrain model, gives valuable information on the forest height structure, which may be used for forest inventory and biomass estimation tasks (Section 6.3). Dense airborne laser scanner data even allow single-tree detection and modelling (Section 6.4). Going well beyond the detection of the first and last pulse echoes, some laser scanner systems offer the option of full-waveform digitisation, generating an echo intensity profile over the entire returned pulse. In forestry applications, these waveforms contain information on the 3D structure of the canopy (Section 6.5).

Conventional procedures for forest inventory data acquisition include rather labour-intensive fieldwork and are restricted to regular visits to a limited number of sample plots, which are used in spatial extrapolation schemes. Among other parameters, data acquisition in these plots includes the counting of trees, the measurement of tree height and diameter at breast height, and the estimation of timber volume. Maltamo *et al.* (2007) report a prediction accuracy of 15–30% in fieldwork-based forest inventory techniques. If it is possible to extract some of these parameters automatically from airborne laser scanner data, the technique represents quite a valuable source for full-field (rather than plot-based) forest inventory data acquisition, supporting or

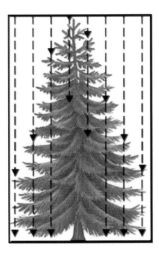

Figure 6.2 First and last echoes.

even replacing conventional techniques in some applications. Due to the 3D nature of the data, airborne laser scanning has a much bigger potential here than airborne or satellite imagery.

The convincing capabilities of airborne laser scanning in forest areas have gained special attention in recent years in the context of global carbon cycle studies. As above ground biomass (the total amount of oven-dried biological material present above the soil surface in a specified area) is approximately 50% carbon, there is a clear need for techniques to efficiently and reliably quantify biomass and biomass changes in forests on both local and global scales. Generally, young (growing) forests depict a carbon sink, while old primary forests are either in carbon equilibrium or depict a carbon source if they reach a state of mortality. Conventional biomass estimation techniques are based on destructive sampling and interpolation schemes. Field studies often use allometric relationships between total biomass and tree height and stem diameter. Airborne or satellite 2D image-based techniques can only capture the 2D structure of forests; they depend on empirical parameters such as leaf area index (LAI), woody area index (WAI) and plant area index (PAI) to extrapolate biomass estimation parameters. Synthetic aperture radar (SAR) has also been used, but tends to saturate in dense forest conditions [Drake *et al.*, 2002]. Airborne laser scanning, combined with sophisticated data-processing schemes, has the potential to overcome these problems and to deliver precise and reliable large full-field 3D forest structure data.

Techniques developed for the determination of forest inventory parameters from airborne laser scanner data are generally differentiated into stand-wise methods, which generate average height values and a statistical canopy height description over forest compartments (stands) with an area of typically 1–10 ha, and tree-wise methods, which attempt to extract, model and analyse individual trees. Early airborne laser scanner systems with pulse rates of 10 000 pulses per second or less, resulting in a point spacing of 2–4 m, were limited to stand-wise approaches. Modern systems with pulse rates of 100 000 pulses per second and more, which may deliver more than one point per square metre, also support tree-wise approaches. While forest management and planning data are often based on stands rather than on individual trees in order to keep the amount of data manageable, some forest inventory and biodiversity analysis tasks require 3D data for individual trees. Generally, stand-wise methods apply tools from geostatistics to extract parameters for larger areas, while tree-wise methods rely heavily on image processing and image analysis techniques applied to point clouds or height data interpolated to a raster representation.

In addition to airborne laser scanning, terrestrial laser scanning provides a powerful technique for the capture of very dense 3D representations of the geometry of trees and local forest structure. Techniques for the automatic extraction of tree geometry parameters and results from terrestrial laser scanner based forest inventory pilot studies are shown in Section 6.7.

6.2   Forest region digital terrain models

A first and obvious application of airborne laser scanning was the determination of digital terrain models in forest regions. Until the advent of airborne laser scanning, terrain models were mainly generated by stereo photogrammetric techniques. Until the late 1970s, analogue stereo plotters were used, giving terrain models represented by contour lines, which were measured by a human operator. Analytical plotters allowed for the generation of digital terrain models in grid, TIN (triangulated irregular network) or breakline representations. In the mid-1980s, stereo image matching techniques became available, which enabled a much more efficient automatic generation of digital terrain models. All these photogrammetric terrain model generation techniques will regularly fail in forested regions. The reason for this is the decorrelation of terrain surface patches in images taken from different perspectives, which is caused by the 3D structure of the vegetation. As terrestrial tacheometry is usually too expensive to be an alternative for data capture in larger areas, forest region terrain model generation was basically an unsolved problem until the advent of airborne laser scanning, and forest region DTMs were often of a relatively poor quality.

Airborne laser scanning as an active 3D measurement technique shows some very clear advantages here. In a typical airborne laser scanner dataset, most pulse first echoes will come from the upper parts of the forest canopy. Many pulses will, however, partly penetrate through the canopy to the ground. If a laser scanner is able to record the first and the last echo of a laser pulse or multiple echoes, the last echo has a good chance of reaching the forest ground. Hoss (1997) reports a laser scanner pulse penetration rate of 30–65% for mid-European forests. Kraus and Pfeifer (1998) showed that this penetration rate is sufficient to reliably eliminate the remaining pulses in an automatic filtering approach and to generate a high quality digital terrain model of the forest ground. Various methods for filtering have been discussed in Chapter 4. Reutebuch *et al.* (2003) report that DTM errors show only a slight increase with canopy density.

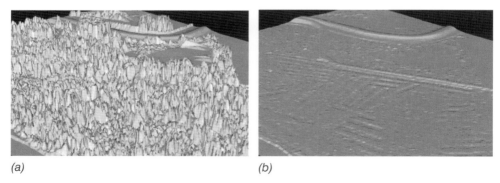

(a)   *(b)*

Figure 6.3 Success of laser scanner data filtering for digital terrain model generation in a forest area. (Vosselman, 2000; courtesy George Vosselman)

Pfeifer *et al.* (1999) show a general degradation of the height accuracy of forest area digital terrain models with increasing terrain slope – an effect which might also be caused by planimetric errors in the laser scanner data. Vosselman (2000) shows the potential of slope-based filtering techniques for the generation of forest area digital terrain models from dense airborne laser scanner data; his results (Figure 6.3) even unveil a drainage channel pattern in a forest. In deciduous forests, the penetration rate can be optimised by acquiring data during the leaf-off season. Only in very dense coniferous forests or in multiple canopy tropical rain forests will the pulse penetration rate be insufficient to generate a digital terrain model of the forest ground.

6.3 Canopy height model and biomass determination

The capability of most airborne laser scanner systems to record the first and last echo of an emitted laser pulse (or multiple echoes) forms a very good basis for the generation of canopy height models and depicts a starting point for powerful forest structure analysis techniques. In some ways, these applications, which try to extract forest geometry information from laser scanner data, can be considered complementary to digital terrain model generation (Chapter 4), which is also termed "virtual deforestation" [Haugeraud and Harding, 2001].

In an idealised case, the first echoes in an airborne laser scanner dataset over a forest region form a surface model of the canopy, while the last echoes, post-processed by filtering to exclude non-terrain points, form the terrain model. A canopy height model (CHM) can then be obtained by subtracting the digital terrain model from the canopy surface model. Such a canopy height model forms a height representation of a forest region in a raster or TIN height data format, which does not specify heights or even geometric models for individual trees. Related to the boundaries of forest stands, the CHM may form a valuable source of information for forest inventory, management and planning tasks.

Airborne laser scanner datasets are usually characterised by their point spacing, which is larger than the actual laser spot size on the ground. As a consequence, laser scanning has to be considered an undersampling of the surface. In the case of forest canopies, this may lead to the average stand height being underestimated, as tree tops are likely to be missed as a consequence of the undersampling nature of airborne laser scanning. Heurich *et al.* (2003) performed a study on undersampling-induced systematic underestimation of tree heights. For a sample of 1000 trees in the Bavarian Forest, they determined a tree height underestimation of 53 cm (79 cm for coniferous trees, 37 cm for deciduous trees) with a tree height standard deviation of 90 cm. The magnitude of the underestimation effect will largely depend on the scanning parameters (mainly point density and beam divergence) and on the specific geometry of different types of trees. The effect will usually be larger for coniferous trees such as fir or spruce. In deciduous forests, seasonal aspects have to be considered. Using this knowledge, a correction term can be developed to be applied to the results of canopy model analysis. In addition to these undersampling-induced effects, canopy height underestimation may also be caused by wind bending, the penetration of the laser pulse into the crown or by dense ground vegetation, which is often difficult to detect in filtering processes.

Using empirical forest research computation models, digital canopy height models can be used to estimate forest stand parameters such as average tree height, basal area (stem area per unit area), timber volume or biomass, which are often used in forest inventory, planning and management schemes. Most of the research work in this field is based on regression methods or other statistical approaches applied to the laser scanner data. MacLean and Krabill (1986) showed, on the basis of laser profiler data, that the cross-sectional area of the forest canopy is directly related to the logarithm of the timber volume. Many researchers have applied quantiles or percentiles of the height distribution in cells of laser scanner data as predictors in regressions models to estimate parameters such as average tree height, basal area and timber volume [Hyyppä *et al.*, 2004]. Early practical tests on the applicability and the information content of airborne laser scanning canopy height models were shown by Næsset (1997), who analysed the accuracy of stand-wise average tree height determination obtained from quantiles of laser scanner height data cells. Holmgren and Jonsson (2004) report on the first large-scale forest inventory laser scanning project in Sweden, covering an area of 50 km^2 including 29 validation stands with ground reference. The point density in the dataset was 1.2 points/m^2. They applied a two-stage procedure, where objectively field-measured sample plots are used for building regression functions with variables extracted from the laser canopy height distribution; these functions are then applied for predictions covering the whole area. Separate function parameters for pine, spruce and deciduous trees were derived from training plots. The results show a relative RMS error at stand level of 5.0% (80 cm) for mean tree height, 9% (1.9 cm) for mean stem diameter, 12% (3.0 m^2/ha) for basal area and 14% (28 m^3/ha) for stem volume estimations. The results are consistent with several other studies on stand-wise forest parameter estimation. They imply that estimations of forest variables based on automatic airborne laser scanner data processing have higher accuracies than traditional methods.

Acquiring and processing multitemporal data, dendrochronology parameters can be determined. Obvious applications are forest growth monitoring or the detection of harvested trees. Gobakken and Næsset (2004) analysed forest growth from two airborne laser scanner datasets, which were acquired two years apart. Using a quantile and mean height based plot-wise analysis technique, they could detect growth with sufficient significance. Generally, the quality of multitemporal laser scanner data analysis will depend on data acquisition with identical or at least similar systems and scanning parameters.

Modelling the relationship between biomass and carbon (or biomass increase and CO_2 absorption), even quantitative statements on the CO_2 storage capacity of forests can be derived. Laser scanning therefore represents an interesting technique for meeting obligations under Article 3.3 of the United Nations Kyoto Protocol, where carbon credits can be earned from net carbon accumulated in forests planted in non-forest land after 1990 ("Kyoto forests") [Stephens *et al.*, 2007]. Article 3.3 requires annual reporting of carbon stock changes arising from land use, land-use change and forestry (LULUCF) activities. Good Practice Guidance for LULUCF activities requires carbon stock changes to be estimated in an unbiased, transparent and consistent manner. Many studies have been performed to verify the agreement between laser scanner data derived parameters and field measurements. Lefsky *et al.* (2001) present a very simple regression model between lidar-derived mean canopy height squared and biomass, which has been tested in three different types of forest and delivered an R^2 of 84% between biomass derived from laser scanner measurements and field measurement schemes. Stephens *et al.* (2007) report investigations to confirm the relationship between lidar variables and forest carbon, height, basal area and age at plot scale. They demonstrate that a simple heuristic height percentile analysis based lidar data processing method is able to provide estimates of total carbon per plot ($R^2 = 0.80$), mean top height ($R^2 = 0.96$), basal area ($R^2 = 0.66$) and age ($R^2 = 0.74$).

St. Onge and Véga (2003) present a sensor fusion approach for canopy height model generation, where the terrain model is obtained from airborne laser scanner data, while the crown model is obtained from aerial stereo image matching. They show that the crown height model obtained from stereo matching is a smoothed version of a laser scanner generated crown height model. An obvious advantage of using images to obtain the crown height model can be seen in the fact that it opens up the possibility of performing long-term retrospective studies by using archived aerial imagery, which may date back to the 1920s. Assuming that there are no significant temporal changes in the terrain model, a terrain model generated from one laser scanner flight can be subtracted from time-resolved crown height model series obtained from (cheaper) aerial imagery. Andersen *et al.* (2003) show results from a comparison of high-density airborne laser scanning and airborne X-band in SAR (synthetic aperture radar) for the generation of canopy- and terrain-level elevation models in a Douglas-fir forest. Their results indicate that both technologies can provide useful terrain models, but airborne laser scanning performs much better in the representation of fine detail in the canopy.

6.4 Single-tree detection and modelling

Beyond raster or TIN structure canopy height models, airborne laser scanner datasets with point densities of 1 point/m^2 or denser offer the possibility of detecting, segmenting and modelling individual tree crowns. Such a tree-wise reconstruction method clearly has greater potential as a basis for forest inventory and forest management tasks, but it also opens up new possibilities in applications such as biodiversity or habitat analysis. Single-tree detection and 3D reconstruction allow counting of trees in a certain area and determination of parameters such as individual tree height, crown diameter, crown area and 3D crown shape. Researchers have attempted to derive other forest inventory relevant parameters such as timber volume, diameter at breast height, basal area or leaf area index (the total area of the leaves, divided by the crown area) from these parameters by applying empirical models. Together with complementary information such as multispectral image data, it is even feasible to automatically recognise tree species and age.

A single-tree detection and segmentation scheme, based on a raster representation of airborne laser scanner data, is shown in Figure 6.4. The raster data are obtained by an interpolation of the first pulse laser scanner data. In a first step, a local maxima search is performed. Proceeding from a local maximum, segmentation is performed by defining height profiles in eight, 16 or 32 directions. The heights on these profiles are analysed for local minima (watersheds to neighbouring tree crowns) or sudden drops (crown boundaries without direct neighbours). Disturbing effects of single branches can be removed by filtering of the raster data or the profile. A polygon through the x,y-coordinates of these profile points defines the 2D projection and the area of the crown. Similarly, a region-growing approach in a height gradient image can be used for segmentation. Region growing can basically also be applied to raw laser scanner data in a TIN structure, but a greater implementational and computational effort is required. The 3D crown shape may be defined by a 3D polyhedron, obtained from the original 3D point cloud within the segmented crown area, or by an ellipsoid or a polynomial function fitted into the point cloud. The success and quality of the results of the approach depend on the laser scanner dataset point density as well as on the size, shape and distribution of trees. It will usually not be able to detect understorey trees or small trees standing close to large trees. As in stand-wise

Figure 6.4 Single-tree detection and segmentation scheme. (Courtesy TU Dresden, Germany)

canopy height determination applications, tree height is likely to be underestimated. Crown diameters will also often be underestimated due to the neglect of crown overlaps and interference in the watershed technique.

Procedures similar to the one described above have been applied in a number of studies. Hyyppä and Inkinen (1999) were among the first to apply single-tree detection to airborne laser scanner data. In practical studies, they reached a standard deviation of about 10% for the determination of stand volume and an RMS of 98 cm for tree height determination, while the proportion of detected trees was only about 40%. This was explained by the multilayered and unmanaged stand structure of the study area. Persson *et al.* (2002) show the results of a study in boreal forest in southern Sweden containing 795 spruce and pine trees with available ground truth. A laser scanner dataset with 5 points/m^2 was used. 71% of all trees and 90% of the trees with a stem diameter >20 cm were detected; the average difference of stem positions was 51 cm. They achieved a correlation of 99% and a standard deviation of 63 cm (with a 1.13 m underestimation) between tree heights obtained from airborne laser scanner data and field measurements. For the crown diameter, a correlation of 76% and a standard deviation of 61 cm were achieved. Somewhat worse results were published in a study by Heurich and Weinacker (2004) based on a total of 2666 spruce and beech trees with available ground reference in 28 plots ranging from simple single-layer single-species spruce stands to very complex multilayered mixed stands. They achieved an overall detection rate of 44%. The detected timber volume was 85% of the total volume measured in the field. The RMS of tree height determination was 1.40 m with an underestimation of 60 cm for coniferous and 23 cm for deciduous trees. The RMS of crown radius determination was 0.93 m. Generally, the results for coniferous trees were better than those for deciduous trees. Wack *et al.* (2003) use airborne laser scanner data for forest inventory in eucalyptus plantations under the aspect of optimised resource management for the paper industry. Due to the rather narrow shape of eucalyptus tree crowns, they limit their study to the detection of trees and tree height determination. In a test area, they achieved 93% correct detection rate and a standard deviation of 1.24 m for tree height determination. Morsdorf *et al.* (2003) use a *k*-means clustering to delineate single trees from laser scanner raw data and validate their approach with a dataset of about 2000 georeferenced trees. They obtained an RMS of 0.60 m for height accuracy with an underestimation of 0.71 m.

Yu *et al.* (2004) address forest growth and harvesting analysis from multitemporal laser surveys. Sixty-one out of 83 field-checked harvested trees could be detected, including all mature trees. Height growth could be determined with a precision of 5 cm at stand level and 10–15 cm at plot level. Yu *et al.* (2005) show that individual tree growth can be determined at an RMS of 0.45 m, which is better than the actual tree height determination accuracy. The fact that the relative accuracy of tree height determination is better than the absolute accuracy corresponds to expectations. It can be explained by deficiencies in the mathematical and stochastic model of tree detection, segmentation and modelling, as well as by effects caused by sensor characteristics, which are correlated and thus partly compensated for when calculating differences.

6.5 Waveform digitisation techniques

So far, most laser scanner systems use discrete return pulse detection techniques delivering either one (first or last) echo, a first and a last echo, or four to five echoes per emitted laser pulse. The pulse detection is performed by signal analysis in the receiver part of the instrument. A straightforward extension of this discrete return detection is the digitisation of the complete echo waveform by a digitiser operating at a very high temporal resolution (Figure 6.5). This waveform digitisation delivers a full intensity profile of the echo of a laser pulse. Besides this full profile information, waveform digitisation may potentially give more user control on range determination.

Waveform digitisation comes with severe hardware requirements. To achieve, for instance, 10 cm vertical resolution in the waveform, the returning pulse echo has to be sampled at 1.5 GHz (0.66 ns sampling interval).[1] As real-time processing will not be possible or desirable in many cases, the full digitised waveform has to be stored, drastically increasing the total amount of data. Assuming 40 m tree height and 10 cm

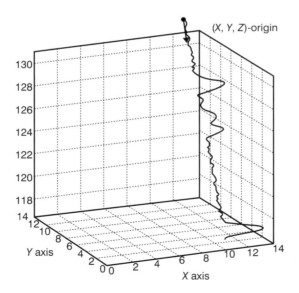

Figure 6.5 128 sample 8-bit laser scanner waveform. (Persson *et al.*, 2005; courtesy the authors)

---

1 Including a factor of two for the round trip time of the laser pulse.

waveform resolution, 400 samples have to be stored for each laser pulse. The first commercial airborne full-waveform laser scanner system (LiteMapper-5600 based on the Riegl LMS-Q560 laser scanner) was introduced in 2004 [Hug *et al.*, 2004]. Its 1 ns waveform sampling offers a vertical profile sampling distance of 15 cm.

While full-waveform digitisation and analysis were first used in hydrographic applications [e.g. Gutelius, 2002], its greatest benefit can be seen in the field of forestry research, as waveform data can provide spatially resolved horizontal and vertical geometric information on trees. It will be especially beneficial in biomass and forest structure estimation techniques, where conventional multipulse methods may be biased due to their limited understorey tree detection capabilities.

Some early efforts to use full-waveform digitisation laser scanner systems in forestry research have been shown by NASA [e.g. Blair *et al.*, 1999]. While most laser scanner systems work with small footprints (narrow beam divergence), their LVIS (laser vegetation imaging sensor) uses a widened beam with a footprint of typically 10–25 m. The entire time history of the outgoing and return pulses is digitised. In combination with the large footprint, the technique shifts the focus from horizontal to vertical resolution in airborne laser scanner data. The return waveform gives a detailed record of the vertical distribution of nadir-intercepted surfaces (i.e. leaves and branches). For surfaces with similar reflectance, the amplitude of the waveform, corrected by attenuation effects, is a measure of the amount of canopy material. Drake *et al.* (2002) developed a processing scheme to determine canopy height, height of median energy, height/median ratio and ground from LVIS waveform data over tropical forest as input into a biomass determination scheme. They achieved a coefficient of determination $R^2 = 0.93$ between biomass derived from the waveform data and field measurements.

Figure 6.6 [Persson *et al.*, 2005] visualises 3D tree structure results for pine and spruce trees. The data were obtained from a TopEye Mark II small footprint waveform digitisation system operating at 50 kHz pulse frequency and 1 GHz 8-bit waveform digitisation. Waveforms were represented with 128 samples and visualised by transforming colour-coded waveform amplitude information into a 3D voxel space at 15 cm discretisation.

Persson *et al.* (2005) process the data by fitting Gaussian distribution functions into the waveform data and introducing the positions of up to eight significant peaks as additional 3D points. A result of the Gaussian decomposition is shown in Figure 6.7.

The approach of Gaussian waveform decomposition, treating the echo of a laser pulse as the sum of a limited number of single echoes reflected from individual scatterers, is also used by Wagner *et al.* (2006). Modelling the waveform by a superimposition of basis functions has clear advantages with respect to data storage aspects. The primary result of the decomposition is the number and position of these scatterers in the digitised waveform. In addition, the decomposition delivers information on the echo width and intensity.

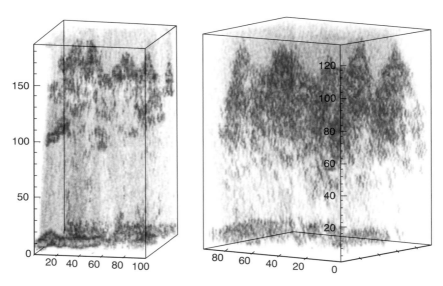

Figure 6.6 128 sample laser scanner waveform visualisation. (Persson *et al.*, 2005; courtesy the authors)

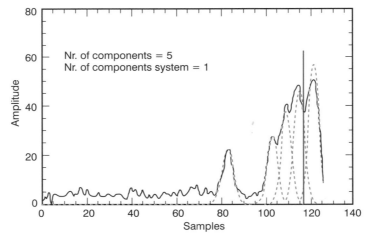

Figure 6.7 Results from waveform analysis: original waveform (black), five echoes from Gaussian decomposition (red), original system last echo (blue). (Persson *et al.*, 2005; courtesy the authors)

Beyond this multi-echo detection capability, full waveforms may also deliver valuable information on low vegetation. Low vegetation can often not be detected in discrete pulse systems due to the finite length of laser pulses. First and last pulses of vegetation lower than about 1.5 m will be detected as a single pulse, leading to errors

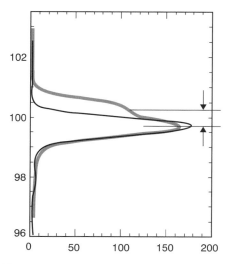

Figure 6.8 Waveform of flat ground (black) and ground with low vegetation (gray). (Hug *et al.*, 2004; courtesy the authors)

in digital terrain models and vegetation height models. Using the return pulse profile information from waveform digitisation, low vegetation can be recognised and measured from the width and shape of the waveform down to a vegetation height of about two to three times the profile sampling distance (see Figure 6.8) [Hug *et al.*, 2004].

The waveform may also deliver information on ground surface roughness. Jutzi and Stilla (2005) use a combination of correlation and Levenberg–Marquardt methods to exploit the shape of a waveform in the determination of surface range and roughness.

6.6 Ecosystem analysis applications

Beyond the more geometry-related approaches of airborne laser scanning in forestry discussed above, a number of studies on using laser scanner data in general ecosystem analysis have been presented. These applications address aspects such as biodiversity or CO_2 efficiency of forests. Biodiversity parameters such as tree species, height and age distribution, plant area index and gap fraction can be determined on the basis of single-tree detection and modelling.

Reitberger *et al.* (2006) used point clouds, which were enriched by additional points obtained from waveform analysis, to distinguish between coniferous and deciduous trees by analysing crown shape parameters. Characteristic shape, structure and intensity measures were used to classify trees. The overall classification accuracy was 80%. Ørka *et al.* (2007) investigate the possibility of tree classification by analysing the mean laser pulse intensity and the standard deviation of intensity. They achieve overall classification accuracies of 68–74% in a study area containing spruce, birch and aspen trees.

Several authors have developed schemes for the determination of canopy closure or canopy gap fraction parameters from airborne laser scanner data. Canopy gap fraction can be used in solar radiation determination for silvicultural analysis with the goal of quantitatively evaluating the growth chances of ground vegetation (e.g. tree regeneration) in forest ecosystems. Here, airborne laser scanning can be considered to be a full-field alternative to local terrestrial hemispheric image analysis techniques (Figure 6.9). Solberg *et al.* (2006) and Morsdorf *et al.* (2006) make use of the ratio of below canopy returns to total returns in a laser scanner dataset and show that it can be used as an indicator for canopy gap fraction. Hopkinson *et al.* (2007) used pulse intensity information to achieve a slight improvement in the precision of canopy gap fraction and fractional coverage.

Laser scanner data may also be used in forest fire fighting by delivering maps of the distribution of canopy fuels and supporting models to predict fire behaviour over the landscape [Andersen *et al.*, 2005]. Besides the total canopy mass in an area, these forest fire fighting relevant parameters may also include spatially explicit forest structure information such as canopy bulk density (the mass of available canopy fuel per canopy volume unit), canopy height and canopy base height.

Bird species diversity is affected by the vertical and horizontal diversity of forest structure. Hashimoto *et al.* (2004) discuss techniques to derive forest structure indices, which can be useful for wildlife habitat assessment, from airborne laser scanner data.

Chasmer *et al.* (2007) use canopy height models obtained from laser scanning to analyse whether an ecosystem is a net sink or source of CO_2. They test the hypothesis that vegetation structural heterogeneity has an influence on CO_2 fluxes. Their results

Figure 6.9 Canopy gap fraction in hemispherical image. (Schwalbe *et al.*, 2006; courtesy the authors)

indicate that the structural variability of the vegetation has a significant though small effect on the variability in CO_2 flux estimates.

6.7   Terrestrial laser scanning applications

While airborne laser scanning is a technique well suited to determining stand- or tree-wise parameters such as height or crown area and to deriving information on timber volume and biomass, it is not very well suited to detailed modelling tasks, due to its point spacing and perspective. As a supplement, terrestrial laser scanning may form a valuable tool for the acquisition of very dense 3D point clouds, which allow for a much more detailed geometric description. Obviously, terrestrial laser scanning is limited to measurement in smaller regions.

Figures 6.10 and 6.11 give an impression of the information content of terrestrial laser scanner point clouds.

(a)   *(b)*

Figure 6.10 Samples of terrestrial laser scanner point clouds of trees: (a) thinned 2D point cloud projection; and (b) laser point intensity detail display. (Courtesy TU Dresden, Germany)

Figure 6.11 Terrestrial laser scanner data of a forest inventory plot (panoramic projection of point cloud intensity). (Courtesy TU Dresden, Germany)

Several research projects on the application of terrestrial laser scanning in forest research, inventory, management and planning tasks have been reported in the literature. These tasks require, besides other parameters, the determination of a number of parameters describing the geometry of a tree or a group of trees. In the simplest case, these parameters are limited to the tree height and diameter at breast height. In some tasks, many more geometry parameters, such as vertical stem profiles, open stem height (the timber-relevant height between the bottom and the first branch), stem ovality, damage or branch diameters are required.

6.7.1 Literature overview

Basically, a laser scanner could be used as a pure 3D point cloud generation tool with the goal of interactive measurement of relevant parameters in the point cloud, thus shifting the survey interpretation task from the field to the office. Much more interesting from a scientific and economic point of view are of course techniques for the automatic derivation of task-relevant parameters from point clouds. Simonse *et al.* (2003) use a 2D Hough transform to detect trees in point clouds and to determine diameters at breast height after height reduction to the digital terrain model. This approach is extended to the determination of diameters at different heights by Aschoff and Spiecker (2004). Gorte and Winterhalder (2004), as well as Gorte and Pfeifer (2004), generate tree topology skeletons by projecting point clouds into a voxel space, where stems and major branches are extracted by morphology operations using 3D structure elements and connectivity analysis. Pfeifer and Winterhalder (2004) model the stem and some major branches of a tree by a sequence of cylinders fitted into the point cloud. Henning and Radtke (2006) show an ICP-based automatic registration of three scans of a group of nine trees using the stem centres detected in the trees at different heights and report a precision of 1–2 cm for stem diameters determined at different heights.

Several other studies show application-specific solutions for automatically deriving parameters beyond the pure stem geometry. Schütt *et al.* (2004) show a technique for the extraction of parameters describing wood quality from laser scanner data. Haala and Reulke (2005) address the recognition of tree species from the fusion of laser scanner and image data. Aschoff *et al.* (2006) used terrestrial laser scanner data to generate 3D forest structure descriptions as a basis for analysing bat habitats by simulating their hunting flight paths. Danson *et al.* (2006) determined canopy gap fracture in spruce forests from terrestrial laser scanner and hemispherical image data. Straatsma *et al.* (2007) analyse the retention potential of flood plain vegetation from density profiles, which were derived from terrestrial laser scanner data.

6.7.2 Forest inventory applications

Forest inventory is a task which has to be fulfilled by forest authorities at regular time intervals to monitor the state of their forests. It includes the determination of a number of parameters characterising the forest. As full area coverage is an unrealistic goal in forest inventory using conventional techniques, inventory schemes based on data acquisition in isolated plots and statistical inference methods have been developed. Some of the most important parameters to be determined in forest inventory are the number of trees in a certain plot, the determination of the diameters at breast height and the tree heights. Terrestrial laser scanning, combined with reliable automatic data-processing techniques, may provide an interesting tool to bridge the gap between conventional inventory techniques and airborne laser scanner data-processing schemes and to facilitate data acquisition for 3D individual tree geometry parameters in large plots.

Maas *et al.* (2008) show the results of four pilot studies on the applicability and success of terrestrial laser scanning in forest inventory. The study areas include coniferous forest, deciduous forest and mixed forest scanned in single panoramic scan mode (Figure 6.12a) as well as multiple scan mode with three instrument positions (Figure 6.12b).

The data-processing chain used in the studies can be outlined as follows:

1. Digital terrain model reduction: the digital terrain model is determined by a simple height histogram analysis searching for minima in predefined *XY*-meshes of the laser scanner data, followed by a neighbourhood consistency check and bilinear interpolation in the meshes [Bienert *et al.*, 2006]. All laser scanner points are reduced to this terrain model.

2. Segmentation and counting of trees: the tree detection process is based on mathematical morphology techniques, extended from raster image analysis to irregularly distributed points in horizontal slices of the laser scanner data.

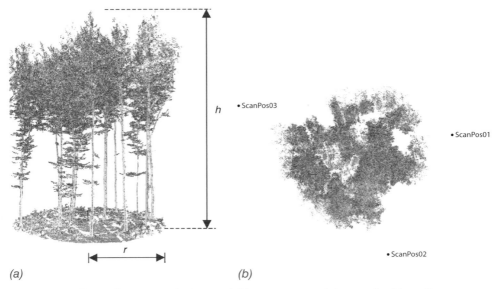

(a) *(b)*

Figure 6.12 Forest inventory plot terrestrial laser scanner data acquired from three instrument positions: (a) point cloud of a single panoramic scan; and (b) point cloud from a multiple scan mode with three instrument positions. (Maas *et al.*, 2008; courtesy TU Dresden, Germany)

3. Determination of diameter at breast height: the diameter at breast height (DBH) is defined as the diameter 1.3 m above the finished grade at the end of the trunk (Figure 6.13). It is determined by fitting a circle into the point cloud of each detected tree and analysing the results using robust estimation techniques.

4. Tree height determination: the tree height is calculated as the height difference between the highest point of the point cloud of a tree and the representative ground point using a similar procedure to that reported by Hopkinson *et al.* (2004). Generally, tree height determination procedures remain rather vague due to undersampling, occlusions and wind effects in the crown.

5. Stem profiles: repeating the technique of stem position and diameter at breast height determination in predefined height intervals at the same (X,Y)-location, stem profiles containing information on shape, uprightness and straightness can be determined. The probability of gross errors in the diameter determination is further reduced by using knowledge of the tapering of the stem in a subsequent filtering operation.

These methods were applied in several pilot studies using data from different scanners in test regions in Germany, Austria and Ireland. The overall success rate

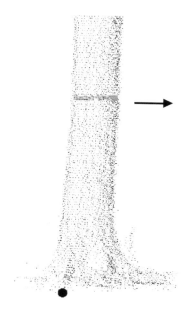

Figure 6.13 Ground point of a tree on sloping terrain; a circle in the data points represents the diameter at breast height. (Courtesy TU Dresden, Germany)

of tree detection was 97.5%. Type I errors (trees which were not detected) were mostly caused by full or partial occlusions of stems in single panoramic scan data. Type II errors (false detections) could largely be eliminated in later processing steps by checking the tree height or by an analysis of the stem profile. From a comparison with conventional calliper measurements, an RMS error of 1.8 cm could be obtained for the automatic determination of diameter at breast height. This value contains the laser scanner instrument precision as well as the precision of calliper measurement and deviations of the stem from circularity. Automatic tree height determination turned out to be much more critical: compared to hand held tachymeter height measurements, an RMS error of 2.07 m was obtained. This error has to be attributed to both occlusions in the point cloud and the limited precision of the reference measurements. Vertical stem profiles were derived by applying the circle fit procedure at multiple height steps. A comparison with harvester data yielded an RMS profile diameter difference of 4.7 cm over the whole tree, with larger deviations occurring in the lower and upper part of the stem. Over the economically most relevant branch and root free part of the stem, the RMS diameter deviation was only 1.0 cm.

References

Andersen, H., McGaughey, R., Carson, W., Reutebuch S., Mercer, B. and Allan, J., 2003. A comparison of forest canopy models derived from LiDAR and InSAR data in a Pacific North West conifer forest. *International Archives of Photogrammetry, Remote Sensing and Spatial Information Sciences*, 34(Part 3/W13), 211–217.

Andersen, H., McGaughey, R. and Reutebuch, S., 2005. Estimating forest canopy fuel parameters using LIDAR data. *Remote Sensing of Environment*, 94(4), 441–449.

Aschoff, T. and Spiecker, H., 2004. Algorithms for the automatic detection of trees in laser-scanner data. *International Archives of Photogrammetry, Remote Sensing and Spatial Information Sciences*, 36(Part 8/W2), 66–70.

Bienert, A., Scheller, S., Keane, E., Mullooly, G. and Mohan, F., 2006. Application of terrestrial laser scanners for the determination of forest inventory parameters. *International Archives of Photogrammetry, Remote Sensing and Spatial Information Sciences*, 36(Part 5) (on CD-ROM).

Blair, B., Rabine, D. and Hofton, M., 1999. The laser vegetation imaging sensor: a medium-altitude, digitisation-only, airborne laser altimeter for mapping vegetation and topography. *ISPRS Journal of Photogrammetry and Remote Sensing*, 54(2–3), 115–122.

Chasmer, L., Barr, L., Black, A., Hopkinson, C., Kljun, N., McCaughey, J. and Treitz, P., 2007. Using airborne lidar for the assessment of canopy structure influences on CO_2 fluxes. *International Archives of Photogrammetry, Remote Sensing and Spatial Information Sciences*, 36(Part 3/W52), 96–101.

Drake, J., Dubayah, R., Clark, D., Knox, R., Blair, B., Hofton, M., Chazdon, R., Weishampel, J. and Prince, S., 2002. Estimation of tropical forest structural characteristics using large-footprint lidar. *Remote Sensing of Environment*, 79(2–3), 305–319.

Gobakken, T. and Naesset, E., 2004. Effects of forest growth on laser derived canopy metrics. *International Archives of Photogrammetry, Remote Sensing and Spatial Information Sciences*, 36(Part 8/W2), 224–227.

Gorte, B. and Pfeifer, N., 2004. Structuring laser-scanned trees using 3D mathematical morphology. *International Archives of Photogrammetry, Remote Sensing and Spatial Information Sciences*, 35(Part B5), 929–933.

Gorte, B. and Winterhalder, D., 2004. Reconstruction of laser-scanned trees using filter operations in the 3D-raster domain. *International Archives of Photogrammetry, Remote Sensing and Spatial Information Sciences*, 36 (Part 8/W2), 39–44.

Gutelius, B., 2002. High performance airborne lidar for terrain and bathymetric mapping technologies. *International Archives of Photogrammetry, Remote Sensing and Spatial Information Sciences*, 34(Part 1); available online from http://www.isprs.org/commission1/proceedings02/paper/00040.pdf.

Hashimoto, H., Imanishi, J., Hagiwara, A., Morimoto, Y. and Kitada, K., 2004. Estimating forest structure indices for evaluation of forest bird habitats by an airborne laser-scanner. *International Archives of Photogrammetry, Remote Sensing and Spatial Information Sciences*, 36(Part 8/W2), 254–258.

Haugerud, R. and Harding, D., 2001. Some algorithms for virtual deforestation (vdf) of lidar topographic survey data. *International Archives of Photogrammetry, Remote Sensing and Spatial Information Sciences*, 34(Part 3/W4), 211–217.

Henning, J. and Radtke, P., 2006. Detailed stem measurements of standing trees from ground-based scanning lidar. *Forest Science*, 52(1), 67–80.

Heurich, M., Schneider, Th. and Kennel, E., 2003. Laser scanning for identification of forest structures in the Bavarian forest national park. *Proceedings ScandLaser Workshop*, 3–4 September 2003, Umeå, Sweden, 98–107.

Heurich, M., Persson, Å., Holmgren, J. and Kennel, E., 2004. Detecting and measuring individual trees with laser-scanning in mixed mountain forest of Central Europe using an algorithm developed for Swedish boreal forest conditions. *International Archives of Photogrammetry, Remote Sensing and Spatial Information Sciences*, 36(Part 8/W2), 307–312.

Holmgren, J. and Jonsson, T., 2004. Large scale airborne laser-scanning of forest resources in Sweden. *International Archives of Photogrammetry, Remote Sensing and Spatial Information Sciences*, 36(Part 8/W2), 157–160.

Hopkinson, C. and Chasmer, L., 2007. Modelling canopy gap fraction from lidar intensity. *International Archives of Photogrammetry, Remote Sensing and Spatial Information Sciences*, 36(Part 3/W52), 190–194.

Hoss, H., 1997. Einsatz des Laser-Scanner-Verfahrens beim Aufbau des Digitalen Geändemodells (DGM) in Baden-Württemberg. *Photogrammetrie, Fernerkundung und Geoinformation*, 1997(2), 131–142.

Hug, C., Ullrich, A. and Grimm, A., 2004. LiteMapper-5600 – a waveform-digitizing LIDAR terrain and vegetation mapping system. *International Archives of Photogrammetry, Remote Sensing and Spatial Information Sciences*, 36(Part 8/W2), 24–29.

Hyyppä, J. and Inkinen, M., 1999. Detecting and estimating attributes for single trees using laser scanner. *The Photogrammetric Journal of Finland*, 16(2), 27–42.

Hyyppä, J., Kelle, O., Lehikoinen, M. and Inkinen, M., 2001. A segmentation-based method to retrieve stem volume estimates from 3-D tree height models produced by laser scanners. *IEEE Transactions on Geoscience and Remote Sensing*, 39(5), 969–975.

Hyyppä, J., Hyyppä, H., Litkey, P., Yu, X., Haggrén, H., Rönnholm, P., Pyysalo, U., Pitkänen, J. and Maltamo, M., 2004. Algorithms and methods of airborne laser-scanning for forest measurements. *International Archives of Photogrammetry, Remote Sensing and Spatial Information Sciences*, 36(Part 8/W2), 82–89.

Jutzi, B. and Stilla, U., 2005. Measuring and processing the waveform of laser pulses. *Optical 3-D Measurement Techniques VII*, Grün, A. and Kahmen, H. (Eds.), vol. I, pp. 194–203.

Kraus, K., and Pfeifer, N., 1998. Determination of terrain models in wooded areas with airborne laser scanner data. *ISPRS Journal of Photogrammetry and Remote Sensing*, 53(4), 193–207.

Lefsky, M., Cohen, W., Harding, D., Parker, G., Acker, S. and Gower, Th., 2001. Lidar remote sensing of aboveground biomass in three biomes. *International Archives of Photogrammetry, Remote Sensing and Spatial Information Sciences*, 34(Part 3/W4), 150–160.

Maas, H.-G., Bienert, A., Scheller, S. and Keane, E, 2008. Automatic forest inventory parameter determination from terrestrial laserscanner data. *International Journal of Remote Sensing*, 29(5), 1579–1593.

Maltamo, M., Packalén, P., Peuhkurinen, J., Suvanto, A., Pesonen, A. and Hyyppä, J., 2007. Experiences and possibilities of ALS based forest inventory in Finland. *International Archives of Photogrammetry, Remote Sensing and Spatial Information Sciences*, 36(Part 3/W52), 270–279.

Morsdorf, F., Meier, E., Allgöwer, B. and Nüesch. D., 2003. Clustering in airborne laser scanning raw data for segmentation of single trees. *International Archives of Photogrammetry, Remote Sensing and Spatial Information Sciences*, 34(Part 3/W13), 27–33.

Morsdorf, F., Kotz, B., Meier, E., Itten, K.I. and Allgower, B., 2006. Estimation of LAI and fractional cover from small footprint airborne laser scanning data based on gap fraction. *Remote Sensing of Environment*, 104(1), 50–61.

Næsset, E., 1997. Estimating timber volume of forest stands using airborne laser scanner data. *Remote Sensing of Environment*, 61(2), 246–253.

Ørka, H., Næsset, O. and Bollandsås, O., 2007. Utilizing airborne laser intensity for tree species classification. *International Archives of Photogrammetry, Remote Sensing and Spatial Information Sciences*, 36(Part 3/W52), 300–304.

Persson, Å., Holmgren, J. and Söderman, U., 2002. Detecting and measuring individual trees using an airborne laser scanner. *Photogrammetric Engineering and Remote Sensing*, 68(9), 925–932.

Persson, Å., Söderman, U., Töpel, J. and Ahlberg, S., 2005. Visualization and analysis of full-waveform airborne laser scanner data. *International Archives of Photogrammetry, Remote Sensing and Spatial Information Sciences*, 36(Part3/W19) 103–108.

Pfeifer N., Reiter T., Briese, C. and Rieger, W., 1999. Interpolation of high quality ground models from laser scanner data in forested areas. *International Archives of Photogrammetry and Remote Sensing*, 32(Part 3/W14), 31–36.

Pfeifer, N. and Winterhalder, D., 2004. Modelling of tree cross sections from terrestrial laser-scanning data with free-form curves. *International Archives of Photogrammetry, Remote Sensing and Spatial Information Sciences*, 36(Part 8/W2), 76–81.

Reitberger, J., Krzystek, P. and Stilla, U., 2006. Analysis of full waveform lidar data for tree species classification. *International Archives of Photogrammetry, Remote Sensing and Spatial Information Sciences*, 36(Part 3), 228–233.

Schwalbe, E., Maas, H.-G., Roscher, M. and Wagner, S., 2006. Profile based sub-pixel-classification of hemispherical images for solar radiation analysis in forest ecosystems. *International Archives of Photogrammetry, Remote Sensing and Spatial Information Sciences*, 36(Part 7) (on CD-ROM).

Simonse, M., Aschoff, T., Spiecker, H. and Thies, M., 2003. Automatic determination of forest inventory parameters using terrestrial laserscanning. *Proceedings of the ScandLaser Workshop*, 3–4 September 2003, Umeå, Sweden, 251–257.

Solberg, S., Næsset E., Hanssen, K. and Christiansen, E., 2006. Mapping defoliation during a severe insect attack on Scots pine using airborne laser scanning. *Remote Sensing of Environment*, 102(3–4), 364–376.

Stephens, P., Watt, P., Loubser, D., Haywood A. and Kimberley, M., 2007. Estimation of carbon stocks in New Zealand planted forests using airborne scanning lidar. *International Archives of Photogrammetry, Remote Sensing and Spatial Information Sciences*, 36(Part 3/W52), 389–394.

St-Onge, B. and Véga, C., 2003. Combining stereo-photogrammetry and LiDAR to map forest canopy height. *International Archives of Photogrammetry, Remote Sensing and Spatial Information Sciences*, 34(Part 3/W13), 205–210.

Vosselman, G., 2000. Slope based filtering of laser altimetry data. *International Archives of Photogrammetry and Remote Sensing*, 33(Part B3/2), 935–942.

Wagner, W., Ullrich, A., Ducic, V., Melzer, T. and Studnicka, N., 2006. Gaussian decomposition and calibration of a novel small-footprint full-waveform digitising airborne laser scanner. *ISPRS Journal of Photogrammetry and Remote Sensing*, 60(2), 100–112.

Wack, R., Schardt, M., Lohr, U., Barrucho, L. and Oliveira, T., 2003. Forest inventory for eucalyptus plantations based on airborne laser scanner data. *International Archives of Photogrammetry, Remote Sensing and Spatial Information Sciences*, 34(Part 3/W13), 40–46.

Yu, X., Hyyppä, J., Kaartinen, H. and Maltamo, M., 2004. Automatic detection of harvested trees and determination of forest growth using airborne laser scanning. *Remote Sensing of Environment*, 90(4), 451–462.

Yu, X., Hyyppä, J., Kaartinen, H., Hyyppä, H., Maltamo, M. and Rönnholm, P., 2005. Measuring the growth of individual trees using multi-temporal airborne laser scanning point clouds. *International Archives of Photogrammetry, Remote Sensing and Spatial Information Sciences*, 34(3/W19), 204–208.

Engineering Applications | 7

Roderik Lindenbergh

This chapter considers the use of both airborne and terrestrial laser scanning for the control and monitoring of newly built and existing infrastructures. In recent years both researchers and companies have explored laser scanning as an alternative to traditional surveying techniques, first for the construction of as-built models of engineering works and second for monitoring their state and safety with time. For traditional techniques like levelling, access to the area of interest is necessary. For areas such as industrial plants, or under hazardous conditions, actual access is often very problematic. Although observations obtained by single point techniques have the advantage of high accuracy, no information is obtained in between the measured points, therefore it is possible that damage on a surface cannot be visualised or modelled, or even that a damage pattern is completely missed.

The current single point precision of laser scanning is of the order of several millimetres for terrestrial scanning to at best several centimetres for airborne scanning. In order to regain part of the accuracy that is somehow lost by using laser scanning, the generally large redundancy in the number of available observations should be exploited. If a series of observations carried out over a long period of time or parameterisable geometric objects such as planes or cylinders are available, parameters can be estimated with accuracies well below the single point precision, provided that systematic errors in raw data or errors introduced by consecutive processing steps do not disturb the signal of interest. Alternatively, spatio-temporal correlation can be exploited in a less explicit way by feature matching or by considering derived parameters as local roughness.

Most of the methodology described below is directly driven by a very specific application. For that reason methods and applications are treated together. For the topic of change detection and deformation analysis the available methods have been organised in an overall framework: first it is considered whether changes occur at all, then changes at single points are quantified and parameterised, and finally changes of complete objects or of homogeneous areas are quantified.

This chapter is organised as follows. Section 7.1 considers how complex industrial sites, such as oil platforms, can be reconstructed from terrestrial laser scanning data. Section 7.2 is devoted to the topic of change detection and deformation analysis. As well as relatively simple, but often very effective, methods a general statistical framework is described that qualifies and quantifies changes but also allows error propagation. Applications mainly consider large constructions such as dams or tunnels but special emphasis is also given to geomorphological hazard management of soft coastal defences (dunes), and landslides and rock fall. In Section 7.3 an overview is given of corridor mapping: in particular, airborne laser scanning from low flying aeroplanes or helicopters turns out to be a successful tool for the monitoring of structures such as electrical power lines and water embankments. The chapter finishes with a short look at expected future developments.

7.1   Reconstruction of industrial sites

Complicated industrial sites such as oil platforms or refineries develop over time spans of decades. As a consequence, the original CAD design model becomes outdated, even if this model already coincided with the actual as-built situation at delivery. For further development and maintenance it is important to produce an accurate up-to-date geometric description. Two examples that illustrate the use of such descriptions are clash detection and retrofitting. If in the design phase of a site extension it is discovered that the planned extension clashes with the current as-built situation, a redesign of the extension can still resolve the clash. For retrofitting planning, i.e. the replacement of existing components, it is essential to determine if a component is still accessible without having to remove other components that were added later.

7.1.1 Industrial site scanning and modelling

Terrestrial laser scanning is a method suited to the capture of 3D data of industrial environments [Langer *et al.*, 2000] possibly in combination with photogrammetry [Markley *et al.*, 2008]. Scanning is fast and often it is not even necessary to access key parts of the installations. For a first clash detection it is sufficient to intersect a CAD model of a proposed extension to a point cloud representing the current as-built situation. If both representations are correctly referenced, clashes can simply be detected by a point in object test.

A more challenging task is to convert the raw 3D point cloud into a complete geometric model of the site, preferably in a fully automatic way. A related issue, which is not considered here, but is still quite important in practice, is how to, again preferably automatically, enrich an updated geometric model with metadata available in a previous version, describing, for example, the age and material composition of individual components of the modelled site. To resolve this issue, the correspondence between identical object parts in different versions of a site model has to be established.

When modelling an ordinary building, it is often enough to consider planar objects for the representation of walls and roofs. Industrial sites are typically composed of slightly more complicated geometric primitives, such as spheres or cylinders representing, for example, pipes and pipe bends. Several software packages nowadays offer the possibility of fitting such primitives to a point cloud in a semi-automatic way, where the user has to select a part of the point cloud or a seed point together with a suitable primitive. Software suites facilitating the conversion of laser scans into 3D models of industrial plants include INOVx PlantLINx, Cyclone Cloudworx, Faro Scene, LFM Modeller, LupoScan and 3Dipsos. In Sternberg and Kersten (2007) the practical performance of complete laser scanner systems, consisting of scanner, targets and off-the-shelf processing software, is discussed in two different industrial as-built documentation applications.

In the remainder of this section methodology is presented which aims at a fully automatic 3D reconstruction of an industrial site. In Section 7.1.2 a segmentation method is discussed which is able to handle curved surfaces. Methods for the automatic identification and parameterisation of geometric objects are discussed in Section 7.1.3. In Section 7.1.4 an integrated method for the simultaneous fitting and registration of industrial site data is described.

7.1.2 Industrial point cloud segmentation

Given a 3D point cloud of an industrial scene as obtained from the alignment of several individual scans, a segmentation is a first step towards a decomposition of the sampled scene into object components. An algorithm that aims at the fully automatic segmentation of such an industrial point cloud should fulfil some specific requirements. It should notably incorporate non-planar, smooth surfaces, as these typically occur in industrial installations; it should be robust in the sense that it can handle noise and also small biases which might be introduced by a registration step; it should be controllable by a minimum number of interpretable parameters; and, finally, it should be able to efficiently handle large datasets.

Rabbani *et al.* (2006) propose such an algorithm. As the measure of similarity that decides whether a point belongs to a segment, the surface normal at that point is used. To ensure robustness against noise and small biases, a suitably large number of nearest points should be incorporated for determining the local normal direction. This can be done in essentially two different ways: either the number of nearest points or the area from which all nearest points are used is kept constant. The first method has the advantage that it adapts automatically to the local point density. Both methods can be efficiently implemented applying existing 3D space partitioning methods such as k-D trees [Goodman and O'Rourke, 1997].

The surface normal determination results in two parameters at each point cloud point: the normal angle itself, and the quality of this angle as indicated by the average

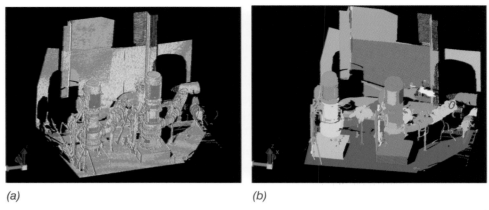

(a) *(b)*

Figure 7.1 Point clouds representing an industrial installation: (a) registered point cloud, coloured by the average size of the residuals resulting from a local planar patch adjustment; and (b) point cloud after segmentation. (From Rabbani *et al.*, 2006; courtesy the authors)

residual of the adjustment to a local planar patch of the neighbouring points. A region-growing approach (see Section 2.3.3) is applied to group neighbouring points with similar normals into segments. As normal variations are only evaluated locally, smoothly varying surface elements can effectively be identified. The procedure is illustrated in Figure 7.1. In Figure 7.1(a) a registered point cloud is shown. Each point is coloured according to the size of the average residual of the local patch adjustment used to determine the surface normal at that point. Thirty points per patch were used. Region growing was applied using an angular tolerance of 15°. Figure 7.1(b) shows the resulting segmentation. It can be seen that point cloud points from cylindrical surfaces were successfully grouped into segments.

The number of segments can be controlled by an average residual size threshold. Based on this threshold it is decided whether a point should serve as a new seed point. If so, it will also lead to a new segment. Allowing large average residuals will therefore lead to a small number of segments. All implementational details can be found in Rabbani *et al.* (2006).

7.1.3 Industrial object detection and parameterisation

Because of the potentially large number of geometric objects in an industrial scene, as represented by a laser scanner point cloud, automatic detection and modelling of geometric objects such as planes and cylinders is not straightforward. How geometric objects can be identified and modelled in an efficient way is discussed here; further details can be found in Rabbani and Van den Heuvel (2005) and Rabbani (2006).

To ensure practical applicability, an object detection and modelling method should be robust against noise and small biases in the point cloud data, should be able to detect and model only partly sampled geometric objects (such as one-half of a cylinder), and should be able to handle many different individual objects at the same time. Moreover, if such a method is to profit from a preceding segmentation step, it should be able to cope with under- and over-segmentation errors.

In Chapter 2 the Hough transformation was discussed. In the case of laser scanner data the Hough transformation basically transforms an *XYZ* point cloud into an alternative space that highlights the features of interest, such as planes or cylinders. The major drawback of using Hough transforms in practice is that the computational efforts increase rapidly with both (1) the total number of (candidate) objects; and (2) the number of parameters needed to describe a geometric object. In an unsorted point cloud of a complicated industrial scene, a Hough transform may detect hundreds of subsets of points that appear merely by accident to be on a geometric object such as a cylinder. This will lead to many false or duplicate hypotheses. Fortunately, this problem can be avoided by first segmenting the point cloud into smooth pieces, using for example the segmentation procedure described in Section 7.1.2. In one smooth segment a few cylinder parts at most will be present. A further way of reducing computational efforts is to split the parameterisation of the geometric objects in the point cloud. Such an approach can be clearly demonstrated in the case of cylinder modelling: at first the orientations of the axes of the possible cylinders in the scenes are considered, and only in the second step are the cylinders fully located (see also Section 2.3.1.3).

The complete resulting procedure is illustrated in Figure 7.2 for a scene representing an industrial plant as scanned by a Cyrax HDS2500 scanner with a reported point accuracy of 6 mm. The point cloud in Figure 7.2(a), consisting of 20 million points, was first segmented using the above procedure, using parameter settings which avoid over-segmentation. For each individual segment the simultaneous method of cylinder

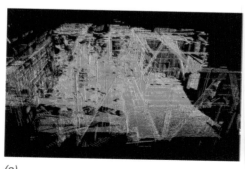

(a)   *(b)*

Figure 7.2 (a) Point cloud from a petrochemical plant; and (b) detected planes and cylinders. (From Rabbani, 2006; courtesy T. Rabbani)

and plane detection was run, including ambiguity testing to distinguish between planes and cylinders. The procedure resulted in a fully automatic detection of 946 planar and 1392 cylinder patches. Part of the results is shown in Figure 7.2(b). Validation of the results showed that more than 50% of the point cloud points are within 1 cm of the closest modelled surface.

7.1.4 Integrated object parameterisation and registration

Standard registration methods typically use either control points or minimise distances between individual points (see Chapter 3). An alternative approach to aligning two scans is to estimate parameters describing objects that are (partly) sampled in both scans and use these parameters for registration. If, for example, three non-parallel cylindrical shapes of different radius are sampled in both scans, the direction of the cylinder axes could be extracted in both scans separately by a Hough transform. If in addition the correspondences between the cylinders in both scans can be established, by using estimates for the radii, the scans can be aligned by matching the three cylinder axes. Such an approach has several advantages. First, many points are used for the estimation of the geometric parameters, therefore these parameters can be estimated with a precision much better than the nominal point precision. Moreover, no actual overlap between the scans is needed as long as sufficient parts of the same geometric object are sampled in each scan. This approach can also be integrated with the industrial modelling method described above. As no separate registration step is needed and registration of several scans in parallel is possible, accumulation of processing errors is minimised.

Rabbani *et al.* (2007) describe two different algorithms that integrate geometric registration with industrial reconstruction. The first method, the *indirect method*, quickly leads to approximate registration parameters. For this purpose points in all scans are labelled as belonging to a certain object, as decided by a segmentation algorithm like the one described above. Then for each object, in each scan, geometric object parameters are estimated using least-squares fitting. Finally the sum of squares of differences between the geometric shape parameters of corresponding objects in the different scans is minimised to obtain parameters describing the transformation that aligns the scans.

In the second method, the *direct method*, the registration and geometric shape parameters are estimated simultaneously. For this approach, good approximate initial registration parameters are needed, which can be obtained from the indirect method. An integrated adjustment step is then performed, which adjusts the observations, i.e. the scan points in all scans, to the model space, which in this case is spanned both by all parameters describing the unique geometric objects in the scene and by the parameters describing the alignment of the different available scans with, say, one reference scan. Note that the direct method uses all point cloud points in its adjustment and therefore is potentially slow.

The direct and indirect methods were tested on four scans representing an industrial site. The scans were obtained using the Cyrax HDS2500 scanner and each contained about 1 million points. Objects in each of the scans were detected, using a method similar to the one described in Section 7.1.3. In total, the four scans contain 88 objects, 72 objects are visible in only one scan, 14 objects in two scans, while two objects are sampled in three scans. First the scans were registered using the indirect method. Then the direct method was applied to estimate a total of 538 parameters of both the objects and the scan transformation simultaneously. The final registration result, obtained without any targets, was compared to a standard ICP registration based on additionally placed artificial targets, which are needed to obtain sufficient overlap between the scans. The mean distance between the ICP registered and the directly registered point cloud was 2.9 mm with a standard deviation of 2.1 mm. Modelling results of four cylinders, once using the direct method and once from ICP followed by a separate modelling step, were compared. For all four cylinders the direct method gave better results. Finally it should be mentioned that the methods described here allow error propagation. The resulting quality description can be used, for example, to evaluate how well the models fit to the point cloud.

7.2   Structural monitoring and change detection

Typically, traditional surveying methods measure the *xyz* coordinates of an apparent reference point or benchmark. After georeferencing of the measurements in a common coordinate system, possible displacements of the reference point through time can be assessed. In general, this procedure is not applicable to laser scan data: the individual laser pulses of repeated scans will not hit exactly the same location. Moreover, scans obtained from different standpoints will possibly sample different parts of objects due to occlusions, while corresponding parts of objects will be sampled at different scan point densities and at different individual point qualities. Combination of scans to a common coordinate system in a registration procedure will inevitably result in local closing errors. As a result, change detection methods tend to find false changes.

On the other hand, the high scanning point densities imply that real changes in the scene are often sampled by many individual scan points. Therefore, at the risk of losing computational feasibility, change detection methods can potentially profit from a large data redundancy. For example, Gordon and Lichti (2007) report results that show that modelled terrestrial laser scanner data achieved an accuracy up to 20 times higher than the single point coordinate precision.

These issues make deformation analysis based on scan data a challenging and interesting problem. In this section the different methodologies developed for the detection and parameterisation of temporal changes in scan data are considered. In most cases a methodology is specially developed for a certain designated application. For this reason, the underlying applications have been given a prominent role.

Some preprocessing issues: Care should be taken when registering datasets that are subject to change. When sufficiently sampled stable objects are available in all scans from different epochs, a suitable surface-based method, such as iterative closest point [Besl and McKay, 1992], TIN least-squares matching [Maas, 2000], or least-squares 3D surface and curve matching [Grün and Akca, 2005] can be used for a registration. Alternatively, a connection to a global reference system can be established using artificial targets which are also independently measured, for example using a total station. Throughout this chapter it is assumed that point clouds from all epochs have been registered in a common Cartesian coordinate system.

In the following sections three different approaches to analysing changes are considered. The topic of Section 7.2.1 is change detection. Here if and where a scene has changed is analysed, but the changes themselves are not necessarily quantified. In Section 7.2.2 methods are given for quantifying deformation at single positions. Such methods enable the monitoring of structural changes when objects in the scene are actually deforming. Section 7.2.3 considers rigid body transformations or the quantification of movement of complete objects. This approach necessarily involves the identification in the scan data of the individual objects that undergo a displacement. Finally, a short look at future developments is given.

7.2.1 Change detection

The primary goal of change detection is to identify changes in a scene over time. Typically it looks at whether particular objects, for example cars, have been added to or removed from the scene in a certain epoch. The result of a change detection procedure is a binary 0/1 map: no change is indicated by 0, change by 1. As it is generally not possible with laser scanning to repeat individual point measurements, changes are mostly considered with respect to an object or at least a certain neighbourhood of a point in the reference situation.

7.2.1.1 Direct DSM comparison

Murakami *et al.* (1999) is one of the first papers to report on application of direct digital surface model (DSM) data based on (airborne) laser scanning for the purpose of change detection. The authors used a laser scanning system with a vertical accuracy of 10–20 cm and a horizontal accuracy of 1 m to update a Japanese GIS database. As the design of many Japanese buildings and houses is based on unit cubes with an edge length of 2 m, the authors use a simple DSM subtraction to detect changes in buildings, followed by an automatic shrinking and expansion filter to remove edges of unchanged features that appear in the difference set due to local misregistration. This example demonstrates that simple DSM comparison is a good solution whenever the needed accuracy is at least an order of magnitude larger than the measurement accuracy.

Bauer *et al.* (2005) also generate georeferenced DSMs for each epoch and use local elevation differences to classify changes into classes such as (in)significant change, unusable measurements, etc. The method was applied, using the long-range Riegl LPM-2k scanner, to the Gries rock fall area in Austria. Deformations of up to 10 m were detected with a reported *xyz* accuracy of 5 cm. Rosser *et al.* (2005) report on the use of terrestrial laser scanning in monitoring hard rock coastal cliff erosion. By comparing direction-dependent triangulated meshes, they report changes in the rock cliffs starting from 10 cm.

7.2.1.2 3D: Octree-based change detection

Girardeau-Montaut *et al.* (2005) suggest the use of an octree, see Section 2.2.2 and Samet (1989), for the purpose of detecting changes in point clouds obtained at different moments by terrestrial laser scanning. The point cloud of each epoch is stored in an octree with the same initial bounding cube. As a consequence, individual octree cells in different epochs are at the same location in the common coordinate system, and therefore the points they contain are in some sense comparable.

A quick overview of possible changes can be obtained by comparing points from different epochs in corresponding octree cells at a suitable octree level: for a fixed octree cell the distance of each point in the cell of a succeeding epoch to the closest point in the same cell in the reference epoch is determined. Average distances above the noise level may indicate change. In particular, octree cells that are empty in, say, the reference epoch but filled in a succeeding epoch may indicate a larger change and give a good indication of where to look in greater detail. Alternatively, the points in an octree cell can be used to determine a local parameter such as a plane normal. A locally changing normal between epochs may give an alternative change indication.

To obtain explicitly for each point from a succeeding epoch the distance to the closest point in the reference epoch is more expensive from a computational point of view. For this purpose it may be necessary to search in several octree cells for the closest point. Nevertheless, exploiting the adjacency structure of the octree subdivision avoids less efficient global searches. Note that this approach is fully 3D.

7.2.1.3 2.5D: Visibility maps

If one considers changes in point clouds as observed from one fixed standpoint, the dimension of the problem can be reduced from full 3D to 2.5D. Three kinds of possible changes may occur: an object might move or deform, an object might appear in the scene, or an object might disappear from the scene. If an object appears, the scene behind the object becomes invisible from the fixed standpoint. Girardeau-Montaut *et al.* (2005) and Zeibak and Filin (2007) both consider the concept of visibility maps for treating these kinds of changes. Changes in the field of view from a fixed scanner position always occur in the range direction. By using a spherical coordinate system

(θ, φ, R), with θ the horizontal angle, φ the vertical angle and R the range, all with respect to the scanner origin, the scanned points in one single scan are behaving functionally: a range panorama is obtained, so for each pair (θ, φ) there is at most one range value, or the obtained surface is 2.5D.

Zeibak and Filin (2007) consider the following question: "Can a newly measured point be seen in a reference scan?". To answer this question one reference scan is fixed. For the sake of simplicity assume that there is only one other scan, the new scan, obtained at a later time and from a different standpoint, but available in the same coordinate system as the reference scan. Three answers are possible: (1) yes, the point is visible in the reference scan, and there is also an object at that location; (2) yes, the point is visible in the reference scan, but there is no object at that location; and (3) no, the point is not visible, because it is occluded. In case (1) no change has occurred, in case (2) a change has occurred, and in case (3) this cannot be decided.

To answer this question in practice, the data from the new scan are transformed into the spherical coordinate system of the reference scan to obtain a range panorama with respect to the same standpoint. After a processing step, which takes care of areas with no data, noise and differences in angular resolution due to different scan positions, it may be assumed that at each (θ, φ) position of a suitable angular grid there is one reference range distance R_{ref} and one range distance R_{new} for the new scan. Now the two range panoramas can be simply subtracted and the three answers above translate, at each angular grid-point location, into

$$\text{no change if } |R_{ref} - R_{new}| \leq \varepsilon$$
$$\text{change if } R_{ref} - R_{new} > \varepsilon$$
$$\text{occlusion if } R_{ref} - R_{new} < -\varepsilon$$

where $\varepsilon > 0$ is chosen to be large enough to incorporate measurement errors. A result of this procedure is shown in Figure 7.3 where a value of $\varepsilon = 5$ cm is used to compare two street scenes. Changes in this case, indicated in red, are mainly caused by by-passers and cars. The advantage of this method is that it is very fast, as only two datasets have to be subtracted from each other.

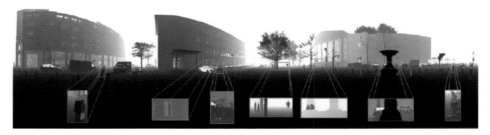

Figure 7.3 Changes (in red) in a road scene detected using visibility maps. (From Zeibak and Filin, 2007; courtesy the authors)

7.2.2 Point-wise deformation analysis

In contrast to change detection, deformation analysis is not primarily aimed at identifying changes but at quantifying and possibly parameterising them. For this purpose models are designed and fitted whose parameters describe deformation over time. Change detection can be considered a special case of deformation: when a model has been adapted that describes a certain type of non-trivial deformation over time, a change has been detected. This section starts with very direct, simple methods that work well in the case of favourable signal-to-noise ratios, and ends with more sophisticated approaches that address either deformation in the same order of magnitude as the accuracy of individual scan points, or differentiate between several deformation scenarios over time. The latter case is only possible when sufficient epochs of measurements are available.

Point-wise comparisons can be divided into three classes:

- **Measurement to measurement.** In this case measurements should be repeated, that is the laser scanner should be kept fixed and range measurements are identified by their horizontal and vertical scan angle (in the case of terrestrial laser scanning).

- **Measurement to surface.** Often a surface is created from the point cloud obtained in the first epoch, for example by creating a triangular mesh. Measurements from a different epoch are compared to the surface. This means of comparison involves only one interpolation step for creating the surface in the first epoch. Advantages of this approach are that it is computationally fast and that differences between a second epoch and the reference epoch are available at the full point density of the second epoch.

- **Surface to surface.** If for further processing purposes a time series of, say, heights is needed at a certain position, measurements of each epoch can be interpolated or adjusted to a common grid. An advantage of this approach is that it enables noise reduction in the case of noisy measurements and data reduction in the case of an excessive amount of observations.

7.2.2.1 Repeated virtual target monitoring

Little (2006) reports on the use of the Riegl LPM-2K long distance laser scanner for the purpose of repeated virtual target monitoring, which is an example of direct measurement to measurement comparison. The term "virtual target" refers to a point on an object whose position is only identified by its coordinates, that is no actual target is available. In the case of terrestrial laser scanning, a virtual target can be identified as a point on the object having fixed horizontal and vertical angles with respect to a spherical coordinate system, originating at the scanner location. Using a rigid and stable measurement setup (see Figure 7.4) allows direct monitoring of a virtual target by

Figure 7.4 Repeated virtual target measurements using a fixed scanner. (From Little, 2006; courtesy M. Little)

evaluating its range coordinate through time. Little (2006) applies this technique to monitoring an open pit mining slope 2 km long and 100 m deep by measuring it at 5 m intervals from a distance of between 500 m and 1 km. The laser scans the points one by one and needs nine hours to measure all the points in the wall at an accuracy between 20 and 50 mm. The monitoring results are used to calculate the volume of the mining excavations, but long-term monitoring trends can also be evaluated to identify high risk areas.

7.2.2.2 Point cloud versus reference surface

In Zogg and Ingensand (2008) an experiment is discussed where deformation due to loads placed on a large, concrete highway viaduct are measured by both terrestrial laser scanning and traditional surveying methods, such as levelling and tacheometry. The same methodology of measurement to mesh comparison is also applied in Alba *et al.* (2006), which describes the first results of a project involving the monitoring of a large concrete dam. The Cancano Lake dam in North Italy presents an arc gravity structure 136 m in height and 381 m in length along the crest. In both cases the use of laser scanning could extend the monitoring procedure from a limited number of points to the whole structure. The first results on the dam project show that the estimated deformation due to seasonal variations in water levels is of the order of a few millimetres, close to the measurement and registration accuracy. Investigating deformation rates close to the noise level for such large structures makes the referencing and registration procedure more critical, as relatively small biases in the target measurements may propagate in closure errors between epochs of a size similar to the expected deformation [Alba *et al.*, 2008].

On a smaller scale, Stone *et al.* (2000) describe an experiment where the amount of sand removed by a bucket of known volume from a sand pile is determined using laser scans before and after the removal using interpolated grids in each epoch. This experiment is interesting because it is an easy test experiment for validating methodology. The authors also describe how the measurement process can be automated using wireless sensor technology.

7.2.2.3 *Landslide analysis by local ICP*

Teza *et al.* (2007, 2008), describe a method where terrestrial laser scanner data from the first epoch are used to determine a reference model of a landslide area. The authors have used an 800 m range OPTECH ILRIS 3D, and a 100 m range Leica HDS 3000. Data from the second epoch, available in the same reference frame, are divided into subareas of 5 × 5 m. For each subarea the local correspondence with the reference epoch is found by applying ICP. This procedure results in a local displacement vector for each subarea (see also Figure 7.5). It follows that the spatial resolution of this method is also 5 m. Displacements between 0 and 25 cm could be detected for an area of approximatley 150 × 150 m. In this case displacements are derived for natural terrain (no buildings).

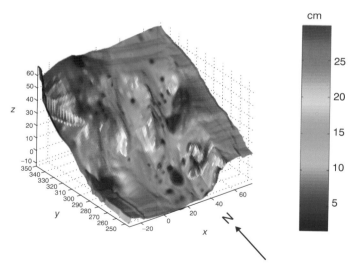

Figure 7.5 Yearly displacements at the Perarolo di Cadore landslide, as obtained by local ICP matching. (Teza *et al.*, 2007)

7.2.2.4 Observation quality

For more critical applications, where the expected signal is closer to the noise level of individual laser scanner observations, insight into the quality of the input data for each epoch is needed. Using a testing theory framework, while incorporating least-squares adjustment together with error propagation [Koch, 1999; Teunissen, 2000b] optimal control is obtained on the detection of deformations. Assume that the object whose deformation is being considered is at least locally 2.5D. This means that locally the object can be expressed in a Cartesian *xyz* coordinate system by specifying at a position *xy* a height *z* to a virtual reference surface $z = 0$. The following errors determine the quality of the input observations.

- **Measurement error.** The measurement error describes the horizontal and vertical accuracy of individual height observations in each epoch. Individual measurements may be of different quality while correlation between measurements may exist.

- **Registration error.** Registration of point clouds of different epochs results in a registration error. The registration error is often a position-dependent systematic error.

- **Interpolation error.** Laser scanning in general does not allow direct measurement of a point at a certain position (x_0, y_0) . If one wants to compare the height z at position (x_0, y_0) as measured in different epochs and maybe even in one epoch from different viewpoints, an interpolation step is needed. This step will result in an interpolated value of a certain quality.

7.2.2.5 Statistical point-wise deformation analysis

In the following the individual deformations at grid points of a suitable grid on the reference surface $z = 0$ are considered. Assume that at a grid point (x, y) heights z_{t_i}, for $i = 1,...m$, are given, with m the number of epochs. Let σ_{t_i} indicate the standard deviation of the height value z_{t_i}. Ideally the height standard deviation is obtained by appropriate propagation of measurement, registration and interpolation error. In practice a workable standard deviation value might be obtained by considering the local noise level of a large enough number of grid points. The m available height observations are used to determine an n-dimensional parameter vector $\hat{x}$ corresponding to a linear deformation model $\mathbf{A}$, that is one considers the system of observation equations

$$E\{\mathbf{y}\} = \mathbf{A} \cdot \mathbf{x}, \qquad D\{\mathbf{y}\} = \mathbf{Q}_y \qquad (7.1)$$

with $E\{.\}$ indicating the expectation of entries in the observation vector $\mathbf{y}$, here containing the temporally varying heights at grid point (x,y), $\mathbf{A}$ the model matrix and $\mathbf{Q}_y$ the variance–covariance matrix describing the dispersion, $D\{.\}$, of the observations. The easiest 1D deformation model corresponds to the case of trivial or no deformation, in this case

$$
\mathbf{y} = \begin{pmatrix} z_{t_1} \\ z_{t_2} \\ \vdots \\ z_{t_m} \end{pmatrix}; \quad \mathbf{A} = \begin{pmatrix} 1 \\ 1 \\ \vdots \\ 1 \end{pmatrix}; \quad \mathbf{x} = (z); \quad \text{and} \quad \mathbf{Q_y} = \begin{pmatrix} \sigma_{t_1}^2 & 0 & \cdots & 0 \\ 0 & \sigma_{t_2}^2 & \ddots & \vdots \\ \vdots & \ddots & \ddots & 0 \\ 0 & \cdots & 0 & \sigma_{t_m}^2 \end{pmatrix} \tag{7.2}
$$

The $\mathbf{A}$-matrix reflects that the expected height in each epoch is the same, that is there is no deformation. The variance–covariance matrix $\mathbf{Q_y}$ expresses that it is assumed that no temporal correlation between the heights z_{t_i} exists, because all entries off the diagonal are equal to zero. System (7.2) is solved using standard adjustment theory [Teunissen, 2000a]. How well model $\mathbf{A}$ describes the ongoing deformation phenomenon is assessed using the framework of testing theory [Teunissen, 2000b], which evaluates the quality of fit relative to both the number and quality of the observations used and the number of parameters in the $\mathbf{x}$-vector.

7.2.2.6 Testing for stability of a tunnel

In Van Gosliga et al. (2006) this approach is applied to detection of an artificial deformation placed on the wall of a concrete tunnel. For the experiment, the second Heinenoord tunnel in Rotterdam was scanned twice using the Leica HDS4500 phase scanner.

Possible deformations are considered in the direction perpendicular to the tunnel wall. For this purpose a coordinate system (y, θ, r) is used, with y the tunnel axis, θ the angle perpendicular to the tunnel axis (for fixed y tracing a profile along the tunnel wall) and r the distance from the tunnel wall to the tunnel axis.

After having scanned the tunnel for the first time, artificial deformations were put against the tunnel wall and the tunnel was scanned a second time. The deformations consisted of (A) several plastic caps and lids of a maximum diameter of 13 cm; (B) a thin wooden plank 10 cm wide and 2 cm thick; and (C) a foam plate $30 \times 40 \times 3$ cm.

The original scan points were interpolated to the same 15 cm grid on the tunnel wall for both epochs, after registration to a common coordinate system. The choice of grid size is a tradeoff: large grid cells in general contain more observations, therefore possible deformations can be detected with a larger probability; on the other hand, the use of large grid cells corresponds to a result with a relatively low spatial resolution. Moreover, simplified local low parameter models of the object may only be valid when applied for small grid cells. Note that in this particular case it is assumed that the cylindrical tunnel wall is to a first approximation flat within patches of 15×15 cm.

As a result of the interpolation procedure, for both epochs 1 and 2, and for each grid cell, range values r_1 and r_2, with associated standard deviations σ_{r_1} and σ_{r_2} respectively, are obtained, depending on both the quality and number of measurements within a grid cell at each epoch. Using testing theory [Teunissen, 2000b] it is formally decided whether the difference between the range values r_1 and r_2 can be attributed to errors caused by the measurement principle and processing or if the difference is so large that actual deformation is the cause. The associated test statistics, basically the quality of fit of the model that expresses stability, are shown for each grid point in Figure 7.6(a). Most test statistics are very small, but at the location of plate C large values are visible, caused by the artificial deformation in the second epoch. A representation such as that in Figure 7.6 is also useful in revealing correlations between neighbouring grid points. Again at plate C it can be seen that the test statistics drop at the boundary of the plate. Here grid cells contain both scan points on and off the plate location. As a consequence, the deformation signal caused by the plate points is damped by the points off the plate.

A division between stable and unstable grid points is shown in Figure 7.6(b). The test statistics were evaluated against an appropriate critical value to decide if a grid point is stable. This particular result is therefore only change detection. The red grid points in Figure 7.6(b) are considered unstable. Foam plate C is considered unstable, as are some grid points near the location of the narrow board B and the lids A. In the two latter cases the results are again less clear because of grid cells containing points both on and off the artificial deformation.

Differences larger than 15 mm were always tested as unstable. Most differences below 15 mm were tested stable, but differences starting from 5 mm were also frequently tested as unstable. This "mixing" is possible because the number of observations

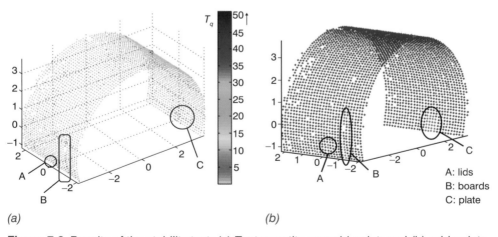

(a)   *(b)*

Figure 7.6 Results of the stability test: (a) Test quantity per grid point; and (b) grid points tested stable (in blue) and unstable (in red). The larger artificial deformations A and B (foam and wooden planks) were found by the testing procedure. Moreover, the procedure finds several other "deformations" near the grip holes, but these are errors caused by registration mismatches. (From Van Gosliga *et al.*, 2006; courtesy the authors)

per grid cell is taken into account: larger numbers of consistent observations will lead to smaller standard deviations per grid cell, which again allows the method to detect smaller deformations.

Apart from the placed artificial deformations, the stability test found "instabilities" at several other grid points. It turns out that these are false alarms which occur near small grip holes in the concrete segments. Due to small registration errors, scan points within a grip hole in one epoch fall within the same grid cell as scan points just outside the grip hole in the other epoch.

These results could be further improved by a thorough segmentation of point clouds per epoch. In this way problems caused by the registration and grid cells containing only parts of objects could be avoided. It should be noted that these results were obtained using only about 1% of the available scan points. The tunnel was scanned using the high-resolution setting of the Leica HDS4500 scanner. For large projects prior determination of the maximal required resolution, by considering how many points of a certain quality are needed to detect deformations of a given size, is recommended.

7.2.2.7 *Morphological deformation*

The same methodology is applied for the analysis of six consecutive yearly airborne laser surveys of a dune area in the north-west Netherlands [Lindenbergh and Hanssen, 2003]. The number of epochs, m, should be larger than the number of parameters, n, of the considered deformation model. As in this case six epochs are available, it is possible to consider several alternative deformation models. As well as stability ($n = 1$), the possibility of a single beach nourishment and linear growth are also considered (both $n = 2$). The case of beach nourishment by means of a sand suppletion was included because it was known that the Dutch Ministry of Public Works maintains the coast line at this location. A suppletion in year $i + 1$ can be be modelled by

$$\mathbf{y} = \begin{pmatrix} z_{t_1} \\ \vdots \\ z_{t_i} \\ z_{t_{i+1}} \\ \vdots \\ z_{t_m} \end{pmatrix}, \qquad \mathbf{A} = \begin{pmatrix} 1 & 0 \\ \vdots & \vdots \\ 1 & 0 \\ 1 & 1 \\ \vdots & \vdots \\ 1 & 1 \end{pmatrix}, \qquad x = \begin{pmatrix} h \\ s \end{pmatrix} \qquad (7.3)$$

where h indicates the height of the beach before the suppletion, and s denotes the height of the suppletion. In this case a grid cell size of 2 m is used. First it is tested, using a stability test, if the height throughout the epochs can be considered stable.

These stable positions, coloured by height, are shown in Figure 7.7(a). From this figure the sea on the left (blue), parts of the beach (yellow) and the first, relatively high, row of dunes can be clearly distinguished. For grid points that fail the stability test, suppletion models, such as Equation (7.3), for suppletions in different years i are considered, as is a constant velocity model. In the latter case, row i of the model matrix equals $(1, t_i)$, corresponding to a parameter vector (h, v), with h the initial height and v the yearly constant increase or decrease. In this case the 2D model which has a test statistic smaller then the critical value for $m - n$ is adopted. If a grid point is found to fit several models, the model with the smallest test statistic is selected.

The results for the non-stable grid points are shown in Figure 7.7(b). Note that this figure is complementary to Figure 7.7(a), as it contains exactly all grid points that are not considered stable. The large, light-blue area corresponds to a beach nourishment in 1999. Directly right of the beach, a red elongated area can be distinguished, corresponding to the top of the first high dune ridge. The prevailing western wind causes a net land-directed sand transport. This effect causes a yearly height increment of the order of 0.5 m at the top of the first dune ridge, which is detected by the constant velocity test.

If only two epochs of data are available, local changes between the epochs can be identified and quantified. The dune application demonstrates that when more than two epochs are available, different deformation regimes can be considered and detected in a data snooping procedure. As well as the deformation parameters themselves, a quality description of these parameters is also obtained as propagated from the variances of, and covariances between, the original observations.

This method could be further refined by incorporating spatial correlation. The whole procedure is now performed for each grid point independently, without considering the behaviour at neighbouring grid points. Nevertheless, the results shown in Figure 7.7(b) demonstrate strong spatial coherence, corresponding to the physical constitution of the terrain.

7.2.3 Object-oriented deformation analysis

In the previous sections changes at individual grid point locations in point clouds obtained by laser scanners were considered. In many applications it is expected that similar displacements occur for a large number of grid points simultaneously, because either the grid points belong to the same moving object or because a homogeneous force triggering the deformation acts on a larger area. The basic advantage of object-oriented methods is that optimal use can be made of the redundancy of the observations. The major challenge is to identify and, if necessary, parameterise points belonging to the same object in the point clouds obtained in different epochs. Before considering changes between objects, first changes relative to an object are considered.

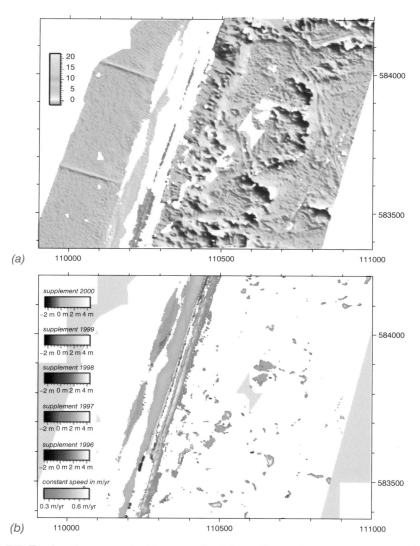

Figure 7.7 Testing theory applied to separate deformation regimes as represented by airborne laser scan data: (a) stable grid points, coloured by height; and (b) non-stable grid points following different deformation patterns, such as instantaneous growth (in blue), and linear growth (in red).

7.2.3.1 Deformations relative to an object

In many applications small changes on the surface of an object are considered, such as damage to a roof or tunnel wall. In order to identify, visualise and isolate such changes, a modelling step can first be applied that allows one to zoom in at the object surface. Such a procedure is illustrated in Figure 7.8 on the doors of a lock used for raising

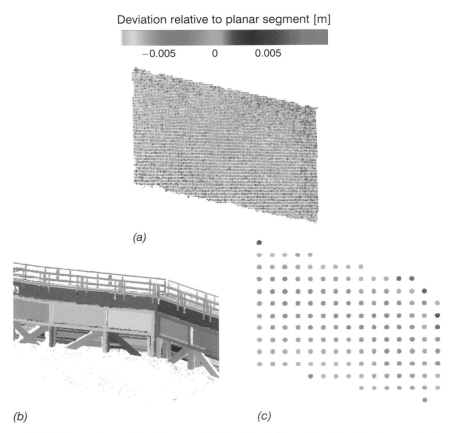

Figure 7.8 Deformation relative to a modelled object: (a) residuals of the laser points after a planar adjustment; (b) segmentation of a point cloud representing a lock; the plate of interest is marked by a cross; and (c) interpolated plate residuals. (From Lindenbergh and Pfeifer, 2005; courtesy the authors)

and lowering boats between stretches of water of different levels on a major canal in The Netherlands. In (b) a segmentation of a point cloud representing the lock doors is given [see also Lindenbergh and Pfeifer, 2005]. The marked segment on the right corresponds to an approximately flat plate. The residuals of the points on the plate after a least-squares adjustment to a planar patch are shown in (c). Note that the size of the residuals is in general less than 5 mm. The sign of the residuals in the middle of the plate is in most cases opposite to the sign near the edges. This observation can be verified by an interpolation of the residuals to a suitable grid (see Figure 7.8c). Now a clear pattern is visible: probably due to water pressure, the middle of the plate is bulging out while the edges of the plate are more constrained in their movement by the surrounding construction. Extraction of point cloud points representing such a plate in different epochs allows the detection of local deformations in the order of millimetres. This is illustrated in Figure 7.9, where deformations (here in the order of centimetres)

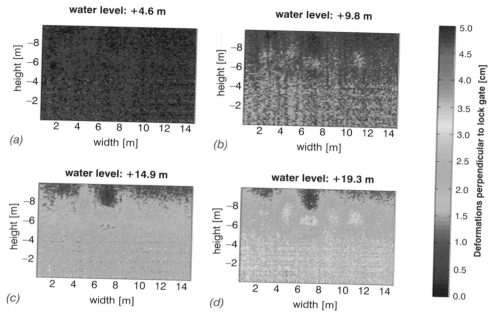

Figure 7.9 Deformation of a lock door relative to a water level of (a) 4.6 m; (b) 9.8 m; (c) 14.9 m; and (d) 19.3 m. (From Schäfer *et al.*, 2004; courtesy Thomas Schäfer)

of a lock door as a function of the water level are shown [Schäfer *et al.*, 2004]. With increasing water level, a regular deformation pattern appears, correlated to the inner framework of the construction. Note that as the point cloud residuals are perpendicular to the model, the applied modelling step directly allows assessment of movement perpendicular to the construction, in many cases the most likely direction of damage.

Another related approach is used in Van Gosliga *et al.* (2006) (see also Section 7.2.2). Here a tunnel is monitored and in order to assess local damage on the tunnel wall basically all data in the available point cloud are used to estimate a cylindrical model through the data, which allows a local coordinate system to be established relative to the tunnel. When the point cloud is connected to an absolute coordinate system, the tunnel model parameters can be compared to the design parameters to determine whether the tunnel is located on the planned position. Similarly, in Schneider (2006), slices of a point cloud representing a television tower are used to determine the axis of the tower, while profiting from a large observational redundancy. In this way a relative movement of 1.3 cm of the top of the tower in a north–south direction could be detected, probably due to heating up of the tower by sunshine.

7.2.3.2 *Local object movement*

Gordon and Lichti (2007) apply a physical model to describe deflections of a deforming concrete beam. In one case, where a force acts at two positions on a horizontal

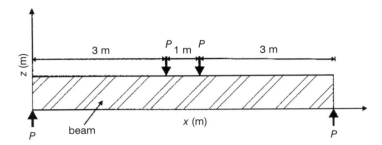

Figure 7.10 Concrete beam deformation experiment. (Gordon and Lichti, 2007)

7 m long beam, which is resting on its end points (see Figure 7.10), this approach results in a deformation model consisting initially of 11 parameters. Using the Riegl LMS-Z210, a total of 12 measurement epochs were acquired with increasing load, until finally the beam failed. In each epoch the 11 parameters describing the shape of the beam were determined using a least-squares adjustment. The resulting models were evaluated against photogrammetric targets placed on top of the beam: the (x,y) positions of the targets were passed into the models in order to obtain adjusted z-coordinates that could be compared to the z-coordinates of the targets. In a slightly simpler second case, this approach resulted in a root mean square error of ±0.3 mm, in the case of a laser point cloud consisting for each epoch of about 7000 points and collected by a Cyrax 2500. Compared to the coordinate precision of the Cyrax 2500 of ±6 mm, this experiment shows an improvement in precision by a factor of about 20. Note that the estimated coordinate precision of the photogrammetric targets was of the order of 0.1–0.2 mm (1σ).

7.2.3.3 Moving rocks

Monserrat and Crosetto (2007) describe an approach that locally estimates rigid body transformation parameters for assessing displacement of relatively small rigid objects, such as rock boulders. First, objects possibly subject to change are identified in both scans, either manually or by an additional automatic step. Unstable areas are masked so as not to negatively influence global registration. After global registration, six parameters describing a rigid body transformation (three for translation, three for rotation) are estimated in a local matching step, following Grün and Akca (2005), for each pair of corresponding objects. The authors suggest identifying these objects (rock boulders in practice) based on intensity differences with the surroundings. This method proved to work well in a simulated deformation case: errors of about 1 cm in each of the three components of the displacement vectors are reported, rotational errors are in general below 1°, for ten artificial targets of about 1 m in size and scanned at 100 m with the Optech Ilris-3D scanner at a sampling density of 0.7 points/cm².

7.2.3.4 Changing roof orientations

Pieraccini *et al.* (2006) compare ground-based SAR (GB-SAR) to terrestrial laser scanning for the purpose of landslide monitoring. The authors used an 800 m range OPTECH ILRIS 3D, and a 100 m range Leica HDS 3000. In this case the village of Lamosano, Belluno, Italy is considered. The village is situated on top of a sliding area. The size of the region of interest is about 200 m × 200 m. In this case deformations are detected by comparing the spatial orientation of surfaces representing the same object, such as roofs or walls, in two georeferenced models. Consistent displacement between 1 and 8 cm were found, which shows good correspondence with the GB-SAR results (see Figure 7.11). The advantage of GB-SAR is that it is very accurate, its spatial coverage is limited though because of its need for coherent scattering behaviour. In this particular case GB-SAR is suitable as a monitoring technique because the many available buildings serve well as coherent scatterers. The authors conclude that displacements of less than 1 cm cannot be found using terrestrial laser scanning.

7.2.3.5 Rock face geometry

Abellán *et al.* (2006) use terrestrial laser scanning to investigate rock falls. They use DTMs derived from laser scans to simulate the trajectories and velocities of falling

Figure 7.11 Landslide displacements in the village of Lamosano, measured by terrestrial laser scanner along the direction normal to the corresponding surface. (Pieraccini *et al.*, 2006)

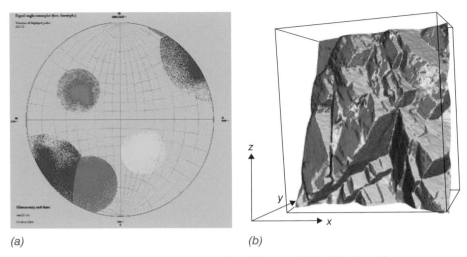

(a)   (b)

Figure 7.12 (a) Polar plot of orientations of rock facets derived by fuzzy k-means clustering followed by a testing procedure; and (b) 3D surface model of a rock face, coloured by orientation. Both images are from Slob *et al.* (2005).

rocks and to reconstruct the joint geometry of past rock fall events at those locations in the rock face where a piece of rock became detached. Slob and Hack (2004) give an overview of terrestrial laser scanning as a monitoring technique for rock face scanning and in particular mention the automatic derivation of rock face properties, such as local orientation or surface roughness. Slob *et al.* (2005) present a method, based on fuzzy k-means clustering, to obtain the orientation and spacing between discontinuity sets in rock faces (see Figure 7.12). This information is directly linked to the size and shape of possibly detaching rocks.

7.2.4 Outlook

In this section an overview is given of the current status of methodology together with some some major applications of laser scanning applied to the detection and quantification of changes. Clearly this topic is still in the pioneering phase: almost all listed applications describe first feasibility studies. It is also clear that it is possible to detect deformations well below the nominal single point accuracy if appropriate modelling in combination with statistical error propagation is used. In cases where such modelling steps are not directly possible, because no rigid objects or no long-term series are available, it is sometimes still possible to profit from the large number of observations that scanning gives by incorporating correlation between nearby scan points, for example by using computer vision techniques such as matching. What is not yet known are the limits of laser scanning for the purpose of deformation analysis. Further research should show the minimum deformation that can be detected in a given setting, and, on the other hand, what resolution and scanner position are needed to be able to detect a change of a certain size at a certain position with a predefined probability. Finally, the techniques described in this section are in general not yet available in standard software. Further spreading of these techniques will profit greatly from professional implementation in surveying software packages.

7.3 Corridor mapping

Airborne laser altimetry turns out to be a useful technique for the monitoring of outdoor engineering constructions such as railroads [Haasnoot, 2001] and their overhead wires, dikes, pipelines [Tao and Hu, 2002], and high-voltage power transmission lines. In these cases not only the condition of the object itself, but also the direct surroundings are of interest: for power lines and railroads vegetation that grows too close will have to be removed, for pipelines terrain parameters can be used to assess the probability and impact of wear and damage to the pipeline and the impact of the damage. Processing of corridor data therefore typically consists of a classification step followed by a modelling step. The classification step is used to identify and separate data representing the railroad or power line from the surrounding vegetation, while the modelling step is used to actually assess the location and/or the shape of the railroad or power line.

7.3.1 Power line monitoring

Power failures in large regions happen due to flashovers in power lines. Such problems can be prevented by monitoring the state and the direct surroundings of the power lines. Traditional ground-based or airborne visual inspection is a difficult, time-consuming and expensive process. Moreover, for a human, even when located in a helicopter, it is very difficult to reliably estimate, for example, the distance between a tree and a transmission cable. The use of airborne laser altimetry can largely automate the monitoring process. In fact, thousands of kilometres of power line have already been monitored in recent years using the FLI-MAP system [FLI-MAP, 2007; Reed *et al.*, 1996], which was introduced on a commercial basis in 1995. Two main issues when monitoring high-voltage transmission lines are first the location of the surrounding vegetation and second the condition of the power poles and notably the power lines themselves (see Figure 7.13). Vegetation close by may eventually damage the power line and should be removed. Power line spans with a high amount of sag should be repaired. Once the after-built positions of power lines under normal conditions are known, better predictions of possible interaction with the 3D environment under more extreme wind or ice loads, or under increased electricity load, can also be obtained.

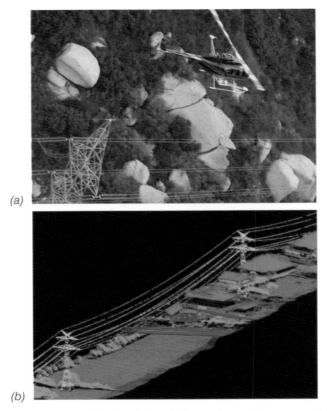

(a)

(b)

Figure 7.13 Power line monitoring: (a) aerial laser data collection; and (b) classified data. (Fugro FLI-MAP; courtesy Fugro Aerial Mapping)

7.3.1.1 Data collection

Power line cables typically have a diameter between 8 and 25 mm. In order to obtain enough laser points on the cables, both the operational height of the laser altimeter and the beam divergence of the laser beam should be low. Note that only a limited corridor around the power line needs to be included in the monitoring process. A helicopter with a low flying height above the power line is therefore most often used for the monitoring (see Figure 7.13). McLaughlin (2006) reports the use of a Saab Topeye Laser system attached to a helicopter operating with a swath width of 76 m and a spatial resolution of 2.5 points/m^2. These settings result in a sampling of the power line cables of about 1 point for every 2 m of cable. As a comparison, for the FLI-MAP 400 system an average point density of 74 points/m^2 is reported, at a flying height of 100 m and a speed of 35 knots [FLI-MAP, 2007].

7.3.1.2 Classification

The first step after data collection and registration is classification of the laser data into relevant classes. McLaughlin (2006) only considers three classes: vegetation, surface and transmission lines. As the data on the cables themselves are relatively sparse, classification is done in elongated, ellipsoidal neighbourhoods, aligned along the direction of the cables coinciding with the direction of flight. Within each neighbourhood the 3×3 covariance matrix of the xyz observations is built. The eigenvectors of the covariance matrix indicate the principal directions of the laser points in the neighbourhood, while the sizes of the corresponding eigenvalues clearly indicate the possible dominance of a direction: for neighbourhoods containing only power line points, the values of the maximal eigenvalue were typically 1000 times larger than the remaining eigenvalues. In this way, power line neighbourhoods are indicated by one large eigenvalue, surface neighbourhoods by two, and vegetation-dominated neighbourhoods by three large eigenvalues. The classification is completed by modelling the three clusters with a Gaussian mixture model: each neighbourhood is mapped to a point in 3D, whose coordinates are given by the ordered eigenvalues. This results in clusters, each corresponding to a cluster distribution. Finally a class is chosen depending on the class-distribution that maximises the likelihood; see also Figure 7.12, where a similar method is applied to classifying the orientation of rock facets. Alternatively, Axelsson (1999) reports the initial classification of power lines based on the reflectance properties and the amount of returned echoes for a multiple return system: power lines almost always give multiple echoes. A more refined classification of the power lines, which is elongated and locally linear, can be obtained using (iterative) Hough transformations [see also Melzer and Briese, 2004].

7.3.1.3 Transmission cable modelling

Given the classification results, it remains to obtain parameters describing the geometry of individual span lines. These can then be applied both to assess the quality of the span lines themselves and to determine the distance of the surrounding vegetation from the span lines. To reconstruct a span line, McLaughlin (2006) starts from an arbitrary neighbourhood, previously classified as containing mainly span line points. Based on a first linear fit of the points in this neighbourhood, a prediction can be made as to where adjacent neighbourhoods should be positioned. The points in such neighbourhoods are then joined to the points already found and a catenary curve is fitted to the point set. A catenary curve is the curve that describes a hanging flexible wire or chain of elasticity α, which is supported at its ends and is acted upon by a uniform gravitational force. The height z at position x of a catenary curve with a minimum at $x = 0$ is given by

$$z = \frac{1}{2} \alpha \left(e^{\frac{x}{\alpha}} + e^{\frac{x}{\alpha}} \right)$$ (7.4)

The algorithm finishes when no additional local models can be added. The main concern with this approach is undersampling of the power lines. Melzer and Briese (2004) describe a related but slightly more involved methodology for separating groups of span lines into individual catenary curves. Both McLaughlin (2006) and Melzer and Briese (2004) report problems using Hough transformations for identifying laser scanner points representing power lines, due to varying sample densities.

7.3.2 Dike and levee inspection

The major reason for setting up dike-monitoring programmes is the large impact of a dike failure. The recent collapse of two minor dikes in The Netherlands resulted in economic damage of about €50 million. The failure of the levees as a result of Hurricane Katrina in New Orleans was catastrophic. In the aftermath of this event the United States Geological Survey used terrestrial laser scanning to visualise the damage and to provide data for investigations of the performance of the levee systems [Seed *et al.*, 2006].

Both airborne and terrestrial laser scanning have proven themselves as methods for the surveying of levees, dikes or embankments. As in the case of power line monitoring, airborne laser scanning is a fast way of recording large distances of dike. Franken and Flos (2006) report a survey speed of 100–150 km a day. Water boards in The Netherlands, for example, have the task of monitoring the dikes in their region on a regular basis. The analysis of the embankments mainly involves the comparison between an ideal, theoretical profile and the real profile as obtained from the laser data (see Figure 7.14).

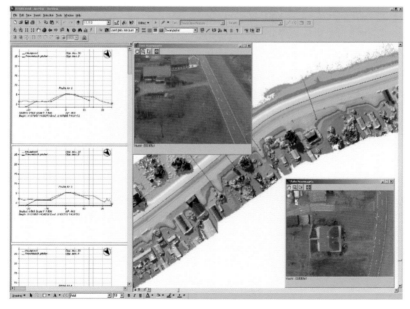

Figure 7.14 FLI-MAP dike monitoring. (Fugro FLI-MAP; courtesy Fugro Aerial Mapping)

Figure 7.15 Leica Scanstation 2 scan of a forced dike collapse. (Courtesy Hansje Brinker)

FLI-MAP samples embankments typically from a helicopter flying 100 m above the ground at a resolution of 10–25 points/m^2.

Nevertheless, dike failure mechanisms and the best way to monitor them are not yet fully understood. In the Dutch IJkdijk experiment [Stichting IJkdijk, 2008], the water pressure in a test dike was gradually increased until it collapsed (see Figure 7.15). Using a Leica Scanstation 2 scanner, more than 100 scans were obtained during the three days of the experiment. A deformation analysis of the scan results is expected to give insight into the signs that can be expected before a possible collapse [Van Leijen *et al.*, 2009].

7.4 Conclusions

In this chapter several applications of laser scanning have been described, focused on the control of both manmade engineering constructions such as locks, tunnels, power lines or dikes, and potentially hazardous natural environments, such as dune areas or landslide regions. These applications have in common that repeated surveys are required to determine if and what intervention is necessary.

Airborne laser scanning is already a fully operational tool for controlling power lines and, to a lesser extent, dikes. It is expected that more and more national and regional bodies will decide to initiate and consecutively maintain a high-resolution archive of airborne laser scanning derived surface elevations. The existence of such archive data will enable assessment of landscape changes after hazardous events such as flooding or landslides by comparing a newly obtained laser dataset with the situation in the archive. For archive update planning, deformation analysis methods will also turn out to be useful. Future methods of deformation analysis should also be able to profit from the enriched full-waveform signal (see Chapter 6).

Methods that profit from the large observational redundancy of terrestrial laser scans are still mainly used and developed by academics. Due to their great potential, it might be expected that more and more engineering companies will start to make such methods operational for everyday use. A partially new area of research would be initiated if laser scanners were used continuously from a fixed standpoint for monitoring of, for example, large construction works. In such cases a Kalman filtering approach could be used to automatically process the resulting near-continuous data streams.

Finally it should be remarked that applications like deformation analysis and corridor monitoring will directly benefit from progress made in the different processing steps treated in other chapters, notably registration, error filtering and methods such as segmentation for structuring point clouds. For statistical methods a reliable quality description of the input point clouds may not be indispensable but is still highly recommended.

References

Abellán, A., Vilaplana, J. and Martínez, J., 2006. Application of a long-range Terrestrial Laser Scanner to a detailed rockfall study at Vall de Núria (Eastern Pyrenees, Spain). *Engineering Geology*, 88(3–4), 136–148.

Alba, M., Fregonese, L., Prandi, F., Scaioni, M. and Valgoi, P., 2006. Structural monitoring of a large dam by terrestrial laser scanning. *International Archives of Photogrammetry, Remote Sensing and Spatial Information Sciences*, 36(Part 5), 1–8.

Alba, M., Roncoroni, F. and Scaioni, M., 2008. Investigations about the accuracy of target measurement for deformation monitoring. *International Archives of Photogrammetry, Remote Sensing and Spatial Information Sciences*, 37(Part B5), 1053–1059.

Axelsson, P., 1999. Processing of laser scanner data, algorithms and applications. *ISPRS Journal of Photogrammetry and Remote Sensing*, 54(2–3), 138–147.

Bauer, A., Paar, G. and Kaltenböck, A., 2005. Mass movement monitoring using terrestrial laser scanner for rock fall management. In *Geo-information for Disaster Management*, Peter van Oosterom, S.Z. and Fendel, E.M. (Eds.). Springer, Berlin and Heidelberg, Germany, 393–406.

Besl, P. and McKay, N., 1992. A method for registration of 3D shapes. *IEEE Transactions on Pattern Analysis and Machine Intelligence*, 14(2), 239–256.

FLI-MAP, 2007. Fli-map transmission lines. http://www.flimap.nl/ (accessed 18 March 2009).

Franken, P. and Flos, S., 2006. Using a helicopter based laser altimetry system (FLI-MAP) to carry out effective dike maintenance and construction policy. In Van Alphen, Van Beek and Taal (Eds.), *Floods, from defence to management: proceedings of the 3rd international symposium on flood defence, Nijmegen, The Netherlands, 25–27 May 2006*, Enschede, The Netherlands. Taylor and Francis Group, London, UK.

Girardeau-Montaut, D., Roux, M., Marc, R. and Thibault, G., 2005. Change detection on points cloud data acquired with a ground laser scanner. *International Archives of Photogrammetry, Remote Sensing and Spatial Information Sciences*, 36(Part 3/W19), 30–35.

Goodman, J.E. and O'Rourke, J., 1997. *Handbook of Discrete and ComputationalGeometry*. CRC Press, Boca Raton, FL, USA.

Gordon, S.J. and Lichti, D.D., 2007. Modeling terrestrial laser scanner data for precise structural deformation measurement. *Journal of Surveying Engineering*, 133(2), 72–80.

Grün, A. and Akca, D., 2005. Least squares 3D surface and curve matching. *ISPRS Journal of Photogrammetry and Remote Sensing*, 59(3), 151–174.

Haasnoot, H., 2001. Aerial survey of fix assets in the right-of way. *Proceedings FIG Working Week 2001*, 6–11 May 2001, Seoul, South-Korea. http://www.fig.net/pub/proceedings/korea/full-papers/session9/haasnoot.htm (accessed 20 March 2009).

Koch, K.-R., 1999. *Parameter Estimation and Hypothesis Testing in Linear Models*. Springer, Berlin and Heidelberg, Germany.

Langer, D., Mettenleiter, M., Härtl, F. and Fröhlich, C., 2000. LADAR for 3-d surveying and cad modeling of real-world environments. *International Journal of Robotics Research*, 19(11), 1075–1088.

Lindenbergh, R. and Hanssen, R., 2003. Eolian deformation detection and modeling using airborne laser altimetry. *Proceedings 2003 IEEE International Geoscience and Remote Sensing Symposium*, Toulouse, France, on CD-ROM.

Lindenbergh, R. and Pfeifer, N., 2005. A statistical deformation analysis of two epochs of terrestrial laser data of a lock. *Proceedings 7th Conference on Optical 3-D Measurement Techniques*, 3–5 October 2005, Vol. II., Vienna, Austria, 61–70.

Little, M., 2006. Slope monitoring strategy at PPRust open pit operation. *Proceedings International Symposium on Stability of Rock Slopes in Open Pit Mining and Civil Engineering*, 3–6 April 2006, Capetown, South Africa, 211–230.

Maas, H.-G., 2000. Least-squares matching with airborne laserscanning data in a tin structure. *International Archives of Photogrammetry, Remote Sensing and Spatial Information Sciences*, 36(Part 3A), 548–555.

Markley, J., Stutzman, J. and Harris, E., 2008. Hybridization of photogrammetry and laser scanning technology for as-built 3D CAD Models. *Proceedings, IEEE Aerospace Conference*, 1–8 March 2008, Big Sky, MT, USA, 1–10.

McLaughlin, R.A., 2006. Extracting transmission lines from airborne LIDAR data. *IEEE Geoscience and Remote Sensing Letters*, 3(2), 222–226.

Melzer, T. and Briese, C., 2004. Extraction and modeling of power lines from als point clouds. *Proceedings of 28th Workshop Austrian Association For Pattern Recognition*, 17–18 June 2004, Hagenberg, Austria, 47–54.

Monserrat, O. and Crosetto, M., 2007. Deformation measurement using terrestrial laser scanning data and least squares 3D surface matching. *ISPRS Journal of Photogrammetry and Remote Sensing*, 63(1), 142–154.

Murakami, H., Nakagawa, K., Hasegawa, H., Shibata, T. and Iwanami, E., 1999. Change detection of buildings using an airborne laser scanner. *ISPRS Journal of Photogrammetry and Remote Sensing*, 54(2–3), 148–152.

Pieraccini, M., Noferini, L., Mecatti, D., Atzeni, C., Teza, G., Galgaro, A. and Zaltron, N., 2006. Integration of radar interferometry and laser scanning for remote monitoring of an urban site built on a sliding slope. *IEEE Transactions on Geoscience and Remote Sensing*, 44(9), 2335–2342.

Rabbani, T., 2006. Automatic reconstruction of industrial installations. Ph.D. thesis, Delft University of Technology.

Rabbani, T., Dijkman, S., van den Heuvel, F. and Vosselman, G., 2007. An integrated approach for modelling and global registration of point clouds. *ISPRS Journal of Photogrammetry and Remote Sensing*, 61(6), 355–370.

Rabbani, T. and Van den Heuvel, F., 2005. Efficient Hough transform for automatic detection of cylinders in point clouds. *International Archives of Photogrammetry, Remote Sensing and Spatial Information Sciences*, 36(Part 3/W19), 60–65.

Rabbani, T., Van den Heuvel, F. and Vosselman, G., 2006. Segmentation of point clouds using smoothness constraints. *International Archives of Photogrammetry, Remote Sensing and Spatial Information Sciences*, 36(5), 248–253.

Reed, M., Landry, C. and Werther, K., 1996. The application of air and ground based laser mapping systems to transmission line corridor surveys. *Proceedings IEEE Position Location and Navigation Symposium*, 22–26 April 1996, Atlanta, GA, 444–457.

Rosser, N., Petley, D., Lim, M., Dunning, S. and Allison, R., 2005. Terrestrial laser scanning for monitoring the process of hard rock coastal cliff erosion. *Quarterly Journal of Engineering Geology and Hydrogeology*, 38(4), 363–375.

Samet, H., 1989. Implementing ray tracing with octrees and neighbor finding. *Computers and Graphics*, 13(4), 445–460.

Schäfer, T., Weber, T., Kyrinovic, P. and Zamecnikova, M., 2004. Deformation measurement using terrestrial laser scanning at the hydropower station of Gabcikovo. *Proceedings INGEO 2004 and FIG Regional Central and Eastern European Conference on Engineering Surveying*, 11–13 November 2004, Bratislava, Slovakia.

Schneider, D., 2006. Terrestrial laser scanning for area based deformation analysis of towers and water damns. *Proceedings 3rd IAG/12th FIG Symposium*, 22–24 May 2006, Baden, Germany.

Seed, R. *et al.*, 2006. *Investigation of the Performance of the New Orleans Flood Protection Systems in Hurricane Katrina on August 29, 2005*. Tech. rep., United States Geological Survey, Independent Levee Investigation Team Final Report.

Slob, S. and Hack, H., 2004. 3D terrestrial laser scanning as a new field measurement and monitoring technique. In *Engineering geology for infrastructure planning in Europe: a European perspective*, Hack, H., Azzam, R. and Charlier, R. (Eds.). Springer, Berlin, Germany, 179–190.

Slob, S., Hack, H., Van Knapen, B., Turner, K. and Kemeny, J., 2005. A method for automated discontinuity analysis of rock slopes with three-dimensional laser scanning. *Transportation Research Record*, 1913, 187–194.

Sternberg, H. and Kersten, T., 2007. Comparison of Terrestrial Laser Scanning systems in industrial as-built-documentation applications. *Proceedings Optical 3-D Measurement Techniques*, *VIII*, 9–12 July 2007, Vol.1., Zurich, Switzerland, 389–397.

Stichting IJkdijk, 2008. (TNO, STOWA, Deltares, NV NOM, IDL Sensor Solustions), in cooperation with Dutch Rijkswaterstaat. http://www.ijkdijk.nl (accessed 16 October 2008).

Stone, W.C., Cheok, G.S. and Lipman, R.R., 2000. Automated earthmoving status determination. *Proceedings ASCE Conference on Robotics for Challenging Environments*, 27 February to 2 March, 2000, Albuquerque, NM, USA, 111–119.

Tao, C. and Hu, Y., 2002. Assessment of airborne LIDAR and imaging technology for pipeline mapping and safety applications. *Proceedings International Joint Conference on Integrat-*

ing Remote Sensing at the Global, Regional, and Local Scale, 10–15 November 2002, Denver, CO, USA, http://www.isprs.org/commission1/proceedings02/paper/00009.pdf (accessed 20 March 2009).

Teunissen, P.J.G., 2000a. *Adjustment Theory*. Delft University Press, Delft.

Teunissen, P.J.G., 2000b. *Testing Theory*. Delft University Press, Delft.

Teza, G., Galgaro, A., Zaltron, N. and Genevois, R., 2007. Terrestrial laser scanner to detect landslide displacement fields: a new approach. *International Journal of Remote Sensing*, 28(16), 3425–3446.

Teza, G., Pesci, A., Genevois, R. and Galgaro, A., 2008. Characterization of landslide ground surface kinematics from terrestrial laser scanning and strain field computation. *Geomorphology*, 97(3–4), 424–437.

Van Gosliga, R., Lindenbergh, R. and Pfeifer, N., 2006. Deformation analysis of a bored tunnel by means of terrestrial laser scanning. *International Archives of Photogrammetry, Remote Sensing and Spatial Information Sciences*, 36(Part 5), 167–172.

Van Leijen, F., Humme, A. and Hanssen, R., 2009. Failure indicators from non-invasive deformation measurements. *Water Resources Research*. In preparation.

Zeibak, R. and Filin, S., 2007. Change detection via terrestrial laser scanning. *International Archives of Photogrammetry, Remote Sensing and Spatial Information Sciences*, 36(Part 3), 430–435.

Zogg, H.-M. and Ingensand, H., 2008. Terrestrial laser scanning for deformation monitoring – load tests on the Felsenau viaduct (CH). *International Archives of Photogrammetry, Remote Sensing and Spatial Information Sciences*, 37(Part B5), 555–561.

Cultural Heritage Applications | 8

Pierre Grussenmeyer and Klaus Hanke

Cultural heritage is a testimony of past human activity and, as such, cultural heritage objects exhibit great variety in their nature, size and complexity, from small artefacts and museum items to cultural landscapes, from historic buildings and ancient monuments to city centres and archaeological sites [Patias *et al.*, 2008]. The importance of cultural heritage documentation is well recognised. ICOMOS (International Council on Monuments and Sites) and the International Society of Photogrammetry and Remote Sensing (ISPRS) combined efforts in 1969 and created the International Committee for Heritage Documentation, CIPA (http://cipa.icomos.org).

Due to the rapidity of data capture and the ability to obtain point clouds instantaneously, laser scanning has become an essential tool along with image-based documentation methods. Total station surveys, on the other hand, require more time on site and usually do not deliver the same level of surface detail.

Cultural heritage objects and sites are often a conglomerate of irregular surface geometries. Although photogrammetry will work for similar structures, laser scanning provides dense 3D information in almost real time with a high capacity for visualisation as a first on-site preresult.

Laser scanning is of interest where documentation is usually a complex task. The variety of applications includes pure visualisation as well as heritage recording and as-built documentation (for example for further restoration, revitalisation or other planning tasks). The value of photographs nevertheless remains inestimable in cultural heritage, and photographs or videos are often used as complementary data sources to colourise point clouds and to texture the final models.

General rules have been set up for site survey planning [English Heritage, 2007] as configuration, accuracy, point density, overlap and instrumental requirements may vary significantly given the numberless applications in the various fields of cultural heritage. To illustrate this variety, a representative but limited selection of five successful projects has been selected for this chapter.

As a first example, an approach to accurate site recording is presented through the 3D reconstruction of a complex historic façade. The generation of a photo-realistic model of a complex shaped object is illustrated using data fusion of image and laser measurements.

A long-distance topographic scanning of an entire UNESCO world heritage site follows. Vectorisation of point clouds leading to CAD models and geometric structural analyses show the potential of as-built documentation for large monuments.

Airborne laser scanning also has high potential for archaeological prospection. The recent full-waveform analysis can be used to find microstructures in relief and faint historic traces hidden by vegetation, even in forested areas.

3D documentation of the excavation progress accompanying fieldwork illustrates the support of laser scanning for the investigation of stratification and finally provides an overall view of the different surface levels and location of finds.

The last example shows different kinds of 3D meshed, photo-textured and solid models from point clouds resulting from accurate post-processing and reconstruction work.

8.1 Accurate site recording: 3D reconstruction of the treasury (Al-Kasneh) in Petra, Jordan

The ancient Nabataean city of Petra has often been called the eighth wonder of the ancient world. Petra city in south-western Jordan (Figure 8.1) prospered as the capital of the Nabataean empire from BC 400 to AD 106. Petra's temples, tombs, theatres and

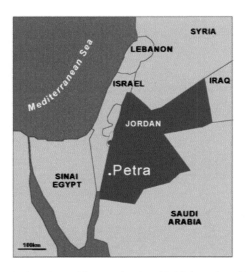

Figure 8.1 Geographical situation of Petra, Jordan (© University of Hawaii).

other buildings are scattered over 400 square miles, carved into rose-coloured sand-stone cliffs. When a visitor enters Petra via Al-Siq, an impressive two-kilometre crack in the mountain, he will first see Al-Khasneh (Figure 8.2), which is probably the most famous object in Petra. The Al-Khasneh façade is 40 m high and remarkably well preserved, probably because the confined space in which it was built has protected it from the effects of erosion. The name Al-Khasneh means "treasury or tax house for passing camel caravans", but others have proposed that the Al-Khasneh monument was a tomb. Behind the impressive façade there are large square rooms that have been carved into the rock.

In this project, the 3D laser scanning system Mensi GS100 was applied. The scanner features a field of view of 360° in the horizontal direction and 60° in the vertical direction, enabling the collection of full panoramic views. The distance measurement is obtained using the time-of-flight measurement principle based on a green laser at 532 nm allowing an object distance between 2 and 100 m. The system is able to measure 5000 points per second. During data collection a calibrated video snapshot of 768 × 576 pixel resolution is additionally captured, which is automatically mapped to the corresponding point measurements. Because of the complex structured sites, multiple scanner stations had to be chosen to avoid most of the occlusions.

For texturing of the 3D model, additional images were collected by a Fuji S1 Pro camera, with a focal length of 20 mm and a sensor format of 1536 × 2034 pixels (Figure 8.3).

Figure 8.2 The Outer Siq at the Khasneh al-Faroun.

Figure 8.3 Images collected using a Fuji camera. (Alshawabkeh and Haala, 2005; courtesy the authors)

The acquired 3D point clouds have been processed using Innovmetric's Poly-Works™ Software. Registration of the scans for both models was done using corresponding points. The models produced have an average point spacing of 5 cm with more than 2 000 000 triangles for the entire model (Figure 8.4).

Figure 8.4. 3D model of Al-Khasneh. (Alshawabkeh and Haala, 2005; courtesy the authors)

Figure 8.5 Final textured model using four images. (Alshawabkeh and Haala, 2005; courtesy the authors)

Distortion-free images and high quality of the registration process require an accurate determination of the camera's interior and exterior orientation parameters. The interior orientation parameters and zero distortion images were computed by the Australis™ software. PhotoModeler™ was applied to calculate the camera position and orientation in the coordinate system of the geometric model.

Performing projective texture mapping to generate photo-realistic models of the complex shaped objects with minimal effort, it was necessary to compute visibility information to map the image only onto the portions of the scene that are visible from its camera viewpoint. For this purpose an efficient algorithm addressing image fusion and visibility has been developed [Alshawabkeh *et al.*, 2004].

The result was a textured 3D model of the site (Figure 8.5) demonstrating the possibilities of the integration of laser scanning and photogrammetry for data capture of historic scenes.

The work was carried out by Yahya Alshawabkeh and Norbert Haala, Institute for Photogrammetry (IFP), University of Stuttgart, Germany, with support from Hashemite University of Jordan, Petra, Regional Authority and Jerash Municipality, Jordan [Alshawabkeh *et al.*, 2004; Alshawabkeh, 2005; Alshawabkeh *et al.*, 2005].

8.2   Archaeological site: scanning the pyramids at Giza, Egypt

The aim of the *Scanning of the Pyramids Project* in 2004 was to apply and test the latest state-of-the-art terrestrial laser scanners combined with a calibrated digital camera for high-accuracy, high-resolution and long-distance topographic scanning in archaeology [Neubauer *et al.*, 2005]. The monuments selected for this campaign were the pyramids and the Sphinx. The data form the basis for a detailed 3D modelling of the monuments to show and to test the instrumentation as a general-purpose tool for the documentation and monitoring of standing monuments. The collected data therefore provide a solid and large database for any further analysis related to scientific problems regarding the construction, destruction and decay of World Heritage Monuments.

A long-range laser scanner Riegl LMS Z420i (Figure 8.6) combined with a calibrated Nikon D100 digital camera was used for this project. The combined sensor of a high-performance long-range laser scanner and a calibrated and orientated high-resolution digital camera provides scan and image data.

The three main objectives of the *Scanning of the Pyramids* project are:

- the collection of high-resolution and high-accuracy topographic data for the creation of a digital elevation model of the Giza plateau (Figure 8.7);
- a 3D documentation of the Cheops Pyramid and surroundings;
- a 3D documentation of the Sphinx (Figure 8.9).

The Riegl Z420i scanner was taken to the top of the Great Pyramid of Cheops for a full ten hours of data collection.

Figure 8.6 LMS-Z420i with calibrated high-resolution digital camera Nikon D100 at the top of the Cheops Pyramid, Egypt. (www.riegl.com; courtesy Riegl)

Figure 8.7 Overview and detail of the digital elevation model of the Giza Plateau created using four single scans from the top of the Cheops Pyramid visualised in ARC GIS 8.2. (Neubauer *et al.*, 2005; courtesy the authors)

Figure 8.8 Single scan from the north and east faces of the Chephren Pyramid visualised as a coloured point cloud. (Neubauer *et al.*, 2005; courtesy the authors)

Figure 8.9 Triangulated point cloud of the Sphinx textured in RiSCAN Pro, combined from seven scanner positions: six from the ground and one from the Cheops Pyramid. (Neubauer *et al.*, 2005; courtesy the authors)

Figure 8.10 Mobile scanner in front of the Sphinx. (Neubauer *et al.*, 2005; courtesy the authors)

Figure 8.11 Single stones can be directly drawn in global 3D coordinates and made visible in the point cloud, superimposed with a photograph using the Microstation application PHIDIAS. (Neubauer *et al.*, 2005; courtesy the authors)

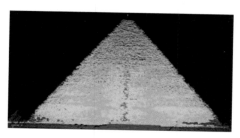

Figure 8.12 Anomalies of the Cheops Pyramid: horizontal projection of the west side with colour-coded deviation from the plane. (Neubauer *et al.*, 2005; courtesy the authors)

These data have been processed automatically with human interaction in some processing steps for the generation of textured triangulated surfaces (Figure 8.8) and orthophotos with depth information (Figure 8.12).

The *Scanning of the Pyramids Project* (2004) was invoked as a scientific project by the Austrian Archaeological Institute, Cairo and VIAS-Vienna Institute for Archaeological Science, University of Vienna, in cooperation with the Egyptian Supreme Council of Antiquities (*Dr Zahi Hawass*). The project was directed by *Professor Dr Manfred Bietak*, Director of the Mission, and *Dr Wolfgang Neubauer*, Field Director. The project description mainly follows Neubauer *et al.* (2005). Riegl LMS GmbH and VIAS-University of Vienna hold the copyright to the pictures.

8.3   Archaeological airborne laser scanning in forested areas

Woodland has a stabilising effect on archaeological remains in relief and therefore protects them from erosion. At the same time, forests and vegetated areas have always caused problems for archaeological investigations: structures in relief are hidden by the vegetation and only larger ones can be easily detected using archaeological prospection methods such as aerial archaeology or field survey. Faint structures can usually only be visualised and detected using detailed and precise digital terrain models (DTM), which can be exaggerated and illuminated from different directions. Over the past few years, therefore, airborne laser scanning has made a significant contribution to archaeological prospection, especially of vegetated areas.

The goal of the project presented here was to evaluate the potential of airborne laser scanning for archaeological prospection in vegetated areas. A 200 km^2 largely forested area (mainly deciduous trees) south east of Vienna was scanned using a full-waveform RIEGL Airborne Laser Scanner LMS-Q560. During the pilot phase, a test scan was made over two selected archaeological sites in April 2006, while the whole area was scanned in spring 2007. The scan was performed by the company Milan-Flug within the short time between melting of the snow and the trees still being leafless. Flight altitude was about 600 m above the ground, which resulted in a laser footprint size of 30 cm on the ground. The area was covered with a scan angle of 45° by parallel flight tracks, which had a width of approximately 500 m and an overlap of 50%. The effective scan rate was 66 kHz, resulting in an overall mean point density of eight measurements per square metre (Doneus *et al.*, 2008).

Areas of extremely dense vegetation and clearance piles could be partly interactively removed from the data before filtering. This resulted in a more reliable classification of the laser data into terrain and off-terrain points in comparison with conventional (discrete echo) lidar [Doneus and Briese, 2006a; Doneus and Briese, 2006b]. Both analysis and georeferencing of the full-waveform data were carried out in cooperation with the Institute of Photogrammetry and Remote Sensing of the Technical University of Vienna. The software package SCOP++ was used for filtering of the data (Chapter 4).

The results are very promising. Most of the forest canopy and brushwood covering the archaeological features could be removed (Figures 8.13 and 8.14). The resulting DTM shows a detailed map of the topography with even faint archaeological structures (Figures 8.15, 8.16 and 8.17) under the forest canopy and therefore clearly demonstrates the potential of full-waveform airborne laser scanning for archaeological prospection of forested areas. A review describing the structure and the physical

(a)

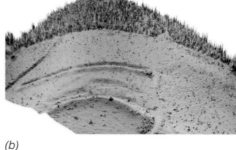

(b)

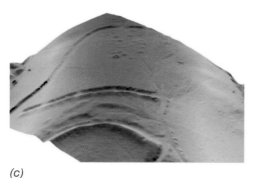

(c)

Figure 8.13 (a) DSM of the first pulse data showing the canopy of the scanned area. In the foreground, the vegetation consists of dense bushes. In the background there is a dense forest with an understorey. (b) DSM resulting from the unfiltered last echo point cloud. There are many points which represent tree trunks, very dense vegetation or narrow vegetation. (c) Filtered DTM showing even faint archaeological traces, for example round barrows with shallow depressions from looting. (Courtesy Michael Doneus)

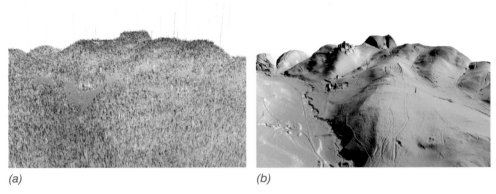

(a)   *(b)*

Figure 8.14 (a) DSM of the first pulse data showing the canopy over the archaeological site of "St. Anna in der Wüste". (b) Filtered DTM showing a former monastery, a ruined castle, remains of several buildings, traces of former fields, hollow ways and round barrows. (Courtesy Michael Doneus)

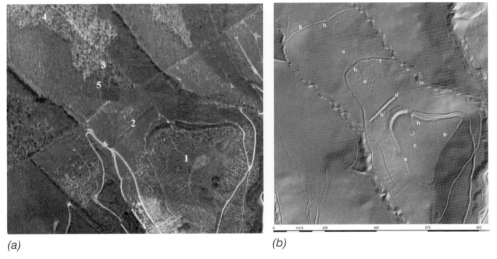

(a)   *(b)*

Figure 8.15 (a) Mosaic of aerial photographs covering the Iron Age hill fort of Purbach. (b) Final DTM after filtering using the theory of robust interpolation with an eccentric and unsymmetrical weight function used within a hierarchical framework. (Courtesy Michael Doneus)

backscattering characteristics of the illuminated surface and the resulting additional information can be found in Mallet and Bretar (2008).

The project "LiDAR supported archaeological prospection of woodland", which was kindly contributed by Michael Doneus, Department of Prehistoric and Mediaeval Archaeology at the University of Vienna, was financed by the Austrian Science Fund (P18674-G02).

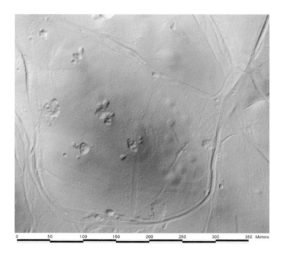

Figure 8.16 Filtered DTM showing round barrows. The barrows are very faint and can hardly be recognised on-site. (Courtesy Michael Doneus)

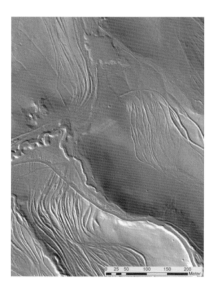

Figure 8.17 Tracks of hollow ways from the mediaeval and post-mediaeval period. (Courtesy Michael Doneus)

8.4 Archaeological site: 3D documentation of an archaeological excavation in Mauken, Austria

Within the framework of the Special Research Program HiMAT (an interdisciplinary research project dedicated to the history of mining in Tyrol and adjacent areas, sponsored by the Austrian Science Foundation), archaeological investigations were conducted in 2007 and 2008 in a prehistoric mining landscape in the lower Inn Valley, Northern Tyrol, Austria. In this region intensive copper ore exploitation took place during the Middle European Late Bronze Age (BC 1200–800). From this period numerous traces of underground mining, mineral processing and ore smelting have been located by archaeological prospection [Hanke, 2007; Hanke, 2008].

Within a former peat-bog, currently drying up due to drainage in the course of agricultural land use, the well-preserved remains of a Late Bronze Age mineral dressing plant was discovered and partially excavated. On this site mechanical crushing and grinding of copper ore from a nearby mine took place, followed by an ore washing process. The finds represent parts of the prehistoric equipment which was in use for mineral treatment, for example a wooden trough, and therefore are of high scientific interest with regard to the reconstruction of the whole metallurgical process chain.

The approach was to put a "permanent care" surveying team in place at the excavation for about five weeks to guarantee 3D documentation of the process at any time. During summer 2007 one field was excavated, followed by the neighbouring four places in 2008. The size of a single excavation field varied from about 4 to 5 m by 3 to 4 m and a depth of 0.5 to almost 2 m. The data acquisition (Figure 8.18) was achieved using a Trimble GX laser scanner with at least four stations for a single field. The average density of the point cloud was about 5 mm at a distance of 4 m. Up to three different archaeological layers were recorded per field.

To provide 3D models (Figure 8.19) with appropriate textures a calibrated Nikon D200 with a 20 mm lens was used for imaging.

The registration was done with Trimble Realworks™ software. All further data processing, filtering, editing and meshing were computed using InnovMetric's PolyWorks™. To texture the models (an example of a textured section is given in Figure 8.20), ShapeTexture software provided by ShapeQuest Inc. was used.

Figure 8.18 Consolidated point cloud of all excavation fields at their lowest level.

Figure 8.19 3D model of all fields; the one from 2007 is already textured.

Figure 8.20 Section cutting the different layers of the 3D model of a single excavation field.

After finishing the excavation of a field it was covered again, so none of the archaeologists ever had an overall view of all the open excavations. The 3D documentation and modelling of the laser scan data finally led to a consolidated common image where comprehensive structures could be identified.

The project was carried out by Klaus Hanke and his group at the Surveying and Geoinformation Unit, University of Innsbruck, Austria within the framework of the Special Research Program HiMAT and was generously supported by the Austrian Science Fund (FWF) under project No. F3114-G02.

8.5 The archaeological project at the Abbey of Niedermunster, France

Located at the bottom of Mount Sainte-Odile, the Abbey of Niedermunster, Alsace, France can be considered as the origin of the sanctuary. Built between AD 1150 and 1180, the Roman-style Abbey was devastated during the War of the Peasants (1525) and by two fires, in 1542 and 1572. The site was then used as a quarry until the nineteenth century. The great western massif of the basilica, still in elevation, and the relics of the crypt allow the beauty of the original buildings to be imaged.

During the first stage, the archive archaeological records were georeferenced in a Geographical Information System after an accurate topographic survey. The second stage of scanning was accomplished in two steps: scanning of the abbey's area in its current state, followed by supplementary scanning site details.

Data acquisition consisted of scanning the complete site using 3D terrestrial laser scanning [Koehl and Grussenmeyer, 2008]. A Trimble GX Advanced scanner was used. This allowed the acquisition of point clouds for which the density was defined by the operator according to the expected degree of detail. A grid density of 1000 points/m² was generally used. The obtained point clouds were georeferenced; the information for every point was thus constituted by X-, Y- and Z-coordinates in the ground coordinate system, completed by intensity or colour. For the complete digitising of the site, about 15 scanner positions were used, which resulted in 40 different scans, giving a total of approximately 30 million points (Figure 8.21).

Figure 8.22 is an illustration of the point cloud of the occidental block which is still standing up to its first level. Interior parts such as stairways and the porch have also been scanned. Eight scanning stations were used for this part of the project.

Figure 8.23 illustrates the process of combination of meshed surfaces computed from the terrestrial laser scanning point clouds and a CAD model used in the case of flat surfaces. The fusion of the two models was used for the creation of the complex and ruined parts of the project.

For the constitution of the 3D model, textures were mapped as shown in Figure 8.24. Some parts have been textured, others were only face-coloured. The inner parts are shown by application of surface transparency.

Figure 8.21 Fusion of point clouds in one project. (Koehl *et al.*, 2009; courtesy the authors)

Figure 8.22 Fusion of point clouds of the occidental block.

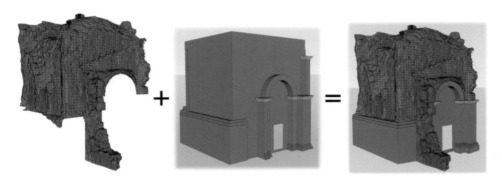

Figure 8.23 Model creation as CAD extrusions and meshed model from point clouds. (Koehl *et al.*, 2009; courtesy the authors)

Figure 8.24 Textured model of the occidental block. (Koehl *et al.*, 2009; courtesy the authors)

Figure 8.25 shows a detailed point cloud of the crypt. This part of the project is composed of two stairways, some pieces of pillar and an inner vaulted wall. As in the other parts of the abbey, the walls remained in a good state of conservation, but other parts were ruined. The resulting model (Figure 8.26) combines meshed point clouds and CAD models. The structure lines of the main parts used in the modelling process result from a manual segmentation. Figure. 8.27 is the result of the texturing process of the crypt model.

Figure 8.28 illustrates the different representation modes in the case of the acquisition of a pillar base. Terrestrial laser scanning offers the possibility of viewing the different point clouds in a colour mode. This is used to show the different segmented clouds. Intensities were also acquired by the laser scanner. The intensity mode can be used to analyse the responses of the scanner according to the surface texture. The true colour rendering mode uses the colour of each point according to its position in an image acquired by the laser scanner camera. The last illustration shows the CAD

Figure 8.25 Point cloud of the crypt.

Figure 8.26 CAD model from terrestrial laser scanning data. (Koehl *et al.*, 2009; courtesy the authors)

Figure 8.27 Textured model of the crypt.

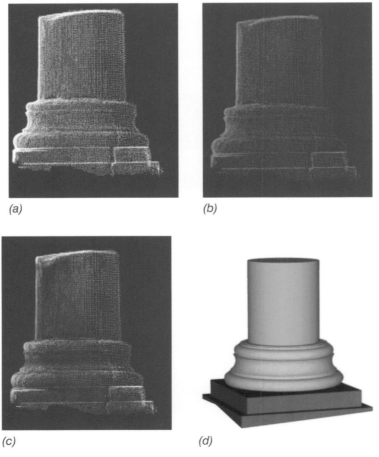

Figure 8.28 Detail of a pillar: (a) white point cloud; (b) intensity cloud; (c) true colour cloud; and (d) CAD model.

Figure 8.29 3D model of the Niedermunster Abbey site. (Koehl *et al.*, 2009; courtesy the authors)

model extracted from the point clouds using a segmentation of the profile. An extrusion process was then used for design of the model.

Figure 8.29 shows the 3D model of the whole site, allowing animations and walkthroughs.

This project was kindly contributed by Mathieu Koehl from the Photogrammetry and Geomatics Group at INSA (Graduate School of Science and Technology), Strasbourg, France.

References

Alshawabkeh, Y. and Haala N., 2004. Laser scanning and photogrammetry: a hybrid approach for heritage documentation. *3rd International Conference on Science and Technology in Archaeology and Conservation*, 7–11 December 2004, Amman, Jordan (available at http://www.ifp.uni-stuttgart.de/publications/2004/Alshawabkeh2004_Hashemite_Uni.pdf).

Alshawabkeh, Y., 2005. Using terrestrial laser scanning for the 3D reconstruction of Petra/Jordan. *Photogrammetric Week '05*, 5–9 September 2005, Stuttgart, Germany, Wichmann, Heidelberg, 39–47.

Alshawabkeh, Y. and Haala N., 2005. Automatic multi-image photo texturing of complex 3D scenes. *International Archives of Photogrammetry, Remote Sensing and Spatial Information Sciences*, '34 (Part 5/C34), 68–73.

Doneus, M. and Briese, C., 2006a. Digital terrain modelling for archaeological interpretation within forested areas using full-waveform laserscanning. *The 7th International Symposium on Virtual Reality, Archaeology and Cultural Heritage VAST*, 30 October to 4 November 2006, Nicosia, Cyprus, Ioannides, M., Arnold, D., Niccolucci , F. and Mania, K. (Eds.) 155–162.

Doneus, M. and Briese, C., 2006b. Full-waveform airborne laser scanning as a tool for archaeological reconnaissance. *From Space to Place. Proceedings of the 2nd International Conference on Remote Sensing in Archaeology*, 4–7 December 2006, Rome, Italy, Campana, S. and Forte, M. (Eds.), BAR International Series, 1568. Archaeopress, Rome, Italy, 99–106.

Doneus, M., Briese, C., Fera, M. and Janner, M., 2008. Archaeological prospection of forested areas using full-waveform airborne laser scanning. *Journal of Archaeological Science*, 35, 882–893.

English Heritage, 2007. *3D Laser Scanning for Heritage: Advice and Guidance to Users on Laser Scanning in Archaeology and Architecture*. Product code 51326. Available at www.heritage3D. org (accessed 13 June 2009).

Hanke, K., 2007. Contribution of laser scanning, photogrammetry and GIS to an interdisciplinary special research program on the history of mining activities (SFB HIMAT). *International Archives of Photogrammetry, Remote Sensing and Spatial Information Science*s, 36(Part 5/C53), 366–370 and *CIPA International Archives for Documentation of Cultural Heritage* 21, 366–370.

Hanke, K., Moser, M., Grimm-Pitzinger, A., Goldenberg, G. and Toechterle, U., 2008. Enhanced potential for the analysis of archaeological finds based on 3D modeling. *International Archives of Photogrammetry, Remote Sensing and Spatial Information Sciences*, 37(Part B5), 187–191.

Koehl, M. and Grussenmeyer, P., 2008. 3D model for historic reconstruction and archaeological knowledge dissemination: the Niedermunster abbey's project (Alsace, France). *International Archives of Photogrammetry, Remote Sensing and Spatial Information Sciences*, 37(Part B5), 325–330.

Koehl, M., Grussenmeyer, P. and Landes, T., 2009. Dokumentation und Denkmalpflege. Von der Kombination von Messungstechniken zum interaktiven 3D-Modell. *15. Internationale Geodätische Woche Obergurgl 2009*. Chesi/Weinold (Eds.), Herbert Wichmann, Heidelberg, 48–57.

Mallet, C. and Bretar, F., 2009. Full-waveform topographic lidar: State-of-the-art. *ISPRS Journal of Photogrammetry and Remote Sensing*, 64, 1–16.

Neubauer, W., Doneus, M., Studnicka N. and Riegl J., 2005. Combined high resolution laser scanning ad photogrammetrical documentation of the pyramids at Giza. XXth CIPA International Symposium, Torino, Italy. *International Archives of Photogrammetry, Remote Sensing and Spatial Information Sciences*, 36(Part 5/C34), 470–475 and *CIPA International Archives for Documentation of Cultural Heritage*, 20, 470–475.

Patias, P., Grussenmeyer, P. and Hanke, K., 2008. Applications in cultural heritage documentation, 2008. *Advances in Photogrammetry, Remote Sensing and Spatial Information Sciences: 2008, ISPRS Congress Book, ISPRS Book Series, Volume 7*, Li, Z., Chen, J., and Baltsavias, E. (Eds.), Taylor and Francis Group, London, UK, 363–383.

Mobile Mapping | 9

Hansjörg Kutterer

9.1 Introduction

Detailed object information based on 3D georeferenced data is of high interest for a broad range of users in science, administration, industry and private life. It is an important basis for geoinformation systems or spatial decision support systems and hence for a variety of applications such as national mapping agency activities, cultural heritage and as-built documentation, 3D city models, facility management, guidance and control in manufacturing and construction, pedestrian or vehicle navigation, and traffic accident or crime scene investigation.

Today, georeferenced object information is derived from highly resolved spatial data which are observed using airborne or terrestrial laser scanners or other imaging devices such as digital cameras. In contrast to most of the other imaging sensors, a terrestrial laser scanner directly provides all three spatial dimensions for quite a flexible range of distances and environments.

Two modes of terrestrial laser scanning procedures can be distinguished: the static mode (fixed mounting of the scanner on one particular station during the total scan) and the kinematic mode (variable scanner position and orientation during the scan time). In both cases dense 3D point clouds are obtained which are parameterised by 3D Cartesian coordinates and possibly a reflectivity value for each observed point (see Chapter 1).

In static mode the point clouds refer to the respective station coordinate systems. In order to describe point clouds from different stations in one unique coordinate system they have to be registered and georeferenced (see Chapter 3). Due to the stable position and orientation during scan time, static scans provide point clouds of good geometric quality. Note that both the data acquisition and the data processing steps are typically rather time consuming.

Efficiency with regard to registration and georeferencing is significantly increased in the kinematic mode, which shows many analogies to airborne laser scanning (see

Chapter 1). The kinematic mode is also known as the dynamic mode; this second notation will not be used in the following. Here, extended objects such as road scenes or building façades are observed much faster than in static mode. Due to the continuous motion of the laser scanner, each observed scan point refers to an individual 3D Cartesian coordinate system. In order to mathematically link all these systems in a unique spatial reference system ("world system"), such as the WGS 84 or some national system, the respective positions and orientations have to be observed using adequate equipment based on GNSS (global navigation satellite systems) and IMU (inertial measurement unit) technology. Moreover, all sensors in use have to be synchronised precisely with a unique time reference.

The so-called stop-and-go mode links the static and kinematic modes. Here, static terrestrial laser scanning is performed from a vehicle-borne platform for the sake of increased efficiency. The spatial distances between the stations for the static scans are rather short to provide overlapping regions for further processing and to avoid gaps in the point clouds.

These descriptions give a rough idea of the design of a mobile mapping system. Such a system is composed of an imaging unit consisting of laser scanners and/or digital cameras, a positioning and navigation unit for spatial referencing and a time referencing unit. Accordingly, mobile mapping is defined as the task of capturing and providing 2D or 3D geometric environmental information using an imaging sensor which is attached to a moving platform. In practice, this platform is typically mounted on a motor vehicle, train or trolley. A competing idea to mobile mapping is rapid mapping, which is considered here as a special case of mobile mapping as it additionally implies the immediate supply of the mapping information.

Mobile mapping systems can be distinguished by the imaging unit in use. Within the scope of this chapter only systems which use terrestrial laser scanners as the main imaging unit are considered. Systems based on photogrammetry such as the KiSS [Sternberg *et al.*, 2001], MoSES [Gräfe, 2003], Visat [El-Sheimy, 1996] and GPS-VAN [Grejner-Brzezinska and Toth, 2002] are not discussed. Note that some of these systems use terrestrial profile laser scanners (i.e. line scanners) as auxiliary sensors. Interested readers are referred to the corresponding publications. Looking at the historical development of mobile mapping technology in the last two decades, mobile mapping systems based on terrestrial laser scanning are the most recent. See Ellum and El-Sheimy (2002) or El-Sheimy (2005) for an overview.

In this chapter a state-of-the-art review of mobile mapping tasks and systems based on terrestrial laser scanning is given. General aspects of mobile mapping are presented in the main as the commercial providers and solutions may change. Given the increasing need for 3D geoinformation and the multitude of efforts in the field of mobile mapping today, the situation will certainly change significantly within the next decade. Note that there is ongoing research on mobile mapping and mobile mapping systems within the framework of several scientific organisations such as the International

Society of Photogrammetry and Remote Sensing (ISPRS, www.isprs.org), the International Association of Geodesy (IAG, www.iag-aig.org) and the International Federation of Surveyors (FIG, www.fig.net), which also organise dedicated joint symposia. Recently, a detailed compilation of advances in mobile mapping technology was published [Tao and Li, 2007].

The contents of this chapter are organised as follows. The observation modes of mobile mapping systems are considered in Section 9.2, their design and typical components as well as the processing flow are described in Section 9.3. The latter is only concerned with the derivation of the final point cloud; further modelling issues are not treated here. In Section 9.4 some present-day systems are shown together with application examples. The validation of a mobile mapping system is addressed in Section 9.5.

9.2 Mobile mapping observation modes

At present, two observation modes are implemented in mobile mapping systems: the stop-and-go mode and the on-the-fly mode. These modes differ with regard to the imaging procedure even if the imaging sensors are identical. Consequently, registration and georeferencing have to be performed in a particular way. Readers who are familiar with the use of the GPS may notice the similarity of the terminology.

9.2.1 Stop-and-go mode

For the stop-and-go mode one or more laser scanners are mounted on a vehicle-borne platform. The choice of the vehicle depends on the particular application. Typically, vans are used for the scanning of road scenes. In the case of indoor applications simple handcarts may be convenient; autonomous vehicles or robots could also be used. Characteristically, the scans are taken in static mode: during scan time the position and orientation of the scanner do not change. The scans can be taken in panoramic or hemispheric mode but sector scans facing only the front or backwards may be more efficient. After one scan, the vehicle changes its position and the next scan is taken (Figure 9.1). In this way, extended or elongated objects are scanned step-by-step.

In stop-and-go mode each point cloud is geometrically consistent regarding the respective local scanner coordinate system, which is defined by the temporary position of the scanner reference point and the orientation of the scanner axes. With respect to a unique coordinate system (world system) there are six (or seven) geometric degrees of freedom: the three coordinates of the scanner reference point, the three rotation angles of the scanner axes (and possibly the scale of the distance measurements).

Because of the formal identity of mobile mapping in stop-and-go mode and the pure static mode, the registration and georeferencing can be based on the same algorithms such as, for example, the ICP (iterative closest point) algorithm [Besl and

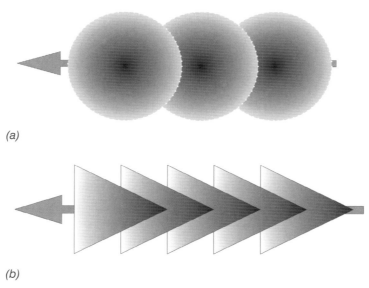

(a)

(b)

Figure 9.1 Mobile mapping in stop-and-go mode: (a) panoramic scans for crossroads scanning; and (b) directed scans on short-term stops along the road. (After Haring, 2007)

McKay, 1992] in combination with a spatial similarity transformation based on points in the object space which are coordinated in the world system (Chapter 3). For this reason system hardware components for positioning and navigation or synchronisation are not required. Obviously, mobile mapping systems in stop-and-go mode provide geometric quality of static scans in terms of accuracy and spatial resolution in an efficient way.

The CityGRID system [Haring, 2007] is an example of a stop-and-go mobile mapping system with respect to terrestrial laser scanning. In pure terrestrial laser scanning mode scan overlaps of at least 50% are required for registration. Typically, sector scans are taken. In order to reduce gaps in the scans the objects can be observed on the way out and back again (two-way mode) or in panorama scans [Haring *et al.*, 2005]. In combination (or exclusively), photographic sequences are taken using two synchronised digital cameras. In the pure photogrammetric case the system can observe in a so-called dynamic mode without stopping the vehicle.

9.2.2 On-the-fly mode

The efficiency of the stop-and-go mode can be exceeded if the scans are taken using the on-the-fly mode. In this case the vehicle is moving along a trajectory without stopping, and the laser scanner is scanning continuously in 2D profile mode. This principle is illustrated most easily if the scan direction is orthogonal to the trajectory of the platform (Figure 9.2).

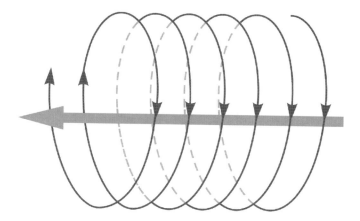

Figure 9.2 Mobile mapping in on-the-fly mode: profile scanning in a cross-direction extended to a helix-shaped point cloud through the motion of the vehicle. (After Hesse and Kutterer, 2007)

Also, in this case a 3D point cloud is obtained as the scanned profiles are fanned out along the trajectory due to the vehicle motion. The spatial resolution of the scan therefore depends mainly on the temporal and spatial resolution of the profile mode and on the velocity of the vehicle.

If, for example, 100 profiles are taken within one second and 10 000 points are observed within a single profile then the spatial resolution along the trajectory is 1.4 cm for a vehicle velocity of 5 km/h (or 5.6 cm for 20 km/h) and in the orthogonal direction 1.3 cm on a distance of 20 m. With the coupled three laser heads system of type HSP 102 of the Fraunhofer Institute for Physical Measurement Techniques (Freiburg, Germany) a profile density of 1.7 cm is obtained at a velocity of 100 km/h (www.ipm. fraunhoferer.de). This system is used with train-borne platforms.

The key problem in the on-the-fly mode is the rectification of the point cloud since each scanned point refers to an individual coordinate system. This is a fundamental difference to the stop-and-go mode and for this reason it is necessary to observe and provide the instantaneous position of the scanner reference point and orientation of the scanner axes with high accuracy and temporal resolution in a world system as well as on a common timescale. The technical issues are presented and discussed in Section 9.3.

The GEOMOBIL system (Alamús *et al.*, 2005) is an example of an on-the-fly mobile mapping system. The position and orientation information is derived from a combined system consisting of an IMU, two GPS receivers and one odometer. Terrestrial laser scanning is performed in profile mode.

At first glance, the on-the-fly case seems to be fully analogous to airborne laser scanning as the position information is derived from GNSS and the orientation information is yielded by an IMU. However, there are some differences due to the characteristics of their respective navigation. This refers mainly to the use of GPS positioning for mobile mapping systems, which is limited in urban environments as the GPS satellites can be partially shaded by buildings or other objects and – in case of indoor applications – even totally shaded.

Note that from the technology viewpoint there are several similarities between on-the-fly mobile mapping systems and the autonomous navigation of robots and vehicles [Thrun *et al.*, 2005]. The main difference is that in autonomous navigation the derived point clouds are only a by-product. They provide essential information for the navigation process which is based on simultaneous localisation and mapping (SLAM) techniques. A thorough discussion of autonomous navigation is therefore outside the scope of this volume.

9.3   Mobile mapping system design

9.3.1 General system design

The design of a mobile mapping system typically comprises the following components:

- imaging;
- referencing
    - spatial: positioning and orientation in relative and absolute mode,
    - temporal: synchronisation;
- communication and process control;
- data processing and product supply.

Depending on the different observation and georeferencing procedures, the design of stop-and-go and on-the-fly mobile mapping systems is in part different. In both cases terrestrial laser scanners are used as imaging components as already indicated in Section 9.1. Note that in addition to the already indicated groups of sensors, other sensors are used, such as digital thermometers which enable compensation for the thermally induced deviations of, for example, the inclination sensors.

9.3.2 Imaging and referencing

In case of stop-and-go mobile mapping systems imaging is performed in full 3D mode as panorama or sector scans of one or more scanners. The spatial referencing is based on object space information. Synchronisation is not required. For communication and process control some computer hardware without any specific properties is needed. Data processing and product supply are principally identical to procedures in the terrestrial laser scanning static mode.

In the case of on-the-fly mobile mapping systems the typical imaging procedure is based on vertical profile scans using one or more scanners. Profile scanners provide the required information directly; 3D scanners have to be used in 2D mode. The profiles can be orthogonal to the direction of the platform trajectory or – if more convenient – in oblique-angled orientations such as 45° and 335° in the case of two forward-viewing line scanners (or 135° and 215° in the case of backward-viewing scanners) with 0° the reference direction aligned with the vehicle motion.

Spatial referencing is mainly based on GNSS and IMU technology. If vans are used, additional car navigation and control sensors such as odometers, accelerometers or laser gyros can be integrated. Direct absolute spatial referencing is possible if the GNSS observations can be linked to a permanent GNSS reference network, such as SAPOS® in Germany, or if additional static GNSS reference equipment mounted on a reference point of the national geodetic network is used.

Temporal referencing requires a real-time operating system (hardware and software) which can handle pulse per second (PPS) signals from the components of the mobile mapping systems for synchronisation. This operating system is also the basis for communication between these components and for process control. Depending on the particular mobile mapping system, the quality of the synchronisation can be crucial. A correctly rectified 3D point cloud is presented in Figure 9.3 whereas in Figure 9.4 the same situation is shown with the point cloud geometrically deformed due to significantly incorrect synchronisation.

Figure 9.3 Rectified 3D point cloud of a road scene obtained by mobile mapping in on-the-fly mode. (From Hesse, 2008; courtesy the author)

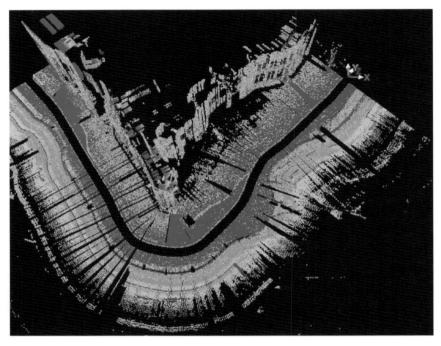

Figure 9.4 3D point cloud of a road scene obtained by mobile mapping in on-the-fly mode, deformed due to incorrect synchronisation. (From Hesse, 2008; courtesy the author)

Note that if object space information is considered, software-based solutions for spatial and temporal referencing are also possible. A software-based synchronisation of the laser scanner and the positioning–navigation unit is described by Vennegeerts *et al.* (2008). The idea is that the mirror of the laser scanner provides its own timescale by its rotation at a stable angular velocity. The shift and drift of this timescale with respect to the GPS timescale are then derived using at least two object points which are coordinated independently, for example with RTK (real time kinematic) GPS.

Note that at present the total cost of a mobile mapping system may be significant if only high-end terrestrial laser scanning, GPS and IMU equipment is used. For this reason several systems have been proposed which allow the reduction or even omission of the IMU, which is typically the most expensive system component. Such an approach has been used for the stop-and-go mobile mapping system City-GRID [Haring, 2007], where the spatial referencing information is completely derived from the object space. Alternatively, Hesse and Kutterer (2007) present a GPS-based on-the-fly mobile mapping system which works without an IMU. The position and orientation information is taken from the GPS trajectory apart from the roll angle, which is observed using an inclination sensor.

9.3.3 Indoor applications

In the case of indoor mobile mapping tasks the spatial referencing cannot be based on GNSS technology. At present there is no off-the-shelf solution. Depending on the particular hardware, positioning accuracies can be obtained in the range of several decimetres to metres [Blankenbach *et al.*, 2007], which is not good enough for mobile mapping tasks. However, there are several promising possibilities for indoor positioning which are based on technologies such as RFID (radio frequency identification), mobile phones or UWB (ultra wide band). The indoor supply of GNSS-like emitters such as pseudolites can be a more accurate alternative in the case of cheap and efficient instrumentation.

Polar surveying techniques such as automatic tachymetry with total stations have already been applied to provide indoor positions as they require only the line of sight between instrument and retro-reflector. The additionally required instantaneous orientation can be derived from inclination sensors or accelerometers.

9.3.4 Communication and control

For data capture all sensors have to be integrated with the real-time operating system. For the control of this process a personal computer (PC) such as a notebook is used, which serves as host server for the real-time computer and as a data storage unit. The communication between the PC, the sensors and the real-time processing system is typically based on library routines provided by the manufacturers. The complete data capture process is divided into three parts: preparation of the mobile mapping system, data capture, and finalisation.

The preparation includes the initialisation of all sensors in use. If, for example, a full 3D scanner is used in profile mode it has to be rotated to the destined orientation with respect to the platform. In addition, the state of the sensors has to be checked. If it is not available from a previous step with sufficient accuracy and reliability, the relative orientation of the sensors has to be determined as the joint use of the imaging units and the positioning and navigation units requires the definition and realisation of a unique body system.

In order to establish such a body system a calibration of the complete mobile mapping system is carried out to determine the relative positions and orientations of the system components before system operation. The system calibration has to be repeated at regular intervals to check and to guarantee the stability of the corresponding parameters. See Chapter 3 for details.

During data capture the measurement data and the time information of the real-time operating system are recorded and stored on the PC. All individual events such as interrupts are managed and processed by the real-time computer. The system integrity is checked before and during data capture. In the finalisation step the system has to be stopped, and, if this has not already been done, the captured data have to be stored on the PC in the measurement database of the system.

9.3.5 Processing flow

The core of the data processing component is a mathematical algorithm which combines the synchronised information of all spatial referencing sensors in an optimal way in order to derive the trajectory of the scanner and the axes orientations in the chosen world system. The trajectory is then described by the time-referenced absolute 3D positions of the laser scanner reference point. The associated tangent vectors comprise the orientations of two laser scanner axes. This information can already be provided in the case of GPS-only spatial referencing, i.e. without the use of an IMU.

However, the inclination of the system in the plane which is orthogonal to the respective tangent vector and the corresponding orientation of the laser beam are undetermined – unless it is observed using, for example, an additional inclination sensor. If an IMU is used, the orientations of all three laser scanner axes are completely determined. Hence, the relation between the temporary, local scanner systems and the unique world system can be described by concatenated 3D similarity transformations based on known positions and orientations of the laser scanner reference points and axes, respectively.

From the algorithmic point of view, recursive state-space filtering and smoothing is well suited to spatio-temporal referencing. For example the Kalman filter can be applied in combination with a smoother such as that described by the Rauch–Tung–Striebel algorithm [Rauch *et al.*, 1965]. If the results are not needed in real-time, other mathematical methods for estimating the required position and orientation parameters can be applied which are based, for example, on functional approximations of the trajectory. For a detailed data processing workflow see Hesse (2008) and Figure 9.5.

9.4 Application examples

In this subsection several existing mobile mapping systems based on terrestrial laser scanning are compiled and briefly described. These systems can be mainly grouped into railroad-track based systems and road-based systems, depending on the vehicle which they are mounted on. A current overview of land-based systems can be found in Barber *et al.* (2008).

9.4.1 Railroad-track based systems

One type of railroad-track based system is the Swiss Trolley [Glaus, 2006], which was developed at the ETH Zürich, Switzerland. Two line scanners (Sick LMS 200) are used as the imaging unit. For spatial referencing RTK GPS is typically applied. In the case of tunnel surveying, a total station with automated target tracking can optionally be used. The GPS information is stabilised by the measurements obtained with two odometers. In order to derive orientation information, the roll angle and the pitch angle are observed using two inclinometers. The pulse per second signal of a GPS receiver is used for synchronisation. According to Glaus (2006) a synchronisation

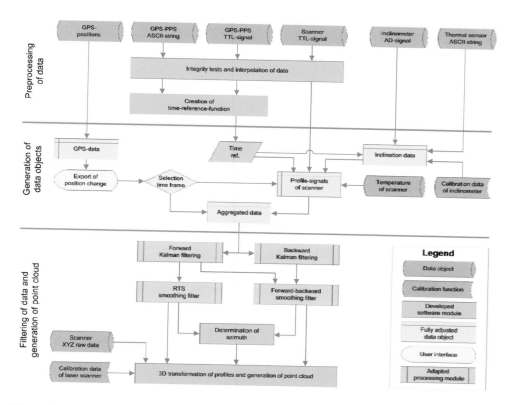

Figure 9.5 Example of data processing workflow of the mobile mapping system RAM-SYS. (From Hesse, 2008; courtesy the author)

accuracy of 1 ms is sufficient as the system is operated at walking speed. See Figure 9.6 for a photograph of this system.

A second railroad-track based system is Amberg Technologies' GRP System FX (GRP 5000) (www.amberg.ch). A hand-driven track-gauging trolley serves as vehicle and sensor platform. A profile scanner (Amberg Profiler 5002) is used as the imaging unit. The spatial referencing can be performed using either RTK GPS or a tracking total station. In addition an odometer, a gauge sensor and a cant sensor are integrated. Typical applications are clearance surveying, track geometry surveying and infrastructure documentation.

The High Speed Profiler HSP 102 of the Fraunhofer Institute for Physical Measurement Techniques (IPM) is a third example (www.ipm.fraunhofer.de). It is a component of a three-scanner head system (HSP 302) which is mounted on a train (inspection car) for clearance surveying. A typical vehicle speed is 100 km/h. Blug *et al.* (2005) report an overall measurement accuracy of this system of below 10 mm on distances up to 10 m. The HSP and its mounting on a train are shown in Figure 9.7.

Figure 9.6 Railroad-track based system Swiss Trolley. (From Glaus, 2006; courtesy the author)

(a)

(b)

Figure 9.7 Railroad-track based system HSP 302 (a) mounted on a train; and (b) scanner head. (Courtesy Fraunhofer IPM)

Typical precision measurements of the railroad-track based systems are in the range of several millimetres to centimetres.

Railroad-track based systems are used for surveying tasks concerned with the planning, construction and monitoring of tracks and the respective structures and infrastructures such as tunnels or signal installations, respectively. Clearance measurements along the tracks and convergence measurements inside tunnels are typical examples.

9.4.2 Road-based systems

At present, there are several road-based MMS (mobile mapping systems) which are mounted on vans. The following examples are typical, although this compilation cannot be exhaustive. The CityGRID system [Haring, 2007], which has already been introduced in Section 9.2.1, represents the stop-and-go mode. All other systems presented in this subsection work in on-the-fly mode. The imaging unit of the CityGRID consists of a 3D laser scanner (Riegl LMS-Z420i), which is relevant in the context of this chapter, and two digital cameras. GPS is used for georeferencing. Other sensors or sensor systems for positioning and navigation are not used. Spatial referencing is mainly based on object space information. It is performed in a combined procedure using the ICP algorithm modified for mobile mapping and control points which are coordinated in the world system.

The StreetMapper system (www.3dlasermapping.com, www.igi.systems.com; see also Figure 9.8) consists of three laser scanners (Riegl) as imaging units and a GPS/IMU positioning and navigation unit (IGI TERRAControl), which is completed by a speed sensor [Redstall and Hunter, 2006]. It can be mounted on a van, a 4 × 4 vehicle, a quad bike or a boat. It is used in several fields of application such as highway surveys, overhead clearance surveys and mining operations. Maximum vehicle speed is 70 km/h. The relative accuracy is 30 mm, the absolute accuracy 50 mm to 1 m, mainly depending on the quality of the spatial referencing.

The VLMS (vehicle-borne laser mapping system) was developed at the Centre for Spatial Science of the University of Tokyo, Japan [Manandhar and Shibasaki, 2002]. Three vertically arranged line scanners (Sick LMS 200) are used as imaging units. Spatial referencing is derived from a GPS/IMU with an integrated odometer. The average vehicle speed is 40 km/h.

Glennie (2007a) has introduced the mobile mapping system TITAN, which can be used on passenger vehicles (Figure 9.9) or small watercrafts. It consists of an array of 3D lidar sensors (at present: Riegl) for imaging and a GPS/IMU for spatial referencing (NovAtel, Honeywell). Rieger *et al.* (2007) describe a car-based system consisting of one 3D laser scanner (Riegl LMS-Z420i) and a GPS/IMU (Applanix Pos LV). The scanner operates in profile mode. The GPS positions are derived within the Austrian service APOS. This system can also be used on marine vehicles.

(a)

Figure 9.8 Road-based system
StreetMapper: (a) mounted on a car;
and (b) sensor platform (Courtesy 3D
Laser Mapping Ltd.)   (b)

The Optech LYNX Mobile Mapper system (www.optech.ca) was presented in 2007. It uses up to four very fast proprietary scanner sensor heads, up to two cameras and a GPS/IMU by Applanix. The rotation rate is 9000 rpm, which leads to a spatial resolution of up to 1 cm at a driving speed of 100 km/h. The absolute accuracy is 5 cm (in good GPS conditions).

The Finnish Geodetic Institute's FGI ROAMER [Kukko *et al.*, 2007] uses a 3D laser scanner (FARO LS) in profile mode together with a GPS/IMU (NovAtel SPAN). The authors state positioning accuracies of a few decimetres and imaging accuracies of a few centimetres.

In contrast, the prototype mobile mapping system RAMSYS [Hesse, 2008] runs without an IMU. Here, the positions and orientations are derived from the GPS

(a)

(b)

Figure 9.9 Road-based system TITAN: (a) system mount; and (b) typical 3D point cloud coloured by elevation. (Courtesy Terrapoint USA Inc.)

trajectory and an additional inclination sensor for cant information. For the GPS positions the German service SA*POS*® is used. For imaging, a 3D laser scanner (Leica HDS 4500) is run in vertical profile mode.

Reulke and Wehr (2004) discuss a mobile mapping system called POSLAS-PANCAM, which additionally contains a 360° panorama camera (rotating CCD line). The 3D-LS is used as the scanning sensor [Wehr, 1999], and positioning and navigation are performed using a GPS/IMU (Applanix).

Road-based systems have been used for a variety of mapping applications. One prominent application area is the capture of 3D information of road surfaces and street furniture for maintenance and asset management. A second area is the supply of data for 3D city models or vegetation databases. Application to clearance measurements of tunnels, bridges or wires is also of interest, for example for oversize cargos.

9.5 Validation of mobile mapping systems

The validation of a mobile mapping system with regard to realistic conditions is a fundamental part of the development of the system. In this subsection three related topics are briefly addressed:

- testing and calibration as a prerequisite for the use of a mobile mapping system;
- theoretical error and accuracy analysis;
- empirical assessment of precision and accuracy.

The testing and calibration procedures generally comprise all system components. Particular emphasis has to be laid on the laser scanner with regard to the scanner reference point and axes, and the positioning and navigation sensors with regard to offset and drift. In addition, temperature effects have to be studied. If applicable, the real-time computer has to be tested with regard to its clock and latency during operation of the system. See Hesse (2008) for detailed considerations and results on this topic.

A theoretical error and accuracy analysis can be based on assessment of the system configuration, the errors of the relevant components and the mathematical processing chain during analysis of the captured data. Lichti *et al.* (2005) discuss error models and error propagation in directly georeferenced terrestrial laser scanning networks. Glennie (2007b) presents a rigorous first-order error analysis of the terrestrial laser scanning georeferencing equations for mobile mapping systems. He considers IMU attitude errors, laser scanner errors, lever-arm offset errors and positioning errors, and discusses various scenarios such as ground-based kinematic lidar. The derived horizontal and vertical errors are distance dependent and mostly below 1 dm. The laser scanner errors are identified as the main impact quantity.

A combination of a theoretical error analysis and empirical tests – in particular for the StreetMapper mobile mapping system – is presented by Barber *et al.* (2008). The authors derived reference data from RTK GPS and short-term static differential GPS observations. As reference points corner points of white road markings were used, which could be identified in the scan data. A thorough statistical analysis revealed a precision of a few centimetres, which was below the values of the theoretical analysis.

Conclusions

Mobile mapping based on terrestrial laser scanning is a rather young technology for the capture and provision of 3D geoinformation. It represents a strongly developing field. The reported number of mobile mapping systems is still increasing as are their fields of application. The main benefit of terrestrial laser scanning as the imaging component compared with photogrammetric systems is the direct acquisition of 3D information. Other such systems, however, allow photo-realistic representations.

Capture of 3D geoinformation based on terrestrial laser scanning can be performed in static and in kinematic mode. The advantage of the static mode is the high accuracy and spatial resolution, whereas the advantage of the kinematic mode lies in the efficiency and hence the fast capture of extended objects. Based on several practical tests Hesse (2008) states reductions of time for data capture and processing (until the derivation of the rectified point cloud) up to 80% in comparison with static terrestrial laser scanning.

Today, direct georeferencing in outdoor environments is solved if high-end GPS/IMU equipment is available. Low-budget or precise indoor solutions are still under development. It can be expected that a refined integrated analysis based on hardware and object space information will further improve the results.

References

Alamús, R., Baron, A., Bosch, E., Casacuberta, J., Miranda, J., Pla, M., Sánchez, S., Serra, A. and Talaya, J., 2005. GEOMOBIL: ICC land based mobile mapping system for cartographic data capture. *Proceedings of XXII International Cartographic Conference of the ICA*, 9–16 July 2005, La Coruna, Spain; available at http://www.icc.es/pdf/common/icc/publicacions_icc/publicacions_tecniques/fotogramefoto/bienni_2005_2006/geomobil.pdf (accessed 18 June 2009).

Barber, D., Mills, J. and Smith-Voysey, S., 2008. Geometric validation of a ground-based mobile laser scanning system. *ISPRS Journal of Photogrammetry and Remote Sensing*, 63(1), 128–141.

Besl, P. J. and McKay, N., 1992. A method for registration of 3-d shapes. *IEEE Transactions on Pattern Analysis and Machine Intelligence*, 14(2), 239–256.

Blankenbach, J., Norrdine, A., Schlemmer, H. and Willert, V., 2007. Indoor-Positionierung aif Basis von Ultra Wide Band. *Allgemeine Vermessungs-Nachrichten*, 2007(5), 169–178.

Blug, A., Baulig, C., Wölfelschneider, H. and Höfler, H., 2005. Fast fiber coupled clearance profile scanner using real time 3D data processing with automatic rail detection. *IEEE Intelligent Vehicles Symposium 2004*, 14–17 June 2004, Parma, Italy, 658–663.

El-Sheimy, N., 1996. The development of VISAT – a mobile survey system for GIS applications. Ph.D. thesis, Department of Geomatic Engineering, University of Calgary.

El-Sheimy, N., 2005. An overview of mobile mapping systems. *Proceedings of the FIG Working Week 2005*, 16–21 April 2005, Cairo, Egypt; http://www.fig.net/pub/cairo/papers/ts_17/ts17_03_elsheimy.pdf (accessed 5 March 2009).

Ellum, C. and El-Sheimy, N., 2002. Land-based mobile mapping systems. *Photogrammetric Engineering and Remote Sensing*, 68(1), 13, 15–17 and 28.

Glaus, R., 2006. Kinematic track surveying by means of a multi-sensor platform. Ph.D. thesis, Eidgenössische Technische Hochschule (ETH), Zürich, No. 16547.

Glennie, C., 2007a. Reign of points clouds – a kinematic terrestrial LIDAR scanning system. *Inside GNSS*, Fall 2007, 22–31.

Glennie, C., 2007b. Rigorous 3D error analysis of kinematic scanning LIDAR systems. *Journal of Applied Geodesy*, 1(4), 147–157.

Gräfe, G., 2003. Mobile mapping with laser scanners using the MoSES. In *Optical 3-D Measurement Techniques VI*, Grün, A. and Kahmen, H. (Eds.), ETH, Zurich, Switzerland, I, 381–388.

Grejner-Brzezinska, D. and Toth, C. (2002): Real-time mobile multisensor system and its applications to highway mapping. *Proceedings of the Second Symposium on Geodesy for Geotechnical and Structural Engineering*, 21–24 May 2002, Berlin, Germany, 202–212.

Haring, A., 2007. Die Orientierung von Laserscanner- und Bilddaten bei der fahrzeuggestützten Objekterfassung. Ph.D. thesis, Technical University Vienna.

Haring, A., Kerschner, M. and Sükar, G., 2005. Vehicle-borne acquisition of urban street space by combined use of 3D laser scanning and close-range photogrammetry. In *Optical 3-D Measurement Techniques VII*, Grün, A. and Kahmen, H. (Eds.), ETH Zürich, Switzerland, I, 275–284.

Hesse, C., 2008. Hochauflösende kinematische Objekterfassung mit terrestrischen Laserscannern. Ph.D. thesis, Leibniz University of Hannover; Wissenschaftliche Arbeiten der Fachrichtung Geodäsie und Geoinformatik der Leibniz Universität Hannover, No. 268.

Hesse, C. and Kutterer, H., 2007. A mobile mapping system using kinematic terrestrial laser scanning (KTLS) for image acquisition. In *Optical 3-D Measurement Techniques VIII*, Grün, A. and Kahmen, H. (Eds.) ETH, Zürich, Switzerland, II, 134 – 141.

Kukko, A., Andrei, C.-O., Salminen, V.-M., Kaartinen, H., Chen, Y., Rönnholm, P., Hyyppä, H., Hyyppä, J., Chen, R., Haggrén, H., Kosonen, I. and Capek, K., 2007. Road environment mapping system of the Finnish Geodetic Institute – FGI ROAMER. *International Archives of Photogrammetry, Remote Sensing and Spatial Information Sciences*, 36(Part 3/W52), 241–247.

Lichti, D.D., Gordon, S.J. and Tipdecho, T., 2005. Error models and propagation in directly georeferenced terrestrial laser scanner networks. *Journal of Surveying Engineering*, 131(4), 135–142.

Manandhar, D. and Shibasaki, R., 2002. Auto-extraction of urban features from vehicle-borne laser data. *International Archives of Photogrammetry, Remote Sensing and Spatial Information Sciences*, 34(Part 4) (on CD-ROM).

Rauch, H.E., Tung, G. and Striebel, C.T., 1965. Maximum likelihood estimates of linear dynamic systems. *American Institute of Aeronautics and Astronautics Journal*, 3(8), 1445–1450.

Redstall, M. and Hunter, G., 2006. Accurate terrestrial laser scanning from a moving platform. *Geomatics World*, July/August, 28–30.

Reulke, R. and Wehr, A., 2004. Mobile panoramic mapping using CCD-line camera and laser scanner with integrated position and orientation system. *International Archives of Photogrammetry, Remote Sensing and Spatial Information Sciences*, 34(Part 5/W16); http://www.commission5.isprs.org/wg1/workshop_pano/papers/PanoWS_Dresden2004_Reulke.pdf (accessed 5 March 2009).

Rieger, P., Studnicka, N. and Ullrich, A., 2007. *"Mobile Laser Scanning" Anwendungen*. http://www.riegl.com/uploads/tx_pxpriegldownloads/_Mobile_Laser_Scanning__Anwendungen.pdf (accessed 5 March 2009).

Sternberg, H., Caspary, W., Heister, H. and Klemm, J., 2001. Mobile data capturing on roads and railways utilizing the Kinematic Survey System KiSS. *Proceedings of the 3rd International Symposium on Mobile Mapping Technology*, 3– 5 January 2001, Cairo, Egypt (on CD-ROM).

Tao, C.V. and Li, J., 2007. *Advances in Mobile Mapping Technology. ISPRS Book Series*, 4, Taylor and Francis, London, UK.

Thrun, S., Burgard, W. and Fox, D., 2005. *Probabilistic Robotics*, MIT Press, Cambridge, MA, USA.

Vennegeerts, H., Martin, J., Becker, M. and Kutterer, H., 2008. Validation of a kinematic laser scanning system. *Journal of Applied Geodesy*, 2(2), 79–84.

Wehr, A., 1999. 3D-Imaging laser scanner for close range metrology. *Proceedings of SPIE*, 6–9 April 1999, Orlando, FL, USA, 3707, 381–389.

Index of advertisers

Index

2.5D representation 156
3D point clouds 274
3D terrain models 156

accuracy 52, 159, 160, 161, 308
    map 159
active shape model 147
adjustment 112, 121, 122
airborne laser scanner 19, 21
    components 21
    error budget 28
    inertial measurement unit (IMU) 23
    processing scheme 33
ambiguity interval 7
amplitude 151, 153
assembly
    boresight 90, 91, 92, 101, 106, 107, 127
    installation 101, 105
    lever arm 90, 91, 101, 105, 125
    misalignment 90, 101, 103, 107, 125

baseline 8
bathymetry 35–6, 37
beam divergence 26
Boolean operations 183
boundary representation (B-rep) 183
breakline 154, 155, 156
building detection 170–7
    CIR image 172
    classification task 170
    in map updating 171
    Laplace operator 171, 174
    morphological filter 171
    normalised difference vegetation index
        (NDVI) 173
building extraction 169–212
    data exchange and file formats *see* modelling
        3D buildings

building reconstruction 182–96
    and generalisation 202
    constraint problem, geometric 197
    constraints, hard/soft 198
    geometric modelling 183–4
    modelling 182
    point splats 183
    proxy objects 183
    regularisation 197
    systems 186–96
    topology 196
building reconstruction systems 186–96
    bottom-up (data-driven) control 186
    convex/concave edges 193
    hypothesise-and-test 189, 196
    Markov chain Monte Carlo
        (RJMCMC) 194–5
    object recognition 186
    rectangle-based 191
    region adjacency graph 192
    semi-automatic 187
    straight skeleton 189, 190
    top-down (model-driven) control 186
    Voronoi diagram 191, 193

CAD 183, 238, 272, 285, 286, 287
calibration 122, 125
    errors 52
cameras 34
canopy height model (CHM) 217–19
change detection 243–61 *see also* tunnel stability,
        morphological deformation
    closure errors 249
    corridor mapping 261–6 *q.v.*
    dam 248
    deformation analysis 243, 247
    dispersion 250
    DSM comparison 244–5

change detection *(continued)*
  Error propagation 250
  highway viaduct 248
  landslide analysis 249
  least squares 3D surface and curve
      matching 244
  least squares adjustment 250
  linear deformation model 250
  local ICP
  measurement to measurement comparison
      247
  measurement to surface comparison 247
  model matrix 250
  noise level 248–9
  object-oriented deformation analysis 254–60 *q.v.*
  occlusion 246
  octree based 245
  range panorama 246
  sand pile 249
  structural monitoring 243
  surface to surface comparison 247
  testing theory 250
  TIN least squares matching 244
  variance-covariance matrix 250–1
  virtual target monitoring 247–8
  visibility maps 245–6
chirp time 7
CityGML 206
classification 137, 170
  of laser data 253
  support vector machines 176
closing (operation) 140
coherent detection 7
COLLADA 206
collimated laser beam 13
computer-aided design (CAD) 183
constant fraction detection 4
constraint problem, geometric 197
constraints 198, 199
  and interactivity 199
  hard/soft 198
  structured introduction 201
constructive solid geometry (CSG) 183
continuous wave modulation 6
  chirp time 7
  coherent detection 7
conversion (of point clouds) to raster
      images 46–53
co-ordinate systems/reference frames
  body frame 89, 92, 105–6, 125
  Earth-centred, Earth-fixed (ECEF) 88
  local-level frame 89–90, 111
  mapping frame 91, 111, 119
  national 86, 91, 111, 119
  object space 86, 111

sensor space/sensor frame 86, 87, 92, 110, 111
  WGS-84 88
correlation 112, 124, 127
corridor mapping 261–6
  catenary curve 263
  classification (of laser data) 253
  dike inspection 264
  dikes 261
  eigenvalues 263
  eigenvectors 263
  forced dike collapse 265
  Gaussian mixture model 263
  Hough transformation 263
  Hurricane Katrina 264
  levee inspection 263
  overhead wires 261
  pipelines 261
  power poles 261
  railroads 261
  sag 261
  span line geometry 263
  transmission lines 261, 262
  vegetation 261
cultural heritage 271–90
  3D point clouds 274
  airborne laser scanning 280
  archaeology 271, 275, 276, 279, 282
  CAD 272, 285, 286, 287
  CIPA 271
  documentation 271, 272, 276, 282
  DEM 277
  DSM 280, 281
  DTM 279, 280, 281, 282
  excavation 272, 284
  façade 272, 273
  full waveform 280
  HiMAT 284
  ICOMOS 271
  mobile scanner 278
  prospection 272, 279
  terrestrial laser scanners 275, 285
  texture 275, 283, 285, 286, 287

data artefacts 97
  blooming 99
  incidence angle 100
  laser beam width 97, 98, 105
  mixed pixels 97, 98
  multipath 100
  walk error 97
data compression 75–8
data-driven control 186
data pyramid 147
data structure 45, 58–63
  3D tree 62

Delaunay triangulation 59
  k-D tree 62–3
  octrees 60–2, 75, 77
Delaunay triangulation 59, 178 *see also* Triangular
    Irregular Network (TIN)
Dempster-Shafer theory 173
density function 195
detection method 4
  constant fraction 4
  leading edge 4
  peak 4
detection of cylinders 67
digital canopy model (DCM) 135
digital elevation model (DEM) 135
digital surface model (DSM) 135, 137,
    280, 281
  comparison 244–5
  nDSM 157
  stripwise 159
digital terrain model (DTM) 135–67, 279, 280,
    281, 282
  accuracy 161
  data reduction 161–3
  data structure 155
  extraction 135–63
  generation 156–63
  model quality 158–61
digital terrain model data reduction 161–3
  mesh decimation 162
  TIN decimation 162
digital terrain model generation 156–63
  grid 157
  hybrid grid model 157
  in forest regions 216–7
  inverse distance weighting 157
  kriging 157
  linear prediction 157
  moving least squares 157
  nDSM 157
  raster 157
  triangulation 157
digital terrain model quality 158–61
  accuracy 159, 160
  data quality 158–60
  point density 158
  point distance 158
  point distance map 158
  precision 160
  dilation 140
Douglas Peucker algorithm 178

echo
  amplitude 151, 153
  multiple 3, 28, 29
  width 151–2, 153

echoes 214
  first and last pulse 214
electronic distance measurement (EDM) 93, 122
engineering applications 237–69
  corridor mapping 261–6 *q.v.*
  industrial sites, reconstruction 238–43 *q.v.*
  see also change detection
erosion 140
error
  analysis 308
  budget 28, 100, 106, 107–9
  GPS/INS 101, 102
  systematic 92
Euler angles 84, 90, 91
eye safety 12

false colour infrared (CIR) image 172
feature-based segmentation 148
field of view (FOV) 96
filter algorithms 138–9, 160
filter comparison 150–1
filtering 137–54
  of point clouds 137–54
first and last pulse echoes 214, 217
first echo 3
flash lidar 19
flight plan 31
flight planning 30–1
footprint extraction 178–82
forest inventory data acquisition 214
forest inventory parameters 215
  stand-wise 215
  tree-wise 215
forest region digital terrain models 216–7
forestry applications 213–35
  biodiversity parameters 225
  biomass determination 217, 218
  canopy closure 226
  canopy height model (CHM) 217–19, 226
  CO_2 storage capacity 218
  inventory 229 *q.v.*
  Kyoto Protocol 218–19
  single-tree detection 219–22
  tree height estimation 217, 220, 221
  using terrestrial laser scanning 227–8
forest inventory 229–31
  data processing chain 229–30
  success rate 230–1
formal grammars 184–5
formline 154
full-waveform 174, 280
  airborne laser scanning 151
  analysis 222–3
  digitisation 28, 222–3
  information 151–4

Gaussian beams 12–13, 104–5
Gaussian decomposition 223, 224
Gaussian mixture model 263
Gauss-Markov model 112, 121
generalisation 202
    map 203
geodata infrastructure 206
geographic information systems (GIS)
        135, 185
geometric constraint problem 197
geometric modelling 183–4
    Boolean operations 183
    boundary representation (B-rep) 183
    computer-aided design (CAD) 183
    constructive solid geometry (CSG) 183
glass fibre scanner 18
global positioning system (GPS) 20, 22,
        24, 32
Google Maps 170
GPS 101, 105, 119, 125, 127
GPS/INS 88, 89–90, 91, 100, 101, 102, 105, 119,
        127
GPS/INS errors 101, 102
grammars
    formal 184–5
    set 185
    split 185

Hough transform 63, 263

industrial point cloud segmentation 239–40
    k-D trees 239
    local point density 239
    region growing 240
    robustness 239
    seed point 240
    surface normal 240
industrial sites 238–43
    see industrial point cloud segmentation
    see object detection
    see reconstruction of
    see registration methods
industry foundation classes (IFC) 206
inertial measurement unit (IMU) 23, 24, 32

Jarvis march algorithm 179

Kalman filter 302
k-D tree 62–3, 239
kriging 157

Laplace operator 171, 174
laser beam propagation 12–14
    Gaussian beams 12–13
    spatial resolution 13

laser scanner properties 25–6
    beam divergence 26
    swath width 26
laser scanners 1
    components 11–19
    light sources 12
    principles 2–11
    terrestrial 36–9
laser scanning
    basics 19
    operational aspects 30–1
    principle 21
    technology 1–42
last echo 3
leading edge detection 4
least squares 157, 244, 250
light propagation 14
light sources 12
    speckle noise 12
Lindenmayer systems 185

machine learning 203
map generalisation 203
map updating 171
Markov chain Monte Carlo (RJMCMC) 194–5
    density function 195
mathematical morphology 140, 141
    operations 140
mesh decimation 162
Microsoft Virtual Earth 170
mirror systems 16
    oscillating 16
mobile mapping 293–311
    accuracy analysis 308
    autonomous navigation 298
    calibration 301, 308
    clearance surveying 303
    error analysis 308
    infrastructure documentation 303
    kinematic mode 293
    mining operations 305
    observation modes 295–8
    on-the-fly mode 296–8, 299
    railroad track surveying 302–5
    rapid 294
    real-time operating system 299
    road-based systems 305–8
    spatial referencing 299
    static mode 293
    stop-and-go mode 294, 295–6, 298
    synchronisation 294, 296
    systems 294, 298, 303
    testing 308
    temporal referencing 299
    validation 308

model-driven control 186
modelling 3D buildings 206
 CityGML 206
 COLLADA 206
 geodata infrastructure 206
 industry foundation classes (IFC) 206
 shapefile format 206
 VRML97, X3D 206
morphological deformation 253–4, 255
 alternative deformation models 253
 data snooping 254
 sand suppletion 253
morphological filtering 139–42
 closing 140
 dilation 140
 erosion 140
 opening 140
morphological operators 171, 172
multiple echo 3, 28, 29
multiple returns 3

navigation
 attitude 89, 90, 91, 103, 107, 120
 GPS 101, 105, 119, 125, 127
 GPS/INS 88, 89–90, 91, 100, 101, 102, 105,
  119, 127
 GPS/INS errors 101, 102
 IMU 92, 101, 106, 107, 125
 Klaman filter 91, 103
 trajectory 88, 90, 91, 101, 102, 119
network design 122, 127
normalised difference vegetation index
  (NDVI) 173
normalised digital surface model (nDSM) 157

object-based segmentation 148
object detection, industrial 240–2
 cylinder modelling 241
 Hough transformation 241
 object modelling 240–1
 over-segmentation 241
object-oriented deformation analysis 254–60
 changes between objects 254
 changes relative to an object 254, 255, 256
 concrete beam 257–8
 fuzzy k-means clustering 260
 ground-based SAR 259
 landslide monitoring 259
 lock door 256, 257
 minimum deformation 260
 moving rocks 258
 rigid body transformation 258
 rock boulders 258
 rock face geometry 259–60
 rock facets orientation 260

roof orientation 259
 television tower 257
 water pressure 256
object recognition 186
observations 83
 elevation angle 85, 95, 124
 encoder angle 90, 126
 horizontal direction 85, 94
 range 85, 90, 92–3, 122, 125
octrees 60–2, 75, 77, 245
opening (operation) 140
optical 3D measurement systems 2
optical triangulation 8, 9
 laser-based 10
orientation 88, 89, 90, 96, 100, 102, 106, 123, 126
oscillating mirror 16
outline simplification 178
outlining of footprints 178, 181

Palmer scanner 18
parameterisation 240–3
peak detection 4
phase measurement 5–8
 continuous wave modulation 6
 techniques 5
photodetection 14
plane detection 65
plane fitting 66, 69
point-based rendering 53–8
point cloud 33, 45
 3D 274
 acquisition 285
 classification 137
 conversion to raster images 46–53
 filtering 137–54
 fusion of 285, 286
 segmentation 63, 148, 239–40
point density 26–7, 51
 images 50
point rendering 55
point splatting 46
point splats 183
polygonal mirror 18
precision 160, 237
primitives, geometric 184
projection mechanisms 16–19
progressive densification 142–4
 progressive TIN 142–4
proxy objects 183
pulse characteristics 4
pulse repetition frequency 26–7
 and point density 26–7
pulse repetition time 4
pulse rise time 4
pulse width 4

quadtree 61
quality control 51–3
quaternion 85, 115

random sample consensus (RANSAC) 67–72,
     180
range uncertainty 5, 6, 7, 9
reconstruction of industrial sites 238–43
     as-built model 238
     CAD model 238
     clash detection 238
     geometric primitives 239
     pipes 239
     retrofitting 239
reflectance 14
region growing 72, 149, 220, 240
registration 109–22
     direct 119
     feature-based 118, 125
     ICP (iterative closest point) 111, 113–18
     strip adjustment 121
     target 111
registration and calibration 83–133
registration methods 242–3
     cylinder axis 242
     direct 242
     error propagation 243
     geometric 242
     ICP 243
     indirect 242
     industrial reconstruction 242
robust interpolation 144, 148, 191
root mean square error (RMSE) 160, 218, 231
rotation matrix 84–5, 89–90, 100, 119

scan line segmentation 75
scanning mechanisms 16–19
scanning mirror 16
     oscillating 16
     polygonal 18
seed detection 72
segmentation 63, 74
     feature-based 148
     Hough transform 63
     object-based 148
     plane detection 65
     scan line 75
segment-based filtering 147–50
semi-automatic building reconstruction 187
sensor integration 23
set grammar 185
shaded height images 48
shapefile format 206
signal-to-noise ratio 4

slope depending filter 142
SNR 4, 10
space partitioning 203
spatial resolution 13, 19
speckle noise 9, 12
split grammar 185
step edge 154
straight skeleton 189, 190
structure line determination 154–6
     breakline 154, 155, 156
     formline 154
     step edge 154
structure line information 154
support vector machines 176
surface-based filtering 144–7
surface growing 72–4
     seed detection 72
surface splatting 55–8
survey flight 31–2
swath width 26
systematic errors 92
     angle encoder/circle 93, 101, 104
     axis wobble 94, 96
     cyclic 93, 124
     eccentricity 94, 96
     index 95
     non-orthogonality 94, 96, 124
     rangefinder offset 92–3, 103–5, 122, 125
     scale 93, 94, 95, 104

terrestrial laser scanners 36–9
terrestrial laser scanning 158, 227–8
time-of-flight measurement 3–5, 213
     detection method 4
translucency 15
Triangular Irregular Network (TIN) 47, 216
     see also Delaunay triangulation
     decimation 162
     progressive 142–4
tunnel stability 251–3
     artificial deformation 251
     false alarms 253
     stability test 252

variance-covariance matrix 250–1
visualisation 45–81
     for quality control 51–3
Voronoi diagram 191, 193
VRML97, X3D 206

waveform 222
     analysis 224
     digitisation techniques 222–5
WGS 84 88, 294